HANDBOOK OF HEAT
VENTILATING AND AIR CON

HANDBOOK OF HEATING VENTILATING AND AIR CONDITIONING

READY-REFERENCE TABLES AND DATA

JOHN PORGES, M.I.Mech.E., M.Inst.F.

revised by
F. PORGES, LL.B., B.Sc. (Eng.), C.Eng.,
M.I.Mech.E., M.I.E.E., F.I.H.V.E.

SEVENTH EDITION

NEWNES – BUTTERWORTHS
LONDON BOSTON
Sydney Wellington Durban Toronto

THE BUTTERWORTH GROUP

UNITED KINGDOM
Butterworth & Co (Publishers) Ltd
London: 88 Kingsway, WC2B 6AB

AUSTRALIA
Butterworths Pty Ltd
Sydney: 586 Pacific Highway, NSW 2067
Also at Melbourne, Brisbane, Adelaide and Perth

CANADA
Butterworth & Co (Canada) Ltd
Toronto: 2265 Midland Avenue, Scarborough,
Ontario, M1P 4S1

NEW ZEALAND
Butterworths of New Zealand Ltd
Wellington: 26–28 Waring Taylor Street, 1

SOUTH AFRICA
Butterworth & Co (South Africa) (Pty) Ltd
Durban: 152–154 Gale Street

USA
Butterworth (Publishers) Inc
Boston: 19 Cummings Park, Woburn, Mass. 01801

First published by George Newnes Ltd, 1942
Second edition 1946
Third edition 1952
Fourth edition 1960
Fifth edition 1964
Sixth edition published by Newnes-Butterworths, 1971
Seventh edition 1976

ISBN 0 408 00233 6

Printed and bound in England by Hazell, Watson & Viney Ltd., Aylesbury, Bucks.

PREFACE TO THE SEVENTH EDITION

Although there are several good books on heating, ventilating and air conditioning there is still a need for a concise manual containing data, charts and tables which may be required by the heating engineer many times a day.

This book is intended to fulfil the need in a concise manner in order to facilitate the work of the heating and ventilating engineer. It is designed for daily use, and a comprehensive bibliography has been included for the benefit of those who wish to pursue the theoretical side of any particular branch.

Practising engineers are frequently concerned with alterations and extensions to systems originally designed in Imperial units and for many years yet will on occasion want to look up data in both Imperial and S.I. units. I have therefore as far as possible retained data in both sets of units.

I am grateful to the Cambridge University Press for permission to reproduce data from "Thermodynamic Tables in S.I. (metric) Units", by Haywood. Some of the data on heat transmission is based on technical data available in German reference books, but the majority of the metric tables I have calculated myself. In doing this I have worked to an accuracy of 2 or 3 per cent, which I am sure is adequate for ordinary design work.

In this edition the list of conversion factors has been improved, new standards on pipes and cylinders have been added, the data on chimney sizes and combustion air requirements have been improved, atomic weights for combustion calculations have been added, the bibliography has been brought up to date and the section on British Standards has been completely rewritten.

The previous edition was the first one incorporating metric units. I am grateful to readers who have drawn attention to errors which had crept in, and I have corrected these and other errors which I have found in checking the text for this edition.

FRED PORGES

CONTENTS

Section I

SIGNS AND SYMBOLS

ABBREVIATIONS

in	inch
ft	foot
yd	yard
in^2	square inch
ft^2	square foot
in^3	cubic inch
ft^3	cubic foot
gal	gallon
gr	grain
lb	pound
cwt	hundredweight
t	ton
B.t.u.	British thermal unit
C.H.U.	Centigrade Heat Unit
kcal	kilocalorie
gcal, cal	gramcalorie
°F	°Fahrenheit
°C	°Celsius
ft lb	foot pound
lb/in^2	pounds per square inch
h	hour
min	minute
s	second
atm	atmosphere
J	Joule
N	Newton
Nm	Newton-metre
N/m^2	Newtons per square metre
kg/s	kilograms per second
W	watt
V	volt
A	ampere
h.p.	horse power
b.h.p.	brake horse power
ft/min	feet per minute
ft^3/min	cubic feet per minute
gal/min	gallons per minute
m^3/s	cubic metres per second
mm	millimetre
m	metre
mm^2	square millimetre
m^2	square metre
m^3	cubic metre
l	litre
g	gram
kg	kilogram
met t	metric tonne
w.g.	water gauge

ABBREVIATIONS USED ON DRAWINGS

BBOE	bottom bottom opposite ends (radiator connections)
BV	butterfly valve
CF	cold feed
CW	cold water
CWM	cold water main
DC	drain cock
EC	emptying cock
EXP	expansion (tank or pipe)
F	flow
FA	from above
FB	from below
FS	fire service
FTA	from and to above
FTB	from and to below
FW	fresh water
GV	gate valve
HTG	heating
HWS	hot water service
LSV	lock shield valve
MF	mixed flow
MV	mixing valve
NB	nominal bore
NTS	not to scale
PRIM	primary (hot water flow)
R	return
SEC	secondary (hot water flow)
TA	to above
TB	to below
TBOE	top bottom opposite ends (radiator connections)
TBSE	top bottom same end
TW	tank water
TWDS	tank water down service

SYMBOLS

l	length (ft, m)
r, R	radius (ft, m)
D, Dia.	diameter (ft, m)
t	thickness (in, mm)
A	area (ft^2, m^2)
V	volume (ft^3, gal, m^3, l)
h, H	height (ft, m)
A	heating surface (ft^2, m^2)
W	weight (force) (lb, N)
H	heat loss (B.t.u./h, W)
D	density (lb/ft^3, kg/m^3)
Q	quantity of water or air (ft^3/min, gal/min, m^3/s)
P	pressure (lb/in^2, N/m^2)
Cp	specific heat at constant pressure (B.t.u./lb °F, J/kg °C)
Cv	specific heat at constant volume (B.t.u./lb °F, J/kg °C)
t	temperature (°F, °C)
T	absolute temperature (°R, °K)
t_o	outside temperature (°F, °C)
t_r	room temperature (°F, °C)
H	heat content (B.t.u./lb, J/kg)
DB	dry-bulb temperature (°F, °C)
WB	wet-bulb temperature (°F, °C)
RH	relative humidity (per cent)
E	efficiency (per cent)
U	coefficient of heat transfer (B.t.u./ft^2 h °F, W/m^2 °C)
a	surface coefficient (B.t.u./ft^2 h, W/m^2)
k	thermal conductance (B.t.u./ft^2 h °F, W/m^2°C)
C	thermal conductivity (B.t.u. in/ft^2 h °F, W/m °C)
R	thermal resistance (ft^2 h °F/B.t.u. in, m °C/W)
x	quality of steam (per cent)
g	accelerating effect of gravity (=32·2 ft/s^2, 9·81 m/s^2)

Length

1 in.	= 25·4 mm. = 0·0254 m.
1 ft.	= 0·3048 m.
1 yd.	= 0·9144 m.
1 m.	= 3·2809 ft. = 1·0936 yd.
1 mm.	= 0·03937 in.

Area

1 in.^{2}	= 6·452 cm.^{2} = 6·452 × 10^{-4} m.^{2}
1 ft.^{2}	= 0·0929 m.^{2}
1 yd.^{2}	= 0·836 m.^{2}
1 mm.^{2}	= 1·55 × 10^{-5} in.^{2}
1 m.^{2}	= 10·764 ft.^{2} = 1·196 yd.^{2}

Volume

1 in.^{3}	= 16·39 cm.^{3} = 1·639 × 10^{-5} m.^{3}
1 ft.^{3}	= 0·0283 m.^{3} = 6·23 gal.
1 yd.^{3}	= 0·7646 m.^{3}
1 gal.	= 4·546 l. = 4·546 × 10^{-3} m.^{3}
	= 0·16 ft.^{3}
1 pint	= 0·568 l.
1 U.S. gal.	= 0·83 Imp. gal.
1 cm.^{3}	= 0·061 in.^{3}
1 m.^{3}	= 35·31 ft.^{3} = 1·308 yd.^{3}

Weight

1 grain	= 0·000143 lb. = 0·0648 g.
1 lb.	= 7000 grains = 0·4536 kg.
1 g.	= 15·43 grains
1 kg.	= 2·205 lb.
1 tonne	= 1000 kg. = 0·984 tons
1 gr./lb.	= 0·143 g./kg.
1 g./kg.	= 7·0 gr./lb.
1 lb./ft.^{3}	= 16·02 kg./m.^{3}
1 kg/l.	= 62·425 lb./ft.^{3}
1 kg./m.^{3}	= 62425 lb./ft.^{3}

Velocity & Volume Flow

1 ft./min.	= 0·00441 m./s.
1 m./s.	= 226·85 ft./min.
1 kg./s. (water)	= 13·23 gal./min.
1 m.^{3}/s.	= 2118·6 ft.^{3}/min.
1 ft.^{3}/min.	= 1·7 m.^{3}/h. = 0·47 l./s.

Pressure

1 atm.	= 1·033 × 10^4 kg./m.^{2}
	= 14·7 lb./in.^{2}
	= 1·013 × 10^5 N/m.^{2} = 1·013 bar
	= 407·69 in. (10·33 m.) water at 62°F.
	= 30 in. (760 mm.) Mercury at 62°F.
1 lb./in.^{2}	= 6895 N/m.^{2} = 6·895 × 10^{-2} bar
	= 27·71 in. (703·6 mm.) water at 62°F
	= 2·0416 in. (51·8 mm.) Mercury at 62°F.
	= 703·6 kg./m.^{2}
	= 0·068 atm.
1 kg./m.^{2}	= 1·422 × 10^{-3} lb./in.^{2}
	= 9·80 N/m.^{2}
	= 0·0394 in water
	= 1 mm. water
	= 0·0736 mm. Mercury
	= 0·9677 × 10^{-4} atm.
1 N/m.^{2}	= 0·1452 × 10^{-3} lb./in.^{2}
	= 1 × 10^{-5} bar
	= 4·03 × 10^{-3} in. (0·1024 mm.) water
	= 0·295 × 10^{-3} in. (7·55 × 10^{-3} mm.) Mercury
	= 0·1024 kg./m.^{2}
	= 0·993 × 10^{-5} atm.
1 in. water	= 0·0361 lb./in.^{2}
	= 249 N/m.^{2}
	= 25·4 kg./m.^{2}
	= 0·0739 in Mercury
1 mm. water	= 1·42 × 10^{-3} lb./in.^{2}
	= 9·80 N/m.^{2}
	= 1 kg./m.^{2}
	= 0·0736 mm. Mercury
	= 0·9677 × 10^{-4} atm.
1 in. Mercury	= 0·49 lb./in.^{2}
	= 3378 N/m.^{2}
	= 12·8 in. water
1 mm. Mercury	= 0·0193 lb./in.^{2}
	= 133 N/m.^{2}
	= 12·8 mm. water
1 bar	= 1 × 10^5 N/m.^{2}

Energy and heat

1 joule	= 1 watt second
	= 0·74 ft. lb.
	= 9·699 × 10^{-9} B.t.u.
1 B.t.u.	= 1·055 × 10^3 Joule
	= 0·252 kcal.
	= 778·5 ft. lb.
	= 0·293 watt hours
	= 3·9683 B.t.u. = 427 kg. m.
	= 4·183 × 10^3 Joule
1 B.t.u./hr.	= 0·293 watts
1 kW.	= 1000 J/s. = 3·6 × 10^6 J/h.
	= 1·358 metric h.p.
	= 737 ft. lb./s.
	= 3412 B.t.u./h.
	= 860 kcal./h.
1 ft. lb.	= 0·1383 kg. m.
	= 0·001286 B.t.u.
	= 1·356 Joule
1 kg. m.	= 7·233 ft. lb.
	= 0·009301 B.t.u.
	= 9·807 Joule
1 B.t.u./ft.^{2}	= 2·713 kcal./m.^{2}
	= 1·136 × 10^4 J/m.^{2}
1 B.t.u./ft.^{2} h.	= 3·155 W/m.^{2}
1 B.t.u./ft.^{3} h.	= 10·35 W/m.^{2}
1 B.t.u./ft.^{2} °F.	= 4·88 kcal./m.^{2} °C.
	= 2·043 × 10^4 J/m.^{2} °C.
1 B.t.u./ft.^{3}	= 8·9 kcal./m.^{3}
	= 3·727 × 10^4 J/m.^{3}
1 B.t.u./lb.	= 0·556 kcal./kg.
	= 2326 J/kg.
1 kcal./m.^{2}	= 0·369 B.t.u./ft.^{2}
1 kcal./m.^{2} °C.	= 0·205 B.t.u./ft.^{2} °F.
1 kcal./m.^{3}	= 0·1125 B.t.u./ft.^{3}
1 kcal./kg.	= 1·800 B.t.u./lb.
1 h.p.	= 550 ft. lb./s. = 33000 ft. lb./min.
	= 1·0139 metric h.p.
	= 746 watts
	= 2545 B.t.u./h.
1 metric h.p.	= 735·56 watts
	= 75 kg. m/s.
	= 0·986 English h.p.
1 ton refrigeration	= 12000 B.t.u./h.
	= 3·516 kW.

Temperatures

°F. = (9/5 × °C.) + 32 °C. = 5/9 × (°F. − 32)

1 deg. F. = 0·555 deg. C. 1 deg. C. = 1·8 deg. F.

Viscosity

1 poise = 0·1 kg./m. s. = 0·1 N s/m.^{2}

1 stoke = 1 × 10^{-4} m.^{2}/s.

Gravity

Acceleration due to gravity

in London = 32·2 ft./s.^{2} = 9·81 m/s.^{2}

at Equator = 32·1 ft./s.^{2} = 9·78 m/s.^{2}

STANDARD SIZES OF DRAWINGS B.S.3429: 1961.
SIZES OF DRAWING SHEETS FOR ENGINEERING

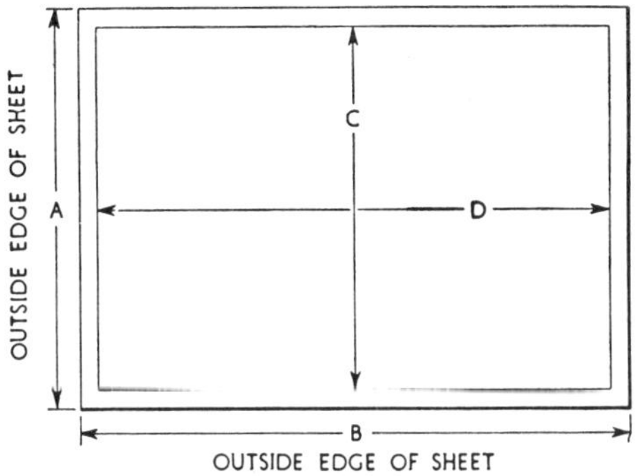

Nominal sizes of drawings and tracings. Overall dimensions across edges of sheet (see diagram above)				*Maximum frame size. Dimensions between border lines (see diagram above)*			
A		B		C		D	
in.	mm.	in.	mm.	in.	mm	in.	mm.
40	1016	72	1829	38	965	70	1778
40	1016	60	1524	38	965	58	1473
30	762	53	1346	$28\frac{1}{2}$	724	$51\frac{1}{2}$	1308
30	762	40	1016	$28\frac{1}{2}$	724	$38\frac{1}{2}$	978
20	508	30	762	19	483	29	737
15	381	20	508	14	356	19	483
10	254	15	381	9	229	14	356
8	203	13	330	$7\frac{1}{4}$	184	$12\frac{1}{4}$	311
8	203	10	254	$7\frac{1}{4}$	184	$9\frac{1}{4}$	239

SIZES OF DRAWING SHEETS OF INTERNATIONAL 'A' SERIES

mm.	in.	Designation
841 × 1189	33·11 × 46·81	A0
594 × 841	23·39 × 33·11	A1
420 × 594	16·54 × 23·39	A2
297 × 420	11·69 × 16·54	A3
210 × 297	8·27 × 11·69	A4

B.S. 308: 1964 ENGINEERING DRAWING PRACTICE
COLOURING TO REPRESENT MATERIALS

Where it is desired to indicate materials in section by colours, the following are recommended:

Material	*Colour*
Cast iron	Payne's grey
Wrought iron	Prussian blue
Steel	Purple
Brass, phosphor bronze and gunmetal ...	Light yellow
Copper	Crimson lake
Aluminium, tin, white metals and light alloys	Light green
Brickwork	Vermilion
Concrete	Light green
Earth, rock	Sepia
Timber	Burnt sienna
Glass	Pale blue wash
Insulation (electrical)	Black

ALTERNATIVE METHODS FOR SHOWING SCREW THREADS

EXTERNAL THREADS

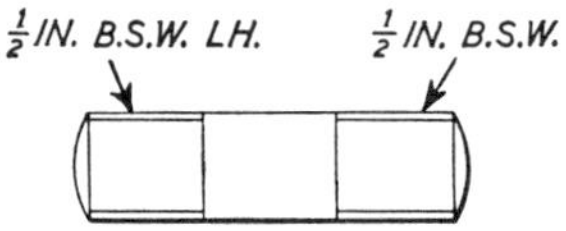

EXTERNAL THREADS

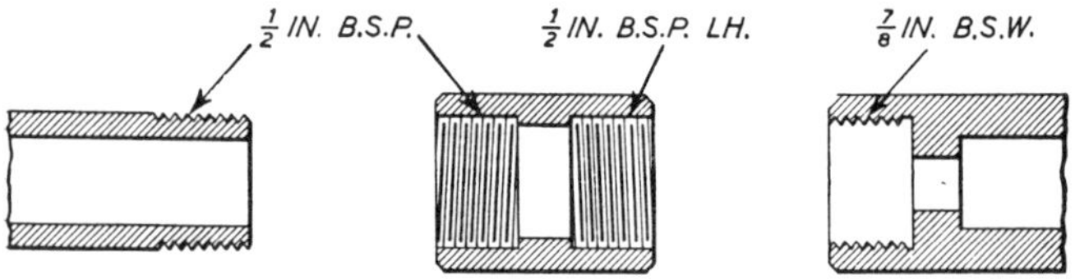

EXTERNAL THREAD (SECTION)

INTERNAL THREADS (SECTION)

FOR REPRESENTING PIPE JUNCTIONS, VALVES AND FITTINGS

- PIPING IN GENERAL
- NON-INTERSECTING PIPES
- STEAM PIPES
- CONDENSATE PIPES
- COLD WATER PIPES
- HOT WATER PIPES
- AIR PIPES
- VACUUM PIPES
- GAS PIPES
- REFRIGERANT PIPES
- OIL PIPES
- LOCK & SHIELD VALVE
- REDUCING VALVE
- DIAPHRAGM VALVE
- THERMOSTAT
- RADIATOR TRAP, PLAN
- RADIATOR TRAP, ELEVATION
- GATE VALVE
- GLOBE VALVE
- ANGLE GLOBE VALVE
- CHECK VALVE
- STOP COCK
- SAFETY VALVE
- QUICK OPENING VALVE
- FLOAT OPERATING VALVE
- MOTOR OPERATED VALVE
- EXPANSION JOINT
- REDUCING FLANGE

- WALL RADIATOR, PLAN
- WALL RADIATOR, ELEVATION
- INDIRECT RADIATOR, PLAN
- INDIRECT RADIATOR, ELEVATION
- SUPPLY DUCT, SECTION
- EXHAUST DUCT, SECTION
- BUTTERFLY DAMPER
- VANES
- AIR SUPPLY OUTLET
- EXHAUST INLET
- JOINT, FLANGES
- JOINT, SCREWED
- JOINT, WELDED
- ELBOW 90°, FLANGED
- ELBOW, TURNED UP
- ELBOW, TURNED DOWN
- ELBOW, LONG RADIUS
- BASE ELBOW
- DOUBLE BRANCH ELBOW
- SINGLE SWEEP TEE
- TEE
- REDUCING ELBOW
- CROSS
- REDUCER
- ECCENTRIC REDUCER

TEMPERATURE CONVERSION TABLE

Degrees Fahrenheit to Degrees Centigrade

(Figures in italics represent negative values on the Centigrade Scale)

Degrees F.	**0**	**1**	**2**	**3**	**4**	**5**	**6**	**7**	**8**	**9**
	°C.	**°C.**	**°C.**	**°C.**	**°C.**	**°C.**	**°C.**	**°C.**	**°C.**	**°C.**
0	*17·8*	*17·2*	*16·7*	*16·1*	*15·6*	*15·0*	*14·4*	*13·9*	*13·3*	*12·8*
10	*12·2*	*11·7*	*11·1*	*10·6*	*10·0*	*9·4*	*8·9*	*8·3*	*7·8*	*7·2*
20	*6·7*	*6·1*	*5·6*	*5·0*	*4·4*	*3·9*	*3·3*	*2·8*	*2·2*	*1·7*
30	*1·1*	*0·6*	—	—	—	—	—	—	—	—
	0	**1**	**2**	**3**	**4**	**5**	**6**	**7**	**8**	**9**
30	—	0	0	0·6	1·1	1·7	2·2	2·8	3·3	3·9
40	4·4	5·0	5·6	6·1	6·7	7·2	7·8	8·3	8·9	9·4
50	10·0	10·6	11·1	11·7	12·2	12·8	13·3	13·9	14·4	15·0
60	15·6	16·1	16·7	17·2	17·8	18·3	18·9	19·4	20·0	20·6
70	21·1	21·7	22·2	22·8	23·3	23·9	24·4	25·0	25·6	26·1
80	26·7	27·2	27·8	28·3	28·9	29·4	30·0	30·6	31·1	31·7
90	32·2	32·8	33·3	33·9	34·4	35·0	35·6	36·1	36·7	37·2
100	37·8	38·3	38·9	39·4	40·0	40·6	41·1	42·7	42·2	42·8
110	43·3	43·9	44·4	45·0	45·6	46·1	46·7	47·2	47·8	48·3
120	48·9	49·4	50·0	50·6	51·1	51·7	52·2	52·8	53·3	53·9
130	54·4	55·0	55·6	56·1	56·7	57·2	57·8	58·3	58·9	59·4
140	60·0	60·6	61·1	61·7	62·2	62·8	63·3	63·9	64·4	65·0
150	65·6	66·1	66·7	67·2	67·8	68·3	68·9	69·4	70·0	70·6
160	71·1	71·7	72·2	72·8	73·3	73·9	74·4	75·0	75·6	76·1
170	76·7	77·2	77·8	78·3	78·9	79·4	80·0	80·6	81·1	81·7
180	82·2	82·8	83·3	83·9	84·4	85·0	85·6	86·1	86·7	87·2
190	87·8	88·3	88·9	89·4	90·0	90·6	91·1	91·7	92·2	92·8
200	93·3	93·9	94·4	95·0	95·6	96·1	96·7	97·2	97·8	98·3
210	98·9	99·4	100·0	100·6	101·1	101·7	102·2	102·8	103·3	103·9
220	104·4	105·0	105·6	106·1	106·7	107·2	107·8	108·3	108·9	109·4
230	110·0	110·6	111·1	111·7	112·2	112·8	113·3	113·9	114·4	115·0
240	115·6	116·1	116·7	117·2	117·8	118·3	118·9	119·4	120·0	120·6
250	121·1	121·7	122·2	122·8	123·3	123·9	124·4	125·0	125·6	126·1

$$F = (C \times 1{\cdot}8) + 32$$

TEMPERATURE CONVERSION TABLE

Degrees Fahrenheit to Degrees Centigrade—*continued*

Degrees F.	0	1	2	3	4	5	6	7	8	9
	°C.	°C.	°C.	°C.	°C.	°C.	°C.	°C.	°C.	°C.
260	126·7	127·2	127·8	128·3	128·9	129·4	130·0	130·6	131·1	131·7
270	132·2	132·8	133·3	133·9	134·4	135·0	135·6	136·1	136·7	137·2
280	137·8	138·3	138·9	139·4	140·0	140·6	141·1	141·7	142·2	142·8
290	143·3	143·9	144·5	145·0	145·6	146·1	146·7	147·2	147·8	148·3
300	148·9	149·4	150·0	150·6	151·1	151·7	152·2	152·8	153·3	153·9
310	154·4	155·0	155·6	156·1	156·7	157·2	157·8	158·3	158·9	159·4
320	160·0	160·6	161·1	161·7	162·2	162·8	163·3	163·9	164·4	165·0
330	165·6	166·1	166·7	167·2	167·8	168·3	168·9	169·4	170·0	170·6
340	171·1	171·7	172·2	172·8	173·2	173·9	174·4	175·0	175·6	176·1
350	176·7	177·2	177·8	178·3	178·9	179·4	180·0	180·6	181·1	181·7
360	182·2	182·8	183·3	183·9	184·4	185·0	185·6	186·1	186·7	187·2
370	187·8	188·3	188·9	189·4	190·0	190·6	191·1	191·7	192·2	192·8
380	193·3	193·9	194·4	195·0	195·6	196·1	196·7	197·2	197·8	198·3
390	198·9	199·4	200·0	200·6	201·1	201·7	202·2	202·8	203·3	203·9
400	204·4	205·0	205·6	206·1	206·7	207·2	207·8	208·3	208·9	209·4
410	210·0	210·6	211·1	211·7	212·2	212·8	213·3	213·9	214·4	215·0
420	215·6	216·1	216·7	217·2	217·8	218·3	218·9	219·4	220·0	220·6
430	221·1	221·7	222·2	222·8	223·3	223·9	224·4	225·0	225·6	226·1
440	226·7	227·2	227·8	228·3	228·9	229·4	230·0	230·6	231·1	231·7
450	232·2	232·8	233·3	233·9	234·4	235·0	235·6	236·1	236·7	237·2
460	237·8	238·3	238·9	239·4	240·0	240·6	241·1	241·7	242·2	242·8
470	243·3	243·9	244·4	245·0	245·6	246·1	246·7	247·2	247·8	248·3
480	248·9	249·4	250·0	250·6	251·1	251·7	252·2	252·8	253·3	253·9
490	254·4	255·0	255·6	256·1	256·7	257·2	257·8	258·3	258·9	259·4
500	260·0	—	—	—	—	—	—	—	—	—

$$F = (C \times 1{\cdot}8) + 32$$

Degrees Centigrade to Degrees Fahrenheit

Degrees C.	**0**	**1**	**2**	**3**	**4**	**5**	**6**	**7**	**8**	**9**
	°F.	°F.	°F.	°F.	°F.	°F.	°F.	°F.	°F.	°F.
0	32·0	33·8	35·6	37·4	39·2	41·0	42·8	44·6	46·4	48·2
10	50·0	51·8	53·6	55·4	57·2	59·0	60·8	62·6	64·4	66·2
20	68·0	69·8	71·6	73·4	75·2	77·0	78·8	80·6	82·4	84·2
30	86·0	87·8	89·6	91·4	93·2	95·0	96·8	98·6	101·4	102·2
40	104·0	105·8	107·6	109·4	111·2	113·0	114·8	116·6	118·4	120·2
50	122·0	123·8	125·6	127·4	129·2	131·0	132·8	134·6	136·4	138·2
60	140·0	141·8	143·6	145·4	147·2	149·0	150·8	152·6	154·4	156·2
70	158·0	159·8	161·6	163·4	165·2	167·0	168·8	170·6	172·4	174·2
80	176·0	177·8	179·6	181·4	183·2	185·0	186·8	188·6	190·4	192·2
90	194·0	195·8	197·6	199·4	201·2	203·0	204·8	206·6	208·4	210·2
100	212·0	213·8	215·6	217·4	219·2	221·0	222·8	224·6	226·4	228·2
110	230·0	231·8	233·6	235·4	237·2	239·0	240·8	242·6	244·4	246·2
120	248·0	249·8	251·6	253·4	255·2	257·0	258·8	260·6	262·4	264·2
130	266·0	267·8	269·6	271·4	273·2	275·0	276·8	278·6	280·4	282·2
140	284·0	285·8	287·6	289·4	291·2	293·0	294·8	296·6	298·4	300·2
150	302·0	303·8	305·6	307·4	309·2	311·0	312·8	314·6	316·4	318·2
160	320·0	321·8	323·6	325·4	327·2	329·0	330·8	332·6	334·4	336·2
170	338·0	339·8	341·6	343·4	345·2	347·0	348·8	350·6	352·4	354·2
180	356·0	357·8	359·6	361·4	363·2	365·0	366·8	368·6	370·4	372·2
190	374·0	375·8	377·6	379·4	381·2	383·0	384·8	386·6	388·4	390·2
200	392·0	393·8	395·6	397·4	399·2	401·0	402·8	404·6	406·4	408·2
210	410·0	411·8	413·6	415·4	417·2	419·0	420·8	422·6	424·4	426·2
220	428·0	429·8	431·6	433·4	435·2	437·0	438·8	440·6	442·4	444·2
230	446·0	447·8	449·6	451·4	453·2	455·0	456·8	458·6	460·4	462·2
240	464·0	465·8	467·6	469·4	471·2	473·0	474·8	476·6	478·4	480·2
250	482·0	483·8	485·6	487·4	489·2	491·0	492·8	494·6	496·4	498·2
260	500·0	501·8	503·6	505·4	507·2	509·0	510·8	512·6	514·4	516·2
270	518·0	519·8	521·6	523·4	525·2	527·0	528·8	530·6	532·4	534·2
280	536·0	537·8	539·6	541·4	543·2	545·0	546·8	548·6	550·4	552·2
290	554·0	555·8	557·6	559·4	561·2	563·0	563·8	566·6	568·4	570·2
300	572·0	573·8	575·6	577·4	579·2	581·0	582·8	584·6	586·4	588·2

$$C=(F-32)\div 1{\cdot}8$$

Fractions of an Inch.

With decimal and metric equivalents.

Fraction	*Decimal*	*Millimetres*	*Fraction*	*Decimal*	*Millimetres*
1/64	0·015625	0·397	33/64	0·515625	13·097
1/32	0·03125	0·794	17/32	0·53125	13·494
3/64	0·046875	1·191	35/64	0·546875	13·891
1/16	0·0625	1·587	9/16	0·5625	14·287
5/64	0·078125	1·984	37/64	0·578125	14·684
3/32	0·09375	2·381	19/32	0·59375	15·081
7/64	0·109375	2·778	39/64	0·609375	15·478
⅛	0·125	3·175	⅝	0·625	15·874
9/64	0·140625	3·572	41/64	0·640625	16·272
5/32	0·15625	3·969	21/32	0·65625	16·669
11/64	0·171875	4·366	43/64	0·671875	17·066
3/16	0·1875	4·762	11/16	0·6875	17·462
13/64	0·203125	5·160	45/64	0·703125	17·859
7/32	0·21875	5·556	23/32	0·71875	18·256
15/64	0·234375	5·953	47/64	0·734375	18·653
¼	0·25	6·349	¾	0·75	19·049
17/64	0·265625	6·747	49/64	0·765625	19·477
9/32	0·28125	7·144	25/32	0·78125	19·844
19/64	0·296875	7·541	51/64	0·796875	20·241
5/16	0·3125	7·937	13/16	0·8125	20·637
21/64	0·328125	8·333	53/64	0·828125	21·034
11/32	0·34375	8·731	27/32	0·84375	21·431
23/64	0·359375	9·128	55/64	0·859375	21·828
⅜	0·375	9·524	⅞	0·875	22·224
25/64	0·390625	9·922	57/64	0·890625	22·622
13/32	0·40625	10·319	29/32	0·90625	23·019
27/64	0·421875	10·716	59/64	0·921875	23·416
7/16	0·4375	11·112	15/16	0·9375	23·812
29/64	0·453125	11·509	61/64	0·953125	24·209
15/32	0·46875	11·906	31/32	0·96875	24·606
31/64	0·484375	12·303	63/64	0·984375	25·003
½	0·50	12·699	1	1·00	25·400

Feet and Inches to Metres

Feet	Inches 0	1	2	3	4	5	6	7	8	9	10	11
	m	m	m	m	m	m	m	m	m	m	m	m
0	—	0·0254	0·0508	0·0762	0·1016	0·1270	0·1524	0·1778	0·2032	0·2286	0·2540	0·2794
1	0·3048	0·3302	0·3556	0·3810	0·4064	0·4318	0·4572	0·4826	0·5080	0·5334	0·5588	0·5842
2	0·6096	0·6350	0·6604	0·6858	0·7112	0·7366	0·7620	0·7874	0·8128	0·8382	0·8636	0·8890
3	0·9144	0·9398	0·9652	0·9906	1·0160	1·0414	1·0668	1·0922	1·1176	1·1430	1·1684	1·1938
4	1·2192	1·2446	1·2700	1·2954	1·3208	1·3462	1·3716	1·3970	1·4224	1·4478	1·4732	1·4986
5	1·5240	1·5494	1·5748	1·6002	1·6256	1·6510	1·6764	1·7018	1·7272	1·7526	1·7780	1·8034
6	1·8288	1·8542	1·8796	1·9050	1·9304	1·9558	1·9812	2·0066	2·0320	2·0574	2·0828	2·1082
7	2·1336	2·1590	2·1844	2·2098	2·2352	2·2606	2·2860	2·3114	2·3368	2·3622	2·3876	2·4130
8	2·4384	2·4638	2·4892	2·5146	2·5400	2·5654	2·5908	2·6162	2·6416	2·6670	2·6924	2·7178
9	2·7432	2·7686	2·7940	2·8194	2·8448	2·8702	2·8956	2·9210	2·9464	2·9718	2·9972	3·0226
10	3·0480	3·0734	3·0988	3·1242	3·1496	3·1750	3·2004	3·2258	3·2512	3·2766	3·3020	3·3274
11	3·3528	3·3782	3·4036	3·4290	3·4544	3·4798	3·5052	3·5306	3·5560	3·5814	3·6068	3·6322
12	3·6576	3·6830	3·7084	3·7338	3·7592	3·7846	3·8100	3·8354	3·8608	3·8862	3·9116	3·9370
13	3·9624	3·9878	4·0132	4·0386	4·0640	4·0894	4·1148	4·1402	4·1656	4·1910	4·2164	4·2418
14	4·2672	4·2926	4·3180	4·3434	4·3688	4·3942	4·4196	4·4450	4·4704	4·4958	4·5212	4·5466
15	4·5720	4·5974	4·6228	4·6482	4·6736	4·6990	4·7244	4·7498	4·7752	4·8006	4·8260	4·8514
16	4·8768	4·9022	4·9276	4·9530	4·9784	5·0038	5·0292	5·0546	5·0800	5·1054	5·1308	5·1562
17	5·1816	5·2070	5·2324	5·2578	5·2832	5·3086	5·3340	5·3594	5·3848	5·4102	5·4356	5·4610
18	5·4864	5·5118	5·5372	5·5626	5·5880	5·6134	5·6388	5·6642	5·6896	5·7150	5·7404	5·7658
19	5·7912	5·8166	5·8420	5·8674	5·8928	5·9182	5·9436	5·9690	5·9944	6·0198	6·0452	6·0706
20	6·0960	6·1214	6·1468	6·1722	6·1976	6·2230	6·2484	6·2738	6·2992	6·3246	6·3500	6·3754

30	9·1440	9·1694	9·1948	9·2202	9·2456	9·2710	9·2964	9·3218	9·3472	9·3726	9·3980	9·4234
40	12·1920	12·2174	12·2428	12·2682	12·2936	12·3190	12·3444	12·3698	12·3952	12·4206	12·4460	12·4714
50	15·2400	15·2654	15·2908	15·3162	15·3416	15·3670	15·3924	15·4178	15·4432	15·4686	15·4940	15·5194
60	18·2880	18·3134	18·3388	18·3642	18·3896	18·4150	18·4404	18·4658	18·4912	18·5166	18·5420	18·5674
70	21·3360	21·3614	21·3868	21·4122	21·4376	21·4630	21·4884	21·5138	21·5392	21·5646	21·5900	21·6154
80	24·3840	24·4094	24·4348	24·4602	24·4856	24·5110	24·5364	24·5618	24·5872	24·6126	24·6380	24·6634
90	27·4320	27·4574	27·4828	27·5082	27·5336	27·5590	27·5844	27·6098	27·6352	27·6606	27·6860	27·7114
100	30·4800											

Metres to Feet

Metres	*0*	*1*	*2*	*3*	*4*	*5*	*6*	*7*	*8*	*9*
0	—	3·281	6·562	9·843	13·123	16·404	19·685	22·966	26·247	29·528
10	32·808	36·089	39·370	42·651	45·932	49·213	52·493	55·774	59·055	62·336
20	65·617	68·898	72·179	74·459	78·740	82·021	85·302	88·583	91·864	95·144
30	98·425	101·706	104·987	108·268	111·549	114·829	118·110	121·391	124·672	127·953
40	131·234	134·515	137·795	141·076	144·357	147·638	150·919	154·200	157·480	160·761
50	164·042	167·323	170·604	173·885	177·166	180·446	183·727	187·008	190·289	193·570
60	196·851	200·131	203·412	206·693	209·974	213·255	216·536	219·816	223·097	226·378
70	229·659	232·940	236·221	239·502	242·782	246·063	249·344	252·625	255·906	259·187
80	262·467	265·748	269·029	272·310	275·591	278·872	282·152	285·433	288·714	291·995
90	295·276	298·557	301·838	305·118	308·399	311·680	314·961	318·242	321·523	324·803
100	328·08									

Cubic Feet to Cubic Metres

cu. ft.	0	1	2	3	4	5	6	7	8	9
	cu. m	cu. m	cu. m	cu. m	cu. m	cu. m	cu. m	cu. m	cu. m	cu. m
0	—	0·0283	0·0566	0·0850	0·1133	0·1416	0·1699	0·1982	0·2265	0·2549
10	0·2832	0·3115	0·3398	0·3681	0·3964	0·4248	0·4531	0·4814	0·5097	0·5380
20	0·5663	0·5947	0·6230	0·6583	0·6796	0·7079	0·7362	0·7646	0·7929	0·8212
30	0·8495	0·8778	0·9061	0·9345	0·9628	0·9911	1·0194	1·0477	1·0760	1·1044
40	1·1327	1·1610	1·1893	1·2176	1·2459	1·2743	1·3026	1·3369	1·3592	1·3875
50	1·4158	1·4442	1·4725	1·5008	1·5291	1·5574	1·5857	1·6141	1·6424	1·6707
60	1·6990	1·7273	1·7556	1·7840	1·8123	1·8406	1·8689	1·8972	1·9255	1·9539
70	1·9822	2·0105	2·0388	2·0671	2·0954	2·1238	2·1521	2·1804	2·2087	2·2370
80	2·2653	2·2937	2·3220	2·3503	2·3786	2·4069	2·4352	2·4636	2·4919	2·5202
90	2·5485	2·5768	2·6051	2·6335	2·6618	2·6901	2·7184	2·7467	2·7750	2·8034
100	2·8317	—	—	—	—	—	—	—	—	—

Cubic Metres to Cubic Feet

cu. m	0	1	2	3	4	5	6	7	8	9
	cu. ft.	cu. ft.	cu. ft.	cu. ft.	cu. ft.	cu. ft.	cu. ft.	cu. ft.	cu. ft.	cu. ft.
0	—	35·3148	70·6295	105·9443	141·2590	176·5738	211·8885	247·2033	282·5181	317·8328
10	353·1476	388·4623	423·7771	459·0918	494·4066	529·7214	565·0361	600·3509	635·6656	670·9804
20	706·2951	741·6099	776·9247	812·2394	847·5542	882·8689	918·1837	953·4984	988·8132	1024·1280
30	1059·4427	1094·7575	1130·0722	1165·3870	1200·7017	236·0165	1271·3313	1306·6460	1341·9608	1377·2755
40	1412·5903	1447·9050	1483·2198	1518·5346	1553·8493	1589·1641	1624·4788	1659·7936	1695·1083	1730·4231
50	1765·7379	1801·0526	1836·3674	1871·6821	1906·9969	1942·3116	1977·6264	2012·9411	2048·2559	2083·5707
60	2118·8854	2154·2002	2189·5149	2224·8297	2260·1444	2295·4592	2330·7740	2366·0887	2401·4035	2436·7182
70	2472·0330	2507·3477	2542·6625	2577·9773	2613·2920	2648·6068	2683·9215	2719·2363	2754·5510	2789·8658
80	2825·1806	2860·4953	2895·8101	2931·1248	2966·4396	3001·7543	3037·0691	3072·3839	3107·6986	3143·0134
90	3178·3281	3213·6429	3248·9576	3284·2724	3319·5872	3354·9019	3390·2167	3425·5314	3460·8462	3496·1609
100	3531·47	—	—	—	—	—	—	—	—	—

Gallons to Litres [1 litre = $10^{-3} m^3$]

gal.	0	1	2	3	4	5	6	7	8	9
	l.	l.	l.	l.	l.	l.	l.	l.	l.	l.
0	—	4·546	9·092	13·638	18·184	22·730	27·276	31·822	36·368	40·914
10	45·460	50·006	54·552	59·098	63·643	68·189	72·735	77·281	81·827	86·373
20	90·919	95·465	100·011	104·557	109·103	113·649	118·195	122·741	127·287	131·833
30	136·379	140·925	145·471	150·017	154·563	159·109	163·655	168·201	172·747	177·293
40	181·839	186·384	190·930	195·476	200·022	204·568	209·114	213·660	218·206	222·752
50	227·298	231·844	236·390	240·936	245·482	250·028	254·574	259·120	263·666	268·212
60	272·758	277·304	281·850	286·396	290·942	295·488	300·034	304·580	309·125	313·671
70	318·217	322·763	327·309	331·855	336·401	340·947	345·493	350·039	354·585	359·131
80	363·677	368·223	372·769	377·315	381·861	386·407	390·953	395·499	400·045	404·591
90	409·137	413·683	418·229	422·775	427·321	431·866	436·412	440·958	445·504	450·050
100	454·596	—	—	—	—	—	—	—	—	—

Litres to Gallons [1 litre = $10^{-3} m^3$]

litres	0	1	2	3	4	5	6	7	8	9
	gal.	gal.	gal.	gal.	gal.	gal.	gal.	gal.	gal.	gal.
0	—	0·2200	0·4400	0·6600	0·8800	1·1000	1·3199	1·5398	1·7598	1·9798
10	2·1998	2·4197	2·6397	2·8597	3·0797	3·2996	3·5196	3·7396	3·9596	4·1795
20	4·3995	4·6195	4·8395	5·0594	5·2794	5·4994	5·7194	5·9393	6·1593	6·3793
30	6·5993	6·8192	7·0392	7·2592	7·4792	7·6991	7·9191	8·1391	8·3591	8·5790
40	8·7990	9·0190	9·2390	9·4589	9·6789	9·8989	10·1189	10·3388	10·5588	10·7788
50	10·9988	11·2187	11·4387	11·6587	11·8787	12·0986	12·3186	12·5386	12·7586	12·9785
60	13·1985	13·4185	13·6385	13·8584	14·0784	14·2984	14·5184	14·7384	14·9583	15·1783
70	15·3983	15·6183	15·8382	16·0582	16·2782	16·4982	16·7181	16·9381	17·1581	17·3781
80	17·5980	17·8180	18·0380	18·2580	18·4779	18·6979	18·9179	19·1379	19·3578	19·5778
90	19·7978	20·0178	20·2377	20·4577	20·6777	20·8977	21·1176	21·3376	21·5576	21·7776
100	21·9975	—	—	—	—	—	—	—	—	—

Pounds to Kilogrammes

lb.	0	1	2	3	4	5	6	7	8	9
	kg.	kg.	kg.	kg.	kg.	kg.	kg.	kg.	kg.	kg.
0	—	0·4535	0·9071	1·3607	1·8143	2·2679	2·7215	3·1751	3·6287	4·0823
10	4·5359	4·9895	5·4431	5·8967	6·3503	6·8039	7·2575	7·7111	8·1647	8·6183
20	9·0718	9·5254	9·9790	10·4326	10·8862	11·3398	11·7934	12·2470	12·7006	13·1542
30	13·6078	14·0614	14·5150	14·9686	15·4221	15·8757	16·3293	16·7829	17·2365	17·6901
40	18·1437	18·5973	19·0509	19·5045	19·9581	20·4117	20·8653	21·3188	21·7724	22·2260
50	22·6796	23·1332	23·5868	24·0404	24·4940	24·9476	25·4012	25·8548	26·3084	26·7620
60	27·2155	27·6691	28·1227	28·5763	29·0299	29·4835	29·9371	30·3907	30·8443	31·2979
70	31·7515	32·2051	32·6587	33·1122	33·5658	34·0194	34·4730	34·9266	35·3802	35·8338
80	36·2874	36·7410	37·1946	37·6482	38·1018	38·5554	39·0089	39·4625	39.9161	40·3697
90	40·8233	41·2769	41·7305	42·1841	42·6377	43·0913	43·5449	43·9985	44·4521	44·9057
100	45·3592	—	—	—	—	—	—	—	—	—

Kilogrammes to Pounds

kg.	0	1	2	3	4	5	6	7	8	9
	lb.	lb.	lb.	lb.	lb.	lb.	lb.	lb.	lb.	lb.
0	—	2·204	4·409	6·613	8·818	11·023	13·227	15·432	17·637	19·841
10	22·0462	24·250	26·455	28·660	30·864	33·069	35·273	37·478	39·683	41·887
20	44·0924	46·297	48·502	50·706	52·911	55·116	57·320	59·525	61·729	63·934
30	66·139	68·343	70·548	72·753	74·957	77·162	79·366	81·571	83·776	85·980
40	88·185	90·389	92·594	94·799	97·003	99·208	101·413	103·617	105·822	108·026
50	110·231	112·436	114·640	116·845	119·050	121·254	123·459	125·663	127·868	130·073
60	132·277	134·482	136·686	138·891	141·096	143·300	145·505	147·710	149·914	152·119
70	154·324	156·528	158·733	160·937	163·142	165·347	167·551	169·756	171·960	174·165
80	176·370	178·574	180·779	182·984	185·188	187·393	189·597	191·802	194·007	196·211
90	198·416	200·620	202·825	205·030	207·234	209·439	211·644	213·848	216·053	218·258
100	220·462	—	—	—	—	—	—	—	—	—

Pounds per Square Inch to Kilonewtons per Square Metre

lb./sq.in.	*0*	*1*	*2*	*3*	*4*	*5*	*6*	*7*	*8*	*9*
0	—	6·895	13·79	20·68	27·58	34·47	41·37	48·25	55·16	62·04
10	68·95	75·85	82·74	89·63	96·53	103·4	110·3	117·2	124·1	131·0
20	137·9	144·8	151·7	158·6	165·5	172·4	179·3	186·2	193·1	199·9
30	206·8	213·7	220·6	227·5	234·4	241·3	248·2	255·1	262·0	268·8
40	275·8	282·7	289·6	296·5	303·4	310·3	317·2	324·1	331·0	337·8
50	344·7	351·6	358·5	365·4	372·3	379·2	386·1	353·0	399·9	406·7
60	413·7	420·6	427·5	434·4	441·3	448·2	455·1	462·0	468·9	475·7
70	505·4	512·3	519·2	526·1	533·0	539·9	546·8	553·7	560·6	567·4
80	551·6	558·5	565·4	572·3	579·0	586·1	593·0	599·9	606·8	613·6
90	620·4	627·3	634·2	641·1	648·0	654·9	661·8	668·7	675·6	682·4
100	689·5	—	—	—	—	—	—	—	—	—

Kilonewtons per Square Metre to Pounds per Square Inch

kN/m.²	*0*	*1*	*2*	*3*	*4*	*5*	*6*	*7*	*8*	*9*
0	—	0·145	0·290	0·435	0·580	0·725	0·871	1·02	1·16	1·31
10	1·45	1·60	1·74	1·89	2·03	2·18	2·32	2·47	2·61	2·76
20	2·90	3·05	3·19	3·34	3·48	3·63	3·77	3·92	4·06	4·21
30	4·35	4·50	4·64	4·79	4·93	5·08	5·22	5·37	5·52	5·66
40	5·80	5·94	6·09	6·23	6·38	6·52	6·67	6·81	6·96	7·10
50	7·25	7·40	7·54	7·69	7·83	7·98	8·12	8·27	8·41	8·56
60	8·71	8·85	9·00	9·14	9·23	9·43	9·58	9·72	9·87	10·01
70	10·15	10·30	10·44	10·59	10·73	10·88	11·02	11·17	11·31	11·46
80	11·61	11·76	11·90	12·05	12·19	12·33	12·48	12·63	12·27	12·92
90	13·05	13·20	13·34	13·49	13·63	13·78	13·92	14·07	14·21	14·36
100	14·51	—	—	—	—	—	—	—	—	—

British Thermal Units to Kilojoules

B.t.u.	*0*	*1*	*2*	*3*	*4*	*5*	*6*	*7*	*8*	*9*
0	—	1·05	2·11	3·17	4·22	5·28	6·33	7·39	8·44	9·50
10	10·55	11·61	12·66	13·72	14·77	15·83	16·88	17·94	18·99	20·05
20	21·10	22·16	23·21	24·27	25·32	26·38	27·43	28·49	29·54	30·60
30	31·65	32·71	33·76	34·82	35·87	36·93	37·98	39·03	40·09	41·15
40	42·21	43·26	44·31	45·37	46·42	47·48	48·53	49·59	50·64	51·70
50	52·76	53·81	54·86	55·92	56·97	58·03	59·08	60·14	61·19	62·25
60	63·31	64·36	65·41	66·46	67·52	68·58	69·63	70·69	71·74	72·80
70	73·86	74·91	75·96	77·02	78·07	79·13	80·18	81·24	82·29	83·35
80	84·41	85·46	86·51	87·57	88·62	89·68	90·73	91·79	92·84	93·90
90	94·95	96·01	97·06	98·12	99·17	100·23	101·28	102·34	103·40	104·45
100	105·5	—	—	—	—	—	—	—	—	—

Kilojoules to British Thermal Units

kJ	*0*	*1*	*2*	*3*	*4*	*5*	*6*	*7*	*8*	*9*
0	—	0·95	1·90	2·84	3·79	4·74	5·69	6·63	7·58	8·53
10	9·48	10·42	11·37	12·32	13·27	14·22	15·16	16·11	17·06	18·01
20	18·95	19·90	20·85	21·80	22·74	23·69	24·64	25·59	26·54	27·48
30	28·43	29·38	30·33	31·27	32·22	33·17	34·12	35·06	36·01	36·96
40	37·91	38·86	39·80	40·75	41·70	42·65	43·59	44·54	45·49	46·44
50	47·39	48·33	49·28	50·23	51·18	52·12	53·07	54·02	54·97	55·91
60	56·86	57·81	58·76	59·71	60·65	61·60	62·55	63·50	64·44	65·39
70	66·34	67·29	68·23	69·18	70·13	71·08	72·03	72·97	73·92	74·87
80	75·82	76·76	77·71	78·66	79·61	80·55	81·50	82·45	83·40	84·35
90	85·29	86·24	87·19	88·14	89·08	90·03	90·98	91·93	92·87	93·82
100	94·77	—	—	—	—	—	—	—	—	—

British Thermal Units per Square Foot to Kilojoules per Square Metre

B.t.u./sq. ft.	0	1	2	3	4	5	6	7	8	9
	kJ/m.²	kJ/m.²	kJ/m.²	kJ/m.²	kJ/m.²	kJ/m.²	kJ/m.²	kJ/m.²	kJ/m.²	kJ/m.²
0	—	11·36	22·72	34·08	45·44	56·80	68·16	75·52	90·88	102·2
10	113·6	125·0	136·3	147·7	159·0	170·4	181·8	193·1	204·5	215·8
20	227·2	238·6	245·0	261·3	272·6	284·0	295·4	306·7	318·1	329·4
30	340·8	352·2	363·5	374·9	386·2	297·6	409·0	420·3	431·7	443·0
40	454·4	465·8	477·1	488·5	499·8	511·2	522·6	533·9	545·3	556·6
50	568·0	579·4	590·7	602·1	613·4	624·8	636·1	647·5	658·9	670·2
60	681·6	693·0	704·3	715·7	727·0	738·4	749·8	761·1	772·5	783·8
70	795·2	806·6	817·9	829·3	840·6	852·0	863·4	874·7	886·1	897·4
80	908·8	920·2	931·5	942·9	954·2	965·6	977·0	988·3	999·7	1011·0
90	1022·4	1033·8	1045·1	1056·5	1067·8	1079·2	1090·6	1101·9	1113·3	1124·6
100	1136·0	—	—	—	—	—	—	—	—	—

Kilojoules per Square Metre to British Thermal Units per Square Foot

kJ/m.²	0	1	2	3	4	5	6	7	8	9
	B.t.u./sq. ft.	B.t.u./sq. ft.	B.t.u./sq. ft.	B.t.u./sq. ft.	B.t.u./sq. ft.	B.t.u./sq. ft.	B.t.u./sq. ft.	B.t.u./sq. ft.	B.t.u./sq. ft.	B.t.u./sq. ft.
0	—	0·088	0·176	0·264	0·352	0·440	0·528	0·616	0·704	0·792
10	0·880	0·968	1·056	1·144	1·232	1·320	1·408	1·496	1·584	1·672
20	1·760	1·848	1·936	2·024	2·113	2·201	2·289	2·377	2·465	2·553
30	2·641	2·729	2·817	2·905	2·993	3·081	3·169	3·257	3·345	3·433
40	3·521	3·609	3·697	3·785	3·873	3·961	4·049	4·137	4·225	4·313
50	4·401	4·489	4·577	4·665	4·753	4·841	4·929	5·017	5·105	5·193
60	5·281	5·369	5·457	5·545	5·633	5·721	5·809	5·897	5·986	6·073
70	6·161	6·249	6·337	6·425	6·514	6·602	6·690	6·778	6·866	6·954
80	7·042	7·130	7·218	7·306	7·394	7·482	7·570	7·658	7·746	7·834
90	7·922	8·010	8·098	8·186	8·274	8·362	8·450	8·538	8·626	8·714
100	8·802	—	—	—	—	—	—	—	—	—

B.t.u/ft^2 °F to kJ/m^2 °C

B.t.u. / sq. ft. °F.	0	1	2	3	4	5	6	7	8	9
	kJ/m.^{2} °C.	kJ/m.^{2} °C.	kJ/m.^{2} °C.	kJ/m.^{2} °C.	kJ/m.^{2} °C.	kJ/m.^{2} °C.	kJ/m.^{2} °C.	kJ/m.^{2} °C.	kJ/m.^{2} °C.	kJ/m.^{2} °C.
0	—	20·44	40·88	61·32	81·76	102·21	122·65	143·09	163·53	183·97
10	204·41	224·85	245·29	265·73	286·17	306·62	327·06	347·50	367·94	388·38
20	408·82	429·26	449·70	470·14	490·58	511·03	531·47	551·91	572·35	592·79
30	613·23	633·67	654·11	674.55	694·99	715·44	735·88	756·32	776·76	797·20
40	817·64	838·08	858·52	878·97	999·40	918·85	940·29	960·73	981·17	1001·6
50	1022·1	1042·5	1062·9	1083·4	1103·8	1124·3	1144·7	1165·2	1185·6	1206·0
60	1226·5	1246·9	1267·3	1287·8	1308·2	1328·7	1349·1	1369·5	1390·0	1410·4
70	1430·9	1451·3	1471·8	1492·2	1512·6	1533·1	1553·5	1574·0	1594·4	1614·9
80	1635·3	1655·7	1676·2	1696·6	1717·0	1737·5	1757·9	1778·4	1798·8	1819·2
90	1839·7	1860·1	1880·6	1901·0	1921·5	1941·9	1962·3	1982·8	2003·2	2023·7
100	2044·1	—	—	—	—	—	—	—	—	—

kJ/m^2 °C to B.t.u/ft^2 °F

kJ / m.^{2} °C.	0	1	2	3	4	5	6	7	8	9
	B.t.u/ft^2 °F	B.t.u/ft^2 °F	B.t.u/ft^2 °F	B.t.u/ft^2 °F	B.t.u/ft^2 °F	B.t.u/ft^2 °F	B.t.u/ft^2 °F	B.t.u/ft^2 °F	B.t.u/ft^2 °F	B.t.u/ft^2 °F
0	—	0·049	0·098	0·147	0·195	0·244	0·293	0·342	0·391	0·440
10	0·489	0·538	0·586	0·635	0·684	0·733	0·782	0·831	0·880	0·929
20	0·977	1·026	1·075	1·124	1·173	1·222	1·271	1·320	1·368	1·417
30	1·466	1·515	1·564	1·613	1·662	1·710	1·760	1·808	1·857	1·906
40	1·959	2·004	2·053	2·101	2·150	2·199	2·248	2·297	2·346	2·395
50	2·444	2·492	2·541	2·590	2·639	2·688	2·737	2·786	2·835	2·883
60	2·932	2·981	3·030	3·079	3·128	3·177	3·225	3·274	3·323	3·372
70	3·421	3·470	3·519	3·568	3·616	3·665	3·714	3·763	3·812	3·861
80	3·910	3·959	4·007	4·056	4·105	4·154	4·203	4·252	4·301	4·349
90	4·398	4·447	4·496	4·545	4·594	4·643	4·692	4·740	4·789	4·838
100	4·887	—	—	—	—	—	—	—	—	—

AREAS AND CIRCUMFERENCES OF CIRCLES

Dia.	*Circum.*	*Area.*	*Dia.*	*Circum.*	*Area.*	*Dia.*	*Circum.*	*Area.*
1/16	0·1963	0·00307	2 5/16	7·2649	4·2	6 1/4	19·63	30·67
1/8	0·3927	0·01227	2 3/8	7·4613	4·4302	6 1/2	20·42	33·18
3/16	0·589	0·02761	2 7/16	7·6576	4·6664	6 3/4	21·20	35·78
1/4	0·7854	0·04909	2 1/2	7·854	4·9087	7	21·99	38·48
5/16	0·9817	0·0767	2 9/16	8·0503	5·1573	7 1/4	22·77	41·28
3/8	1·1781	0·1104	2 5/8	8·2467	5·4119	7 1/2	23·56	44·17
7/16	1·3744	0·1503	2 11/16	8·443	5·6723	7 3/4	24·34	47·17
1/2	1·5708	0·1963	2 3/4	8·6394	5·9395	8	25·13	50·26
9/16	1·7771	0·2485	2 13/16	8·8357	6·2126	8 1/4	25·91	53·45
5/8	1·9635	0·3068	2 7/8	9·0321	6·4918	8 1/2	26·70	56·74
11/16	2·1598	0·3712	2 15/16	9·2284	6·7772	8 3/4	27·49	60·13
3/4	2·3562	0·4417	3	9·4248	7·0686	9	28·27	63·62
13/16	2·5525	0·5185	3 1/8	9·8175	7·6699	9 1/4	29·06	67·20
7/8	2·7489	0·6013	3 1/4	10·21	8·2957	9 1/2	29·84	70·88
15/16	2·9452	0·6903	3 3/8	10·602	8·9462	9 3/4	30·63	74·66
1	3·1416	0·7854	3 1/2	10·995	9·6211	10	31·41	78·54
1 1/16	3·3379	0·8866	3 5/8	11·388	10·32	10 1/2	32·98	86·59
1 1/8	3·5343	0·994	3 3/4	11·781	11·044	11	34·56	95·03
1 3/16	3·7306	1·1075	3 7/8	12·173	11·793	11 1/2	36·13	103·87
1 1/4	3·927	1·2271	4	12·566	12·566	12	37·69	113·09
1 5/16	4·1233	1·353	4 1/8	12·959	13·364	12 1/2	39·27	122·71
1 3/8	4·3197	1·4848	4 1/4	13·351	14·186	13	40·84	132·73
1 7/16	4·516	1·6229	4 3/8	13·744	15·033	13 1/2	42·41	143·13
1 1/2	4·7124	1·7671	4 1/2	14·137	15·904	14	43·98	153·93
1 9/16	4·9087	1·9175	4 5/8	14·529	16·8	14 1/2	45·55	165·13
1 5/8	5·1051	2·0739	4 3/4	14·922	17·72	15	47·12	176·71
1 11/16	5·3014	2·2365	4 7/8	15·315	18·665	15 1/2	48·69	188·69
1 3/4	5·4978	2·4052	5	15·708	19·635	16	50·26	201·06
1 13/16	5·6941	2·58	5 1/8	16·1	20·629	16 1/2	51·83	213·82
1 7/8	5·8905	2·7611	5 1/4	16·493	21·647	17	53·40	226·98
1 15/16	6·0868	2·9483	5 3/8	16·886	22·69	17 1/2	54·97	240·52
2	6·2832	3·1416	5 1/2	17·278	23·758	18	56·54	254·47
2 1/16	6·4795	3·3410	5 5/8	17·671	24·85	18 1/2	58·12	268·80
2 1/8	6·6759	3·5465	5 3/4	18·064	25·967	19	59·69	283·53
2 3/16	6·8722	3·7584	5 7/8	18·457	27·108	19 1/2	61·26	298·64
2 1/4	7·0686	3·976	6	18·85	28·27	20	62·83	314·16

Section II

DIMENSIONS OF RADIATORS, TUBES AND FITTINGS

FEED AND EXPANSION CISTERNS TO B.S.417:1964

Ref. No.	Length		Width		Depth		Capacity		Thickness			
									Body		Loose Cover	
	in.	mm.	in.	mm.	in.	mm.	gal.	lit.	B.G.	mm.	B.G.	mm.
SC 10	18	460	12	310	12	310	4	18				
15	24	610	12	310	15	380	8	36	16	1·59	20	0·996
20	24	610	16	410	15	380	12	55				
25	24	610	17	430	17	430	15	68				
30	24	610	18	460	19	480	19	86	16	1·59	20	0·996
40	27	690	20	510	20	510	25	114				
50	29	740	22	560	22	560	35	159				
60	30	760	23	580	24	610	42	191	14	1·99	20	0·996
70	36	910	24	610	23	580	50	227				
80	36	910	26	660	24	610	58	264	14	1·99	20	0·996
100/2	38	970	27	690	27	690	74	336	14	1·99	20	0·996
125	38	970	30	760	31	790	93	423	12	2·52	18	1·257
150	43	1090	34	860	29	740	108	491				
200	46	1170	35	890	35	890	156	709	12	2·52	18	1·257
250	60	1520	36	910	32	810	185	841				
350	60	1520	45	1140	36	910	270	1250	$\frac{1}{8}$ in.	3·175		
500	72	1830	48	1220	40	1020	380	1727	$\frac{1}{8}$ in.	3·175	16	1·588
600	72	1830	48	1220	48	1220	470	2137	$\frac{1}{8}$ in.	3·175		
1000	96	2440	60	1520	48	1220	740	3364	$\frac{3}{16}$ in.	4·762	16	1·588

CLOSED TANKS TO B.S.417:1964

Ref. No.	Length		Width		Depth		Capacity		Thickness	
	in.	mm.	in.	mm.	in.	mm.	gal.	lit.	in.	mm.
T25/1	24	610	17	430	17	430	21	95		
25/2	24	610	24	610	12	310	21	95	$\frac{1}{8}$	3·175
30/1	24	610	18	460	19	480	25	114		
30/2	24	610	24	610	15	380	27	123	$\frac{1}{8}$	3·175
40	27	690	20	510	20	510	34	15		

COPPER INDIRECT CYLINDERS TO B.S.1566:1972

Ref. No.	Dia. mm.	Height mm.	Capacity litres	Heating Surface	
				Annular type m.^{2}	Coil type m.^{2}
1	350	900	72	0·40	0·27
2	400	900	96	0·52	0·35
3	400	1050	114	0·63	0·42
4	450	675	84	0·46	0·31
5	450	750	95	0·52	0·35
6	450	825	106	0·60	0·40
7	450	900	117	0·66	0·44
8	450	1050	140	0·78	0·52
9	450	1200	162	0·91	0·61
10	500	1200	190	1·13	0·75
11	500	1500	245	1·30	0·87
12	600	1200	280	1·60	1·10
13	600	1500	360	2·10	1·40
14	600	1800	440	2·50	1·70

COPPER DIRECT CYLINDERS TO B.S. 699:1972

Ref. No.	Dia. mm.	Height mm.	Capacity litres
1	350	900	74
2	400	900	98
3	400	1050	116
4	450	675	86
5	450	750	98
6	450	825	109
7	450	900	120
8	450	1050	144
9	450	1200	166
10	500	1200	200
11	500	1500	255
12	600	1200	290
13	600	1500	370
14	600	1800	450

Standard Wire Gauge No.	Birmingham Gauge No.	German Sheet Gauge No. (DIN 1541)	ISO Metric R20 Preferred Series mm.	Thickness or Diameter		Weight of Sheet	
				in.	mm.	lb./sq.ft.	kg/m.²
30	—	—	0·315	0·0124	0·315	0·48	2·5
—	—	27	—	0·0126	0·32	0·52	2·5
29	—	—	—	0·0136	0·345	0·52	2·7
—	29	—	—	0·0139	0·354	0·56	2·8
—	—	—	0·355	0·140	0·355	0·56	2·8
28	—	—	—	0·0148	0·376	0·56	2·9
—	28	—	—	0·0156	0·397	0·63	3·1
—	—	26	—	0·0150	0·38	0·62	3·0
—	—	—	0·400	0·0158	0·400	0·64	3·1
27	—	—	—	0·0164	0·417	0·64	3·2
—	27	—	—	0·0175	0·443	0·71	3·5
—	—	25	—	0·0172	0·44	0·70	3·5
—	—	—	0·450	0·0177	0·450	0·72	3·5
26	—	—	—	0·018	0·457	0·72	3·6
—	26	—	—	0·0196	0·498	0·79	3·9
—	—	24	0·500	0·0197	0·500	0·80	3·9
25	—	—	—	0·020	0·508	0·80	4·0
24	—	—	—	0·022	0·559	0·88	4·4
—	25	—	—	0·022	0·560	0·89	4·4
—	—	23	0·560	0·0221	0·560	0·91	4·4
23	—	—	—	0·024	0·610	1·00	4·8
—	24	—	—	0·025	0·629	1·00	4·9
—	—	22	0·630	0·0248	0·630	1·02	4·9
—	23	—	—	0·028	0·707	1·13	5·5
—	—	—	0·710	0·0280	0·710	1·14	5·6
22	—	—	—	0·028	0·711	1·12	5·6
—	—	21	—	0·0295	0·75	1·21	5·9
—	22	—	—	0·031	0·794	1·27	6·2
—	—	—	0·800	0·0315	0·800	1·28	6·3
21	—	—	—	0·032	0·813	1·28	6·3
—	—	20	—	0·0346	0·88	1·41	6·9
—	21	—	—	0·035	0·887	1·41	7·0
—	—	—	0·900	0·0354	0·900	1·42	7·1
20	—	—	—	0·036	0·914	1·42	7·2
—	20	—	—	0·039	0·996	1·59	7·8
—	—	19	1·000	0·0394	1·000	1·61	7·8
19	—	—	—	0·040	1·016	1·68	8·0
—	19	—	—	0·044	1·12	1·78	8·8
—	—	—	1·12	0·0441	1·12	1·80	8·8
—	—	18	—	0·0443	1·13	1·81	8·9

Standard Wire Gauge No.	Birmingham Gauge No.	German Sheet Gauge No. (DIN 1541)	ISO Metric R20 Preferred Series mm.	Thickness or Diameter		Weight of Sheet	
				in.	mm.	lb./sq.ft.	kg/m.2
18	—	—	—	0·048	1·219	1·96	9·6
—	—	17	1·25	0·0492	1·25	2·00	9·8
—	18	—	—	0·050	1·26	2·00	9·9
—	—	16	—	0·0543	1·38	2·22	10·8
—	—	—	1·40	0·0551	1·40	2·25	11·0
—	17	—	—	0·056	1·41	2·25	11·1
17	—	—	—	0·056	1·422	2·32	11·1
—	—	15	—	0·0591	1·50	2·42	11·7
—	16	—	—	0·063	1·59	2·53	12·4
—	—	—	1·60	0·0630	1·60	2·58	12·5
16	—	—	—	0·064	1·626	2·60	12·7
—	—	14	—	0·0689	1·75	2·82	13·7
—	15	—	—	0·070	1·78	2·83	13·9
—	—	—	1·80	0·0709	1·80	2·90	14·1
15	—	—	—	0·072	1·829	2·94	14·3
—	14	—	—	0·079	1·99	3·18	15·6
—	—	13	2·00	0·0787	2·00	3·18	15·7
14	—	—	—	0·080	2·032	3·32	15·9
—	—	—	—	—	—	—	—
—	13	—	2·24	0·088	2·24	3·57	17·6
—	—	12	—	0·0886	2·25	3·59	17·6
13	—	—	—	0·092	2·337	3·80	18·3
—	—	11	2·50	0·0984	2·50	3·98	19·6
—	12	—	—	0·099	2·52	4·01	19·7
12	—	—	—	0·104	2·642	4·36	20·7
—	—	10	—	0·1083	2·75	4·38	21·6
—	—	—	2·80	0·1102	2·80	4·46	22·0
—	11	—	—	0·111	2·83	4·51	22·2
11	—	—	—	0·116	2·946	4·80	23·1
—	—	9	—	0·1181	3·00	4·56	23·5
—	—	—	3·15	0·1240	3·15	5·02	24·7
—	10	—	—	0·125	3·18	5·06	24·8
—	—	8	—	0·1279	3·25	5·18	25·5
10	—	—	—	0·128	3·251	5·36	25·4
—	—	7	—	0·1378	3·50	5·58	27·4
—	9	—	3·55	0·140	3·55	5·66	27·8
9	—	—	—	0·144	3·658	5·92	28·7
—	—	6	—	0·1476	3·75	5·98	29·4
—	8	—	—	0·157	3·99	6·36	31·3

Standard Wire Gauge No.	Birmingham Gauge No.	German Sheet Gauge No. (DIN 1541)	ISO Metric R20 Preferred Series mm.	Thickness or Diameter		Weight of Sheet	
				in.	mm.	lb./sq.ft.	kg/m.2
—	—	5	4·0	0·1575	4·0	6·38	31·4
8	—	—	—	0·160	4·064	6·60	31·9
—	—	4	—	0·1673	4·25	6·77	33·3
7	—	—	—	0·176	4·470	7·12	35·1
—	7	—	—	0·176	4·48	7·14	35·1
—	—	3	4·5	0·1772	4·50	7·17	35·3
6	—	—	—	0·192	4·877	7·80	38·2
—	—	2	5·0	0·1969	5·00	7·97	39·2
—	6	—	—	0·198	5·032	8·02	39·5
5	—	—	—	0·212	5·385	8·80	42·2
—	—	1	—	0·2165	5·50	8·77	43·1
—	—	—	5·6	0·2205	5·6	8·93	43·9
—	5	—	—	0·222	5·66	9·01	44·4
4	—	—	—	0·232	5·893	9·52	46·2
—	—	—	6·30	0·2480	6·30	10·04	49·4
—	4	—	—	0·250	6·35	10·12	49·9
3	—	—	—	0·252	6·401	10·36	50·2
2	—	—	—	0·276	7·010	11·17	55·0
—	—	—	7·10	0·2795	7·10	11·32	55·7
—	3	—	—	0·280	7·13	11·34	55·9
1	—	—	—	0·300	7·620	12·0	59·7
—	2	—	—	0·315	8·00	12·74	62·7
—	—	—	8·00	0·3150	8·00	12·74	62·7
0	—	—	—	0·324	8·229	13·1	63·9
2/0	—	—	—	0·348	8·839	13·9	69·3
—	1	—	—	0·353	8·98	14·30	70·4
—	—	—	9·00	0·3543	9·00	14·3	70·6
3/0	—	—	—	0·372	9·449	14·9	74·1
—	—	—	10·00	0·3937	10·00	15·9	78·4
—	0	—	—	0·396	10·07	16·0	78·9
4/0	—	—	—	0·400	10·160	16·0	79·7
5/0	—	—	—	0·432	10·973	17·3	86·0
—	—	—	11·2	0·4409	11·2	17·8	87·8
—	2/0	—	—	0·445	11·3	18·0	88·6
6/0	—	—	—	0·464	11·785	18·6	92·4
—	—	—	12·5	0·4921	12·5	19·9	98·0
7/0	3/0	—	—	0·500	12·700	20·0	99·5

Thickness or Dia. mm.	Weight in kg. of		
	Sheet per m.^{2}	Square per m.	Round per m.
5	39·25	0·196	0·154
6	47·10	0·283	0·222
8	62·80	0·502	0·395
10	78·50	0·785	0·617
12	94·20	1·130	0·888
14	109·90	1·539	1·208
16	125·60	2·010	1·578
18	141·30	2·543	1·998
20	157·00	3·140	2·466
22	172·70	3·799	2·984
24	188·40	4·522	3·551
26	204·10	5·307	4·168
28	219·80	6·154	4·834
30	235·50	7·065	5·549
32	251·20	8·038	6·313
34	266·90	9·075	7·127
36	282·60	10·174	7·990
38	298·30	11·335	8·903
40	314·00	12·560	9·865
42	329·70	13·847	10·876
44	345·40	15·198	11·936
46	361·10	16·611	13·046
48	376·80	18·086	14·205
50	392·50	19·625	15·413
52	408·20	21·226	16·671
54	423·90	22·891	17·978
56	439·60	24·618	19·335
58	455·30	26·407	20·740
60	471·00	28·260	22·195
62	486·70	30·175	23·700
64	502·40	32·154	25·253
66	518·10	34·195	26·856

Thickness or Dia. mm.	Weight in kg. of		
	Sheet per m.^{2}	Square per m.	Round per m.
68	533·80	36·298	28·509
70	569·50	36·465	30·210
72	585·20	40·694	31·961
74	600·90	42·987	33·762
76	616·60	45·342	35·611
78	632·30	47·759	37·510
80	628·00	50·240	39·458
85	667·25	56·716	44·545
90	706·50	63·585	49·940
95	745·75	70·846	55·643
100	785·00	78·500	61·654
105	824·25	86·546	67·973
110	863·5	94·985	74·601
115	902·75	103·816	81·537
120	942·0	113·040	88·781
125	981·2	122·656	96·334
130	1020	132·665	104·195
135	1060	143·006	112·364
140	1099	153·860	120·841
145	1138	165·046	129·627
150	1178	176·625	138·721
155	1217	188·596	148·123
160	1256	200·960	157·834
165	1295	213·716	167·852
170	1355	226·865	178·179
175	1394	240·406	188·815
180	1413	254·340	199·758
185	1452	268·666	211·010
190	1492	283·385	222·570
195	1511	298·496	234·438
200	1570	314·000	246·615

Thickness or Dia. in.	Weight in lb. of		
	Sheet per sq. ft.	Square per ft.	Round per ft.
$\frac{1}{8}$	5·10	0·053	0·042
$\frac{3}{16}$	7·65	0·120	0·094
$\frac{1}{4}$	10·20	0·213	0·167
$\frac{5}{16}$	12·75	0·332	0·261
$\frac{3}{8}$	15·30	0·479	0·376
$\frac{7}{16}$	17·85	0·651	0·511
$\frac{1}{2}$	20·40	0·851	0·658
$\frac{9}{16}$	22·95	1·08	0·845
$\frac{5}{8}$	22·50	1·33	1·04
$\frac{11}{16}$	28·05	1·61	1·29
$\frac{3}{4}$	30·60	1·91	1·50
$\frac{13}{16}$	33·15	2·25	1·77
$\frac{7}{8}$	35·70	2·61	2·04
$\frac{15}{16}$	38·25	2·99	2·35

Thickness or Dia. in.	Weight in lb. of		
	Sheet per sq. ft.	Square per ft.	Round per ft.
1	40·80	3·40	2·68
$1\frac{1}{8}$	45·9	4·31	3·38
$1\frac{1}{4}$	51·0	5·32	4·17
$1\frac{3}{8}$	56·1	6·43	5·05
$1\frac{1}{2}$	61·2	7·71	6·01
$1\frac{5}{8}$	66·3	8·99	7·05
$1\frac{3}{4}$	71·4	10·4	8·19
$1\frac{7}{8}$	76·5	12·0	9·39
2	81·6	13·6	10·7
$2\frac{1}{2}$	102·2	21·3	16·8
3	122·4	30·6	24·1
4	163·2	54·4	42·8
5	204·0	85·1	66·9
6	324·8	122·5	96·2

FACING OF FLANGES

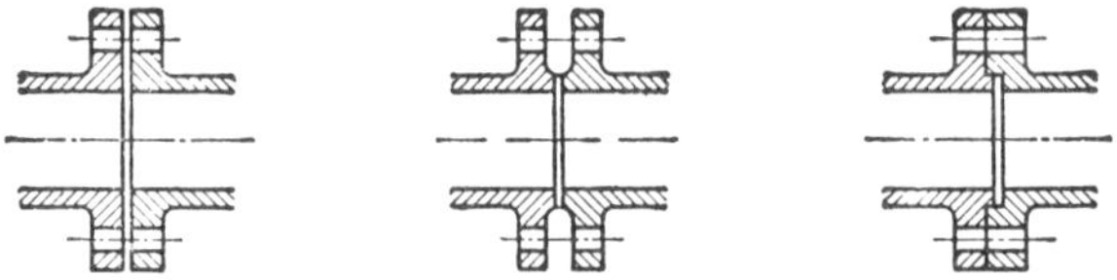

PLAIN FACING *FACING STRIPS* *SINGLE SPIGOT & FAUCET FACING*

DIMENSIONS OF BRITISH STANDARD PIPE FLANGES—SEE TABLES II. 7–10

Water pipes in which the water temperature does not exceed 450°F. may be fitted with flanges to the next lower Table than that required for steam at the same pressure and a maximum temperature of 800°F.

WELDED PIPE JOINTS

BUTT WELDED JOINT SQUARE ENDS *BUTT WELDED JOINT CHAMFERED ENDS* *STEWARTS LONG SLEEVE WELDED JOINT* *STEWARTS SHORT SLEEVE WELDED JOINT*

HIGH PRESSURE FLANGES, METHODS OF FIXING

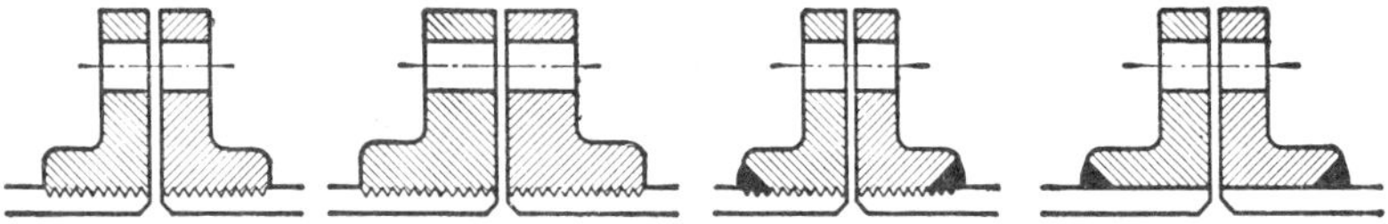

LIGHT SCREWED AND EXPANDED *HEAVY SCREWED AND EXPANDED* *SCREWED AND WELDED OVER* *EXPANDED AND WELDED OVER*

PIPE HANGERS (Usual Dimensions in inches)

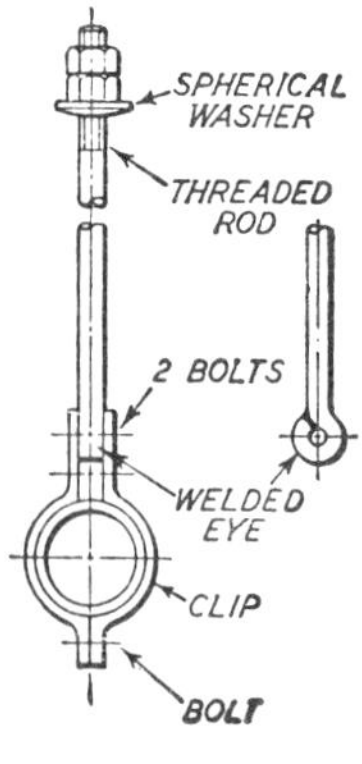

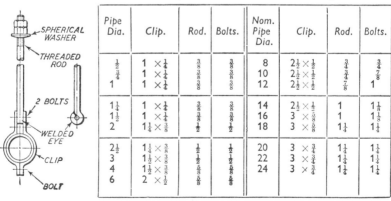

Pipe Dia.	Clip.	Rod.	Bolts.	Nom. Pipe Dia.	Clip.	Rod.	Bolts.
$\frac{1}{2}$	$1 \times \frac{1}{4}$	$\frac{3}{8}$	$\frac{3}{8}$	8	$2\frac{1}{2} \times \frac{1}{2}$	$\frac{3}{4}$	$\frac{3}{4}$
$\frac{3}{4}$	$1 \times \frac{1}{4}$	$\frac{3}{8}$	$\frac{3}{8}$	10	$2\frac{1}{2} \times \frac{1}{2}$	$\frac{3}{4}$	$\frac{7}{8}$
1	$1 \times \frac{1}{4}$	$\frac{3}{8}$	$\frac{3}{8}$	12	$2\frac{1}{2} \times \frac{1}{2}$	$\frac{7}{8}$	1
$1\frac{1}{4}$	$1 \times \frac{1}{4}$	$\frac{3}{8}$	$\frac{3}{8}$	14	$2\frac{1}{2} \times \frac{1}{2}$	1	$1\frac{1}{8}$
$1\frac{1}{2}$	$1 \times \frac{1}{4}$	$\frac{3}{8}$	$\frac{3}{8}$	16	$3 \times \frac{5}{8}$	1	$1\frac{1}{8}$
2	$1\frac{1}{4} \times \frac{3}{8}$	$\frac{1}{2}$	$\frac{1}{2}$	18	$3 \times \frac{5}{8}$	$1\frac{1}{4}$	$1\frac{1}{4}$
$2\frac{1}{2}$	$1\frac{1}{4} \times \frac{3}{8}$	$\frac{1}{2}$	$\frac{1}{2}$	20	$3 \times \frac{3}{4}$	$1\frac{1}{4}$	$1\frac{1}{4}$
3	$1\frac{1}{2} \times \frac{3}{8}$	$\frac{1}{2}$	$\frac{1}{2}$	22	$3 \times \frac{3}{4}$	$1\frac{1}{4}$	$1\frac{1}{4}$
4	$1\frac{1}{2} \times \frac{3}{8}$	$\frac{5}{8}$	$\frac{5}{8}$	24	$3 \times \frac{3}{4}$	$1\frac{1}{4}$	$1\frac{1}{4}$
6	$2 \times \frac{1}{2}$	$\frac{5}{8}$	$\frac{5}{8}$				

BRITISH STANDARD TABLES OF PIPE FLANGES (FOR LAND USE).

TABLE D (superseding Table I of Report No. 10-1904). Flanges for Pipes, Valves, and Fittings for working steam pressures up to 50 lb. per sq. in.

1	1a	2	3	4	5	6(a)	6(b)	6(c)
Nominal Pipe Size	*Actual Outside Dia. of Wrought Pipe*	*Dia. of Flange*	*Dia. of Bolt Circle*	*Number of Bolts*	*Diam. of Bolts*	*Thickness of Flange*		
						Cast Iron	*Cast Steel*	*Iron or Steel (Stamped or Forged) Screwed or riveted on with boss or welded on with fillet*
in.	*in.*	*in.*	*in.*		*in.*	*in.*	*in.*	*in.*
$\frac{1}{2}$	$\frac{27}{32}$	$3\frac{3}{4}$	$2\frac{5}{8}$	4	$\frac{1}{2}$	$\frac{1}{2}$	$\frac{3}{8}$	$\frac{3}{16}$
$\frac{3}{4}$	$1\frac{1}{16}$	4	$2\frac{7}{8}$	4	$\frac{1}{2}$	$\frac{1}{2}$	$\frac{3}{8}$	$\frac{3}{16}$
1	$1\frac{11}{32}$	$4\frac{1}{2}$	$3\frac{1}{4}$	4	$\frac{1}{2}$	$\frac{1}{2}$	$\frac{3}{8}$	$\frac{3}{16}$
$1\frac{1}{4}$	$1\frac{11}{16}$	$4\frac{3}{4}$	$3\frac{7}{16}$	4	$\frac{1}{2}$	$\frac{5}{8}$	$\frac{1}{2}$	$\frac{1}{4}$
$1\frac{1}{2}$	$1\frac{29}{32}$	$5\frac{1}{4}$	$3\frac{7}{8}$	4	$\frac{1}{2}$	$\frac{5}{8}$	$\frac{1}{2}$	$\frac{1}{4}$
2	$2\frac{3}{8}$	6	$4\frac{1}{2}$	4	$\frac{5}{8}$	$\frac{11}{16}$	$\frac{9}{16}$	$\frac{5}{16}$
$2\frac{1}{2}$	3	$6\frac{1}{2}$	5	4	$\frac{5}{8}$	$\frac{11}{16}$	$\frac{9}{16}$	$\frac{5}{16}$
3	$3\frac{1}{2}$	$7\frac{1}{4}$	$5\frac{3}{4}$	4	$\frac{5}{8}$	$\frac{3}{4}$	$\frac{9}{16}$	$\frac{3}{8}$
$3\frac{1}{2}$	4	8	$6\frac{1}{2}$	4	$\frac{5}{8}$	$\frac{3}{4}$	$\frac{9}{16}$	$\frac{3}{8}$
4	$4\frac{1}{2}$	$8\frac{1}{2}$	7	4	$\frac{5}{8}$	$\frac{3}{4}$	$\frac{11}{16}$	$\frac{3}{8}$
5	$5\frac{1}{2}$	10	$8\frac{1}{4}$	8	$\frac{5}{8}$	$\frac{13}{16}$	$\frac{11}{16}$	$\frac{1}{2}$
6	$6\frac{1}{2}$	11	$9\frac{1}{4}$	8	$\frac{5}{8}$	$\frac{3}{16}$	$\frac{11}{16}$	$\frac{1}{2}$
7	$7\frac{5}{8}$	12	$10\frac{1}{4}$	8	$\frac{5}{8}$	$\frac{7}{8}$	$\frac{3}{4}$	$\frac{1}{2}$
8	$8\frac{5}{8}$	$13\frac{1}{4}$	$11\frac{1}{2}$	8	$\frac{5}{8}$	$\frac{7}{8}$	$\frac{3}{4}$	$\frac{1}{2}$
9	$9\frac{5}{8}$	$14\frac{1}{2}$	$12\frac{3}{4}$	8	$\frac{5}{8}$	$\frac{7}{8}$	$\frac{3}{4}$	$\frac{5}{8}$
10	$10\frac{3}{4}$	16	14	8	$\frac{3}{4}$	1	$\frac{3}{4}$	$\frac{5}{8}$
12	$12\frac{3}{4}$	18	16	12	$\frac{3}{4}$	1	$\frac{7}{8}$	$\frac{3}{4}$
*13	14	$19\frac{1}{4}$	$17\frac{1}{4}$	12	$\frac{3}{4}$	1	$\frac{7}{8}$	$\frac{3}{4}$
14	15	$20\frac{3}{4}$	$18\frac{1}{2}$	12	$\frac{7}{8}$	$1\frac{1}{8}$	1	$\frac{7}{8}$
15	16	$21\frac{3}{4}$	$19\frac{1}{2}$	12	$\frac{7}{8}$	$1\frac{1}{8}$	1	$\frac{7}{8}$
16	17	$22\frac{3}{4}$	$20\frac{1}{2}$	12	$\frac{7}{8}$	$1\frac{1}{8}$	1	$\frac{7}{8}$
*17	18	24	$21\frac{3}{4}$	12	$\frac{7}{8}$	$1\frac{1}{8}$	$1\frac{1}{8}$	1
18	19	$25\frac{1}{4}$	23	12	$\frac{7}{8}$	$1\frac{1}{4}$	$1\frac{1}{8}$	1
*19	20	$26\frac{1}{2}$	24	12	$\frac{7}{8}$	$1\frac{1}{4}$	$1\frac{1}{8}$	1
20	21	$27\frac{3}{4}$	$25\frac{1}{4}$	16	$\frac{7}{8}$	$1\frac{1}{4}$	$1\frac{1}{4}$	$1\frac{1}{8}$
21	22	29	$26\frac{1}{2}$	16	$\frac{7}{8}$	$1\frac{3}{8}$	$1\frac{1}{4}$	$1\frac{1}{8}$
*22	23	30	$27\frac{1}{2}$	16	1	$1\frac{3}{8}$	$1\frac{1}{4}$	$1\frac{1}{8}$
*23	24	31	$28\frac{1}{2}$	16	1	$1\frac{3}{8}$	$1\frac{3}{8}$	$1\frac{1}{8}$
24	25	$32\frac{1}{2}$	$29\frac{3}{4}$	16	1	$1\frac{3}{8}$	$1\frac{3}{8}$	$1\frac{1}{4}$

* The Institution recommends that the use of these sizes be avoided.

Thicknesses.—The thicknesses given in this Table include a raised face of not more than $\frac{1}{16}$ in. high if such is used. **Bolt Holes.**—For $\frac{1}{2}$ in. and $\frac{5}{8}$ in. Bolts the diameter of the holes to be $\frac{1}{16}$ in. larger than the diameter of the Bolts, and for larger sizes of Bolts, $\frac{1}{8}$ in. Bolt Holes to be drilled off centre lines.

BRITISH STANDARD TABLES OF PIPE FLANGES (FOR LAND USE)

TABLE E (superseding Table 2 of Report No. 10-1904). Flanges for Pipes, Valves, and Fittings, for working steam pressure above 50 lb. and up to 100 lb. per sq. in.

1	1a	2	3	4	5	6(a)	6(b)	6(c)
Nominal Pipe Size	*Actual Outside Dia. of Wrought Pipe*	*Dia. of Flange*	*Dia. of Bolt Circle*	*Number of Bolts*	*Diam. of Bolts*	*Thickness of Flange*		
						Cast Iron	*Cast Steel & Bronze*	*Iron or Steel (Stamped or Forged) Screwed or Riveted on with boss, or welded on with fillet*
in.	*in.*	*in.*	*in.*		*in.*	*in.*	*in.*	*in.*
1/2	27/32	3 3/4	2 5/8	4	1/2	1/2	3/8	1/4
3/4	1 1/16	4	2 7/8	4	1/2	1/2	3/8	1/4
1	1 11/32	4 1/2	3 1/4	4	1/2	1/2	3/8	9/32
1 1/4	1 11/16	4 3/4	3 7/16	4	1/2	5/8	1/2	5/16
1 1/2	1 29/32	5 1/4	3 7/8	4	1/2	5/8	1/2	11/32
2	2 3/8	6	4 1/2	4	5/8	3/4	9/16	3/8
2 1/2	3	6 1/2	5	4	5/8	3/4	9/16	13/32
3	3 1/2	7 1/4	5 3/4	4	5/8	3/4	9/16	7/16
3 1/2	4	8	6 1/2	8	5/8	3/4	9/16	15/32
4	4 1/2	8 1/2	7	8	5/8	7/8	11/16	1/2
5	5 1/2	10	8 1/4	8	5/8	7/8	11/16	9/16
6	6 1/2	11	9 1/4	8	3/4	7/8	11/16	11/16
7	7 5/8	12	10 1/4	8	3/4	1	3/4	3/4
8	8 5/8	13 1/4	11 1/2	8	3/4	1	3/4	3/4
9	9 5/8	14 1/2	12 3/4	12	3/4	1	13/16	13/16
10	10 3/4	16	14	12	3/4	1	7/8	7/8
12	12 3/4	18	16	12	7/8	1 1/8	1	1
*13	14	19 1/4	17 1/4	12	7/8	1 1/8	1	1
14	15	20 3/4	18 1/2	12	7/8	1 1/4	1	1 1/8
15	16	21 3/4	19 1/2	12	7/8	1 1/4	1	1 1/4
16	17	22 3/4	20 1/2	12	7/8	1 1/4	1	1 1/4
*17	18	24	21 3/4	12	7/8	1 3/8	1 1/8	1 3/8
18	19	25 1/4	23	16	7/8	1 3/8	1 1/8	1 3/8
*19	20	26 1/2	24	16	7/8	1 3/8	1 1/4	1 1/2
20	21	27 3/4	25 1/4	16	7/8	1 1/2	1 1/4	1 1/2
21	22	29	26 1/2	16	1	1 1/2	1 3/8	1 5/8
*22	23	30	27 1/2	16	1	1 1/2	1 3/8	1 3/4
*23	24	31	28 1/2	16	1	1 5/8	1 3/8	1 3/4
24	25	32 1/2	29 3/4	16	1 1/8	1 5/8	1 1/2	1 7/8

* The Institution recommends that the use of these sizes be avoided.

Thicknesses.—The thicknesses given in this Table include a raised face of not more than 1/16 in. high if such be used. **Bolt Holes.**—For 1/2 in. and 5/8 in. Bolts the diameters of the holes to be 1/16 in. larger than the diameters of the Bolts, and for larger sizes 1/8 in. Bolt Holes to be drilled off centre lines.

BRITISH STANDARD TABLES OF PIPE FLANGES (FOR LAND USE)

TABLE F (superseding Table 2 of Report No. 10-1904). Flanges for Pipes, Valves, and Fittings, for working steam pressure above 100 lb. and up to 150 lb. per sq. in.

1	1(a)	2	3	4	5	6(a)	6(b)
Nominal Pipe Size	*Actual Outside Dia. of Wrought Pipe*	*Dia. of Flange*	*Dia. of Bolt Circle*	*Number of Bolts*	*Diam. of Bolts*	*Thickness of Flange*	
						Cast Iron	*Cast Steel Bronze, Iron or Steel (Stamped or Forged) Screwed or Riveted on with boss or welded on with fillet.*
in.	*in.*	*in.*	*in.*		*in.*	*in.*	*in.*
½	27/32	3¾	2⅝	4	½	½	⅜
¾	1 1/16	4	2⅞	4	½	½	⅜
1	1 11/32	4¾	3 7/16	4	⅝	½	⅜
1¼		5¼	3⅞	4	⅝	⅝	½
1½	1 29/32	5½	4⅛	4	⅝	⅝	½
2	2⅜	6½	5	4	⅝	¾	⅝
2½	3	7¼	5¾	8	⅝	¾	⅝
3	3½	8	6½	8	⅝	¾	⅝
3½	4	8½	7	8	⅝	⅞	¾
4	4½	9	7½	8	⅝	⅞	¾
5	5½	11	9¼	8	¾	1	⅞
6	6½	12	10¼	12	¾	1	⅞
7	7⅝	13¼	11½	12	¾	1	⅞
8	8⅝	14½	12¾	12	¾	1⅛	1
9	9⅝	16	14	12	⅞	1⅛	1⅛
10	10¾	17	15	12	⅞	1⅛	1⅛
12	12¾	19¼	17¼	16	⅞	1¼	1¼
*13	14	20¾	18½	16	1	1¼	1⅜
14	15	21¾	19½	16	1	1⅜	1⅜
15	16	22¾	20½	16	1	1⅜	1⅛
16	17	24	21¾	20	1	1⅜	1⅝
*17	18	25¼	23	20	1	1½	1¾
18	19	26½	24	20	1⅛	1½	1¾
*19	20	27¾	25¼	20	1⅛	1½	1¾
20	21	29	26½	24	1⅛	1⅝	2
21	22	30	27½	24	1⅛	1⅝	2
*22	23	31	28½	24	1⅛	1⅝	2⅛
*23	24	32½	29¾	24	1¼	1¾	2¼
24	25	33½	30¾	24	1¼	1¾	2¼

* The Institution recommends that the use of these sizes be avoided.

Thicknesses.—The thicknesses given in this Table include a raised face of not more than 1/16 in. high if such be used. **Bolt Holes.**—For ½ in. and ⅝ in. Bolts the diameters of the holes to be 1/16 in. larger than the diameters of the Bolts, and for larger sizes of Bolts ⅛ in. Bolt Holes to be drilled off centre lines.

BRITISH STANDARD TABLES OF PIPE FLANGES (FOR LAND USE).

TABLE H (superseding Table 2 of Report No. 10-1904). Flanges for Pipes, Valves, and Fittings, for working steam pressure above 150 lb. and up to 250 lb. per sq. in.

1	1(a)	2	3	4	5	6
Nominal Dia. of Pipe	*Actual Outside Dia. of Wrought Pipe*	*Diameter of Flange*	*Diameter of Bolt Circle*	*Number of Bolts*	*Diameter of Bolts*	*Thickness of Flange* — *Cast Steel and Bronze, Steel (Stamped or Forged) Screwed or Riveted on with Boss, or welded on with fillet*
in.	*in.*	*in.*	*in.*		*in.*	*in.*
$\frac{1}{2}$	$\frac{27}{32}$	$4\frac{1}{2}$	$3\frac{1}{4}$	4	$\frac{5}{8}$	$\frac{1}{2}$
$\frac{3}{4}$	$1\frac{1}{16}$	$4\frac{1}{2}$	$3\frac{1}{4}$	4	$\frac{5}{8}$	$\frac{1}{2}$
1	$1\frac{11}{32}$	$4\frac{3}{4}$	$3\frac{7}{16}$	4	$\frac{5}{8}$	$\frac{9}{16}$
$1\frac{1}{4}$	$1\frac{11}{16}$	$5\frac{1}{4}$	$3\frac{7}{8}$	4	$\frac{5}{8}$	$\frac{11}{16}$
$1\frac{1}{2}$	$1\frac{29}{32}$	$5\frac{1}{2}$	$4\frac{1}{8}$	4	$\frac{5}{8}$	$\frac{11}{16}$
2	$2\frac{3}{8}$	$6\frac{1}{2}$	5	4	$\frac{5}{8}$	$\frac{3}{4}$
$2\frac{1}{2}$	3	$7\frac{1}{4}$	$5\frac{3}{4}$	8	$\frac{5}{8}$	$\frac{3}{4}$
3	$3\frac{1}{2}$	8	$6\frac{1}{2}$	8	$\frac{5}{8}$	$\frac{7}{8}$
$3\frac{1}{2}$	4	$8\frac{1}{2}$	7	8	$\frac{5}{8}$	$\frac{7}{8}$
4	$4\frac{1}{2}$	9	$7\frac{1}{2}$	8	$\frac{5}{8}$	1
5	$5\frac{1}{2}$	11	$9\frac{1}{4}$	8	$\frac{3}{4}$	$1\frac{1}{8}$
6	$6\frac{1}{2}$	12	$10\frac{1}{4}$	12	$\frac{3}{4}$	$1\frac{1}{8}$
7	$7\frac{5}{8}$	$13\frac{1}{4}$	$11\frac{1}{2}$	12	$\frac{3}{4}$	$1\frac{1}{4}$
8	$8\frac{5}{8}$	$14\frac{1}{2}$	$12\frac{3}{4}$	12	$\frac{3}{4}$	$1\frac{1}{4}$
9	$9\frac{5}{8}$	16	14	12	$\frac{7}{8}$	$1\frac{3}{8}$
10	$10\frac{3}{4}$	17	15	12	$\frac{7}{8}$	$1\frac{3}{8}$
12	$12\frac{3}{4}$	$19\frac{1}{4}$	$17\frac{1}{4}$	16	$\frac{7}{8}$	$1\frac{5}{8}$
*13	14	$20\frac{3}{4}$	$18\frac{1}{2}$	16	1	$1\frac{3}{4}$
14	15	$21\frac{3}{4}$	$19\frac{1}{2}$	16	1	$1\frac{7}{8}$
15	16	$22\frac{3}{4}$	$20\frac{1}{2}$	16	1	2
16	17	24	$21\frac{3}{4}$	20	1	$2\frac{1}{8}$
*17	18	$25\frac{1}{4}$	23	20	1	$2\frac{1}{4}$
18	19	$26\frac{1}{2}$	24	20	$1\frac{1}{8}$	$2\frac{3}{8}$
*19	20	$27\frac{3}{4}$	$25\frac{1}{4}$	20	$1\frac{1}{8}$	$2\frac{1}{2}$
20	21	29	$26\frac{1}{2}$	24	$1\frac{1}{8}$	$2\frac{5}{8}$
21	22	30	$27\frac{1}{2}$	24	$1\frac{1}{8}$	$2\frac{3}{4}$
*22	23	31	$28\frac{1}{2}$	24	$1\frac{1}{8}$	$2\frac{3}{4}$
*23	24	$32\frac{1}{2}$	$29\frac{3}{4}$	24	$1\frac{1}{4}$	3
24	25	$33\frac{1}{2}$	$30\frac{3}{4}$	24	$1\frac{1}{4}$	3

* The Institution recommends that the use of these sizes be avoided.

Thicknesses.—The thicknesses given in this Table include a raised face of not more than $\frac{1}{16}$ in. high if such be used. **Bolt Holes.**—For $\frac{1}{2}$ in. and $\frac{5}{8}$ in. Bolts the diameters of the Holes to be $\frac{1}{16}$ in. larger than the diameter of the Bolts, and for larger sizes of Bolts $\frac{1}{8}$ in. Bolt Holes to be drilled off centre lines.

MALLEABLE IRON PIPE FITTINGS, BANDED AND BEADED, BLACK AND GALVANIZED.

Air Tested 100 lb. under water, sizes up to 2 in.; fittings $2\frac{1}{2}$ in. size and over tested 300 lb. hydraulic.

	Fitting.	*Crane*	*Ideal "P"*	*G.F.*
	M & F. Bend	192	P80A	1
	Female Bend	193	P80	2
	Elbow, M. & F.	152	P25	92
	Elbow	151	P20	90
	Close Return Bend	211		61
	Open Return Bend	213	P61	70
	Wide Return Bend	214		60
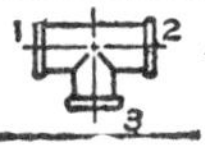	Tee	161	P21	130
	Twin Elbow	197	P81	132
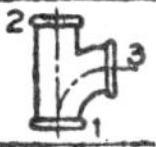	Pitcher Tee	199	P82	131
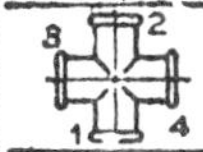	Cross	171	P22	180
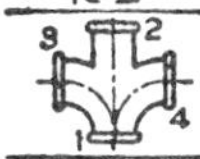	Pitcher Cross	201	P83	181

	Fitting.	*Crane*	*Ideal "P"*	*G.F.*
	Parallel Thread Socket	176	P30	270
	Socket R. & L.	178		271
	Hexagonal Nipple Equal	144	P115	281
	Hexagonal Bush	140	P110	241
	Reducing Socket	179	P27	240
	Eccentric Socket	180	P28	260
	Beaded Plug	146	P114	290
	Cap	185	P31	300
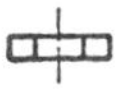	Back Nut	150	P6	312
	Union Elbow Female	246	P70	95
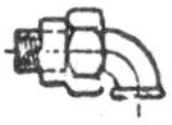	Union Elbow M. & F.	247	P71	97
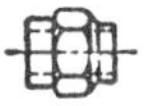	Standard Union F.	241	P90	330
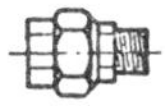	Standard Union M. & F.	242	P92	331

BRITISH STANDARD CAST IRON PIPES FOR WATER, GAS, AND SEWAGE. Standard thickness and external diameters for Spigot and Socket and Flange Straight Pipes. Brit. Stand. Spec. No. 78:1961 (Abstract)

DIMENSIONS IN INCHES

Nom. Internal dia. of Pipe.	GAS		WATER AND SEWAGE						Nom. Internal dia. of Pipe.
	CLASS A. Test Pressure 200 ft. head.		CLASS B. Test Pressure 400 ft. head.		CLASS C. Test Pressure 600 ft. head.		CLASS D. Test Pressure 800 ft. head.		
	Thick.	Ext. dia.	Thick.	Ext. dia.	Thick.	Ext. dia.	Thick.	Ext. dia.	
3	0·38	3·76	0·38	3·76	0·38	3·76	0·40	3·76	3
4	0·39	4·80	0·39	4·80	0·40	4·38	0·46	4·80	4
5	0·41	5·90	0·41	5·90	0·45	5·90	0·52	5·90	5
6	0·43	6·98	0·43	6·98	0·49	6·98	0·57	6·98	6
7	0·45	8·06	0·45	8·06	0·53	8·06	0·61	8·06	7
8	0·47	9·14	0·47	9·14	0·57	9·14	0·65	9·14	8
9	0·49	10·20	0·49	10·20	0·60	10·20	0·69	10·20	9
10	0·52	11·26	0·52	11·26	0·63	11·26	0·73	11·26	10
12	0·55	13·14	0·57	13·14	0·69	13·60	0·80	13·60	12

BASIC SIZES FOR B.S. PIPE THREADS

Abstract from Table 1, B.S.21 : 1973

1	2	3	4	5	6	7	8
				Diameters at Gauge Plane (Basic)			
B.S.P. Size (Nom. Bore of Tube	No. of Threads per inch	Pitch	Depth of Thread	Major (Gauge Diam.)	Effective	Minor	Gauge Length
$\frac{1}{8}$	28	0·3571	0·2209	0·383	0·3601	0·3372	0·1563
$\frac{1}{4}$	19	0·05263	0·0337	0·518	0·4843	0·4506	0·2367
$\frac{3}{8}$	19	0·05263	0·0337	0·656	0·6223	0·5886	0·2500
$\frac{1}{2}$	14	0·07143	0·0457	0·825	0·7793	0·7336	0·3214
$\frac{3}{4}$	14	0·07143	0·0457	1·041	0·9953	0·9496	0·3750
1	11	0·09091	0·0582	1·309	1·2508	1·1926	0·4091
$1\frac{1}{4}$	11	0·09091	0·0582	1·650	1·5918	1·5336	0·5000
$1\frac{1}{2}$	11	0·09091	0·0582	1·882	1·8238	1·7656	0·5000
2	11	0·09091	0·0582	2·347	2·2888	2·2306	0·6250
$2\frac{1}{2}$	11	0·09091	0·0582	2·960	2·9018	2·8436	0·6875
3	11	0·09091	0·0582	3·460	3·4018	3·3436	0·8125
$3\frac{1}{2}$	11	0·09091	0·0582	3·950	3·8918	3·8336	0·8750
4	11	0·09091	0·0582	4·450	4·3336	4·3336	1·0000
5	11	0·09091	0·0582	5·450	5·3918	5·3336	1·1250
6	11	0·09091	0·0582	6·450	6·3918	6·3336	1 1250

GENERAL DIMENSIONS OF STEEL TUBES TO B.S.1387: 1967

(Subject to standard tolerances and usual working allowances)

Nominal Bore		Outside Diameter				Thickness					
		Light		Heavy & Med		Light		Medium		Heavy	
in.	mm.	in.	mm.	in.	mm.	in.	mm.	in.	mm.	in.	mm.
1/8	6	0·396	10·1	0·411	10·4	0·072	1·8	0·080	2·0	0·104	2·65
1/4	8	0·532	13·6	0·547	13·9	0·072	1·8	0·092	2·35	0·116	2·90
3/8	10	0·671	17·1	0·685	17·4	0·072	1·8	0·092	2·35	0·116	2·90
1/2	15	0·871	21·4	0·856	21·7	0·080	2·0	0·104	2·65	0·128	3·25
3/4	20	1·059	26·9	1·072	27·2	0·092	2·35	0·104	2·65	0·128	3·25
1	25	1·328	33·8	1·346	34·2	0·104	2·65	0·128	3·25	0·160	4·05
1¼	32	1·670	42·5	1·687	42·9	0·104	2·65	0·128	3·25	0·160	4·05
1½	40	1·903	48·4	1·919	48·8	0·116	2·9	0·128	3·25	0·160	4·05
2	50	2·370	60·2	2·394	60·8	0·116	2·9	0·114	3·65	0·176	4·50
2½	65	2·991	76·0	3·014	76·6	0·128	3·25	0·144	3·65	0·176	4·50
3	80	3·491	88·7	3·524	89·5	0·128	3·25	0·160	4·05	0·192	4·85
4	100	4·481	113·9	4·524	114·9	0·144	3·65	0·176	4·5	0·212	5·40
5	125	—	—	5·534	140·6	—	—	0·192	4·85	0·212	5·40
6	150	—	—	6·539	166·1	—	—	0·192	4·85	0·212	5·40

Nominal Bore		Weight of Black Tube											
		Light				Medium				Heavy			
		Plain		Screwed		Plain		Screwed		Plain		Screwed	
in.	mm.	lb/ft	kg/m	lb/ft	kg/m	lb/ft	kg/m	lb/ft	kg/m	lb/ft	kg/m	lb/ft	kg/m
1/8	6	0·24	0·36	0·25	0·36	0·27	0·41	0·28	0·41	0·33	0·49	0·33	0·50
1/4	8	0·35	0·52	0·35	0·52	0·44	0·65	0·44	0·65	0·52	0·77	0·52	0·77
3/8	10	0·45	0·67	0·46	0·68	0·57	0·85	0·58	0·85	0·69	1·02	0·69	1·03
1/2	15	0·64	0·95	0·64	0·96	0·82	1·22	0·83	1·23	0·98	1·45	0·98	1·46
3/4	20	0·94	1·41	0·95	1·42	1·06	1·58	1·07	1·59	1·27	1·90	1·28	1·91
1	25	1·35	2·01	1·36	2·03	1·64	2·44	1·65	2·46	2·00	2·97	2·01	2·99
1¼	32	1·73	2·58	1·75	2·61	2·11	3·14	2·13	3·17	2·58	3·84	2·60	3·87
1½	40	2·19	3·25	2·22	3·29	2·43	3·61	2·46	3·65	2·98	4·43	3·01	4·47
2	50	2·76	4·11	2·81	4·18	3·42	5·10	3·47	5·17	4·14	6·17	4·19	6·24
2½	65	3·90	5·80	3·98	5·92	4·38	6·51	4·46	6·63	5·31	7·90	5·39	8·02
3	80	4·58	6·81	4·69	6·98	5·69	8·47	5·80	8·64	6·76	10·1	6·87	10·3
4	100	6·64	9·89	6·84	10·2	8·14	12·1	8·34	12·4	9·71	14·4	9·41	14·7
5	125	—	—	—	—	10·9	16·2	11·2	16·7	12·0	17·8	12·3	18·3
6	150	—	—	—	—	12·9	19·2	13·3	19·8	14·3	21·2	14·7	21·8

SURFACES AREAS AND CONTENTS OF TUBES

Nominal Bore		*Surface*		*Area of inside*		*Contents*	
in.	*mm.*	*sq. ft./ft.*	*m^2/m.*	*sq. in.*	*sq. mm.*	*gal./ft.*	*l./m.*
$\frac{1}{8}$	6	0·11	0·036	0·086	55	0·0037	0·055
$\frac{1}{4}$	8	0·14	0·048	0·163	105	0·0071	0·105
$\frac{3}{8}$	10	0·18	0·060	0·277	178	0·0120	0·178
$\frac{1}{2}$	15	0·22	0·074	0·44	285	0·0196	0·285
$\frac{3}{4}$	20	0·28	0·093	0·75	472	0·0323	0·472
1	25	0·35	0·117	1·17	753	0·0506	0·753
$1\frac{1}{4}$	32	0·44	0·147	1·91	1,447	0·0828	1·45
$1\frac{1}{2}$	40	0·50	0·167	2·52	1,872	0·110	1·87
2	50	0·62	0·208	3·98	2,910	0·173	2·91
$2\frac{1}{2}$	65	0·78	0·262	6·47	4,610	0·281	4·61
3	80	1·05	0·306	8·90	5,740	0·386	5·74
4	100	1·18	0·393	14·9	9,580	0·645	9·58
5	125	1·44	0·481	22·5	14,500	0·973	14·5
6	150	1·70	0·568	31·7	20,460	1·37	20·5
7	175	1·96	0·60	38·5	25,590	1·67	25·6
8	200	2·23	0·68	50·3	32,690	2·37	32·7
9	225	2·49	0·76	63·6	41,000	2·75	41·0
10	250	2·74	0·84	77·5	50,700	3·35	50·7
11	275	3·01	0·92	91·2	60,700	3·95	60·7
12	300	3·27	1·00	110·7	72,100	4·77	72·1

SUGGESTED MAXIMUM WORKING PRESSURES

The pressures given below can be taken as conservative estimates for tubes screwed taper with sockets tapped parallel under normal (non-shock) conditions

	Grade	*Nom. Bore*	$\frac{1}{8}$ *to 1 in.*	$1\frac{1}{4}$ & $1\frac{1}{2}$ *in.*	2 & $2\frac{1}{2}$ *in.*	*3 in.*	*4 in.*	*5 in.*	*6 in.*
Water	*light*	*lb/in^2*	150	125	100	100	80	—	—
		kN/m^2	1000	850	700	700	550	—	—
	medium	*lb/in^2*	300	250	200	200	150	150	125
		kN/m^2	2000	1750	1400	1400	1000	1000	850
	heavy	*lb/in^2*	350	300	250	250	200	200	150
		kN/m^2	2400	2000	1750	1750	1400	1400	1000
Steam or air	*medium*	*lb/in^2*	150	125	100	100	80	80	60
		kN/m^2	1000	850	700	700	550	550	400
	heavy	*lb/in^2*	175	150	125	125	100	100	80
		kN/m^2	1200	1000	850	850	700	700	550

The following allowed for plain end tubes end-to-end welded for steam or compressed air.

	medium	*lb/in^2*	250	200	200	150	150	150	125
		kN/m^2	1750	1400	1400	1000	1000	1000	850
	heavy	*lb/in^2*	300	300	300	200	200	200	175
		kN/m^2	2000	2000	2000	1400	1400	1400	1200

HEAVY GAUGE COPPER TUBES: B.S.61 : Part 1 : 1947

NOTE: B.S.61 has not been republished with metric dimensions. Metric standards are contained in B.S.2871 :1972 (see page II. 16).

Nominal Bore	*Low Pressure Tubes*		*Medium Pressure Tubes*		*High Pressure Tubes*		*Nominal Bore*
in.	*Thickness S.W.G.*	*Weight lb./ft.*	*Thickness S.W.G.*	*Weight lb./ft.*	*Thickness S.W.G.*	*Weight lb./ft.*	*in.*
½	15	0·50	14	0·53	12	0·91	½
¾	15	0·72	13	0·89	11	1·30	¾
1	14	1·04	12	1·33	10	1·84	1
1¼	14	1·29	12	1·64	9	2·63	1¼
1½	14	1·52	12	1·96	9	3·01	1½
2	13	2·33	12	2·61	9	3·85	2
2½	13	2·88	11	3·60	7	5·94	2½
3	12	3·90	10	4·77	6	7·60	3
3½	11	5·07	9	6·25	5	9·60	3½
4	10	6·39	8	7·93	4	11·85	4

LIGHT GAUGE COPPER TUBES: B.S. 659:1967

Nominal Bore in.	*Outside Dia.*		*Thickness*		*Weight*	
	in.	*mm.*	*in.*	*mm.*	*lb./ft.*	*kg./m.*
⅛	0·205	5·21	0·022	0·56	0·05	0·08
3/16	0·283	7·19	0·024	0·61	0·08	0·12
¼	0·346	8·79	0·024	0·61	0·10	0·15
⅜	0·471	11·96	0·026	0·66	0·15	0·22
½	0·596	15·14	0·028	0·71	0.20	0·30
¾	0·846	21·49	0·034	0·86	0·35	0·52
1	1·112	28·25	0·036	0·91	0·48	0·72
1¼	1·362	34·59	0·044	1·12	0·72	1·1
1½	1·612	40·94	0·044	1·12	0·86	1·3
2	2·128	54·05	0·048	1·22	1·2	1·8
2½	2·628	66·75	0·048	1·22	1·5	2·3
3	3·144	79·86	0·056	1·42	2·1	3·2
3½	3·660	92·96	0·056	1·42	2·5	3·7
4	4·184	106·27	0·064	1·63	3·2	4·8
5	5·184	131·67	0·064	1·63	4·0	5·9
6	6·208	157·68	0·072	1·83	5·4	8·0

COPPER, SOIL, WASTE & VENTILATING PIPES

(G.L.C. BYE-LAWS)

Internal dia. in.	*Thickness S.W.G.*	*Weight lb./yard*
1¼	18	2·25
1½	18	2·70
2	17	4·17
2½	17	5·19
3	16	7·11
3½	15	9·33
4	14	11·85
4½	14	13·29
5	14	14·76
6	14	17·64

COPPER TUBE TO B.S.2871:1972

Nominal Bore mm.	Outside Dia. mm.	Table X Half Hard Light Gauge Tube		Table Y Half Hard and Annealed Tube		Table Z Hard Drawn Thin Wall Tube	
		Thickness mm.	Maximum Working Pressure N/mm.^{2}	Thickness mm.	Maximum Working Pressure N/mm.^{2}	Thickness mm.	Maximum Working Pressure N/mm.^{2}
6	6	0·6	13·3	0·8	14·4	0·5	11·3
8	8	0·6	9·7	0·8	10·5	0·5	9·8
10	10	0·6	7·7	0·8	8·2	0·5	7·8
12	12	0·6	6·3	0·8	6·7	0·5	6·4
15	15	0·7	5·8	1·0	6·7	0·5	5·0
18	18	0·8	5·6	1·0	5·5	0·6	5·0
22	22	0·9	5·1	1·2	5·7	0·6	4·1
28	28	0·9	4·0	1·2	4·2	0·6	3·2
35	35	1·2	4·2	1·5	4·1	0·7	3·0
42	42	1·2	3·5	1·5	3·4	0·8	2·8
54	54	1·2	2·7	2·0	3·6	0·9	2·5
67	67	1·2	2·0	2·0	2·8	1·0	2·0
76·1	76·2	1·5	2·4	2·0	2·5	1·2	1·9
108	108·1	1·5	1·7	2·5	2·2	1·2	1·7
133	133·4	1·5	1·4	—	—	1·5	1·6
159	159·4	2·0	1·5	—	—	1·5	1·5

STANDARD DIMENSIONS OF FLANGED FITTINGS

Nominal Bore	A	B	C	D	E	F	G	H	J
ins.	ins.	ins.	ins.	ins.	ins.	ins.	ft. ins.	ft. ins.	ft. ins.
1	4	5½	3	6	3	4	1 5	1 6	0 9
2	5	6½	3½	9½	6	5	2 0	2 1	1 3
2½	5½	7½	4	11½	7½	5½	2 3	2 6	1 6
3	6	8½	4½	13	9	6	2 6	2 9	1 9
3½	6½	9½	4½	15½	10½	6½	2 9	3 3	2 0
4	7	10½	5	17	12	7	3 1	3 7	2 3
5	8	11½	5½	21	15	8	4 2	4 8	3 0
6	9	12½	6	25	18	9	4 6	5 3	3 6
7	10	13½	6½	31½	24½	10	5 6	6 6	4 1
8	11	15	7	36	28	11	6 6	7 5	4 8
9	12	16½	7½	39½	31½	12	7 8	8 4	5 3
10	13	18	8	49	40	13	9 7	10 6	6 8
12	15	21	9	58	48	15	12 1	12 10	8 0

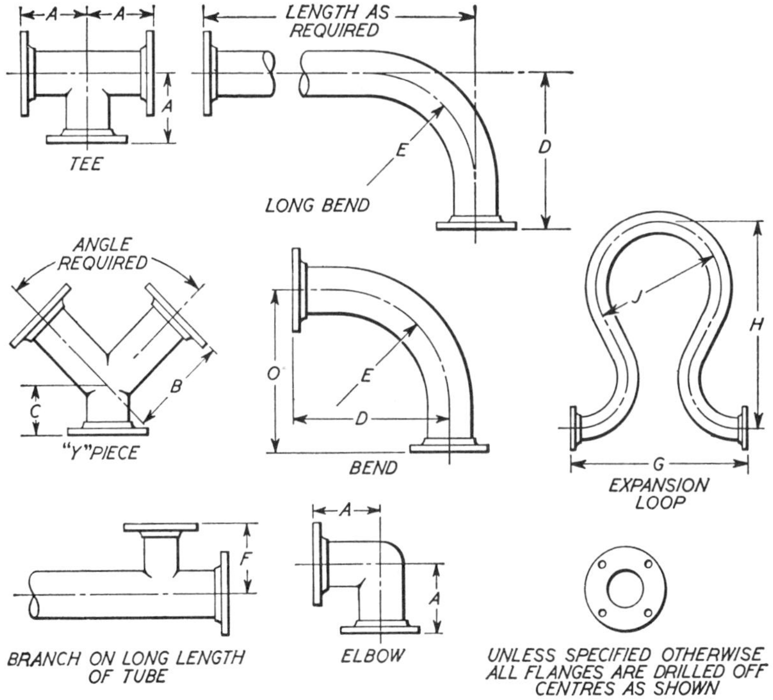

FLANGED FITTINGS

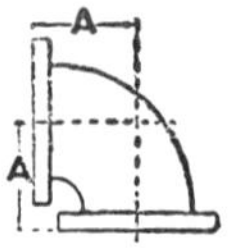

No. A51. Equal
Reducing.

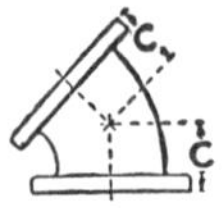

No. A54

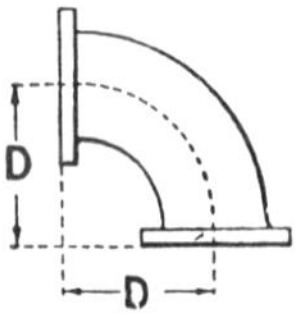

No. A56

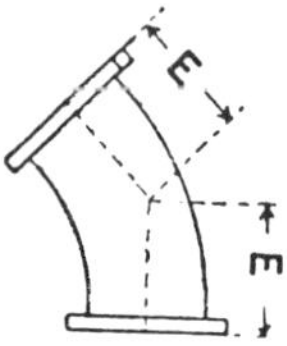

No. A57

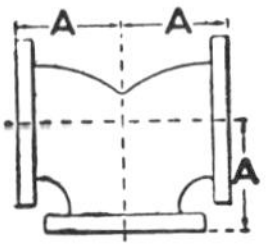

No. A59. Equal.
Reducing.

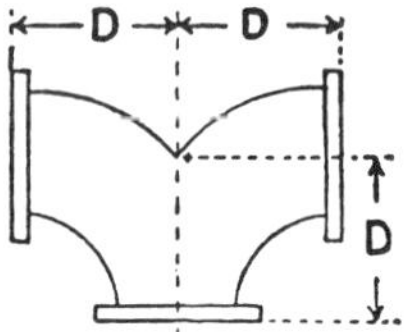

No. A61

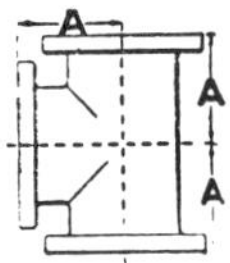

No. A63

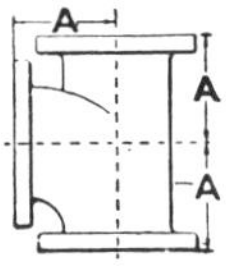

No. A68

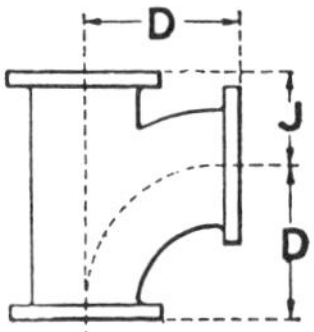

No. A70

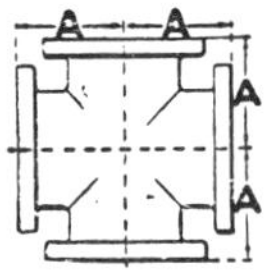

No. A72. Equal.
Reducing.

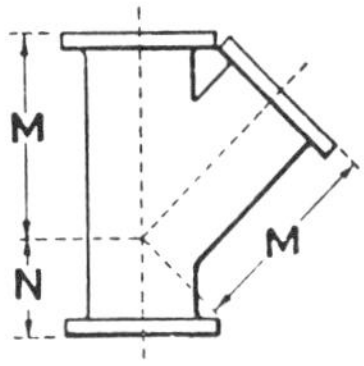

No. A74

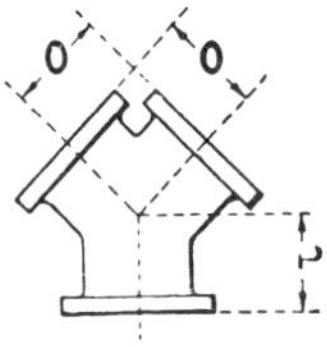

No. A79

CRANE CAST IRON FLANGED FITTINGS

Flanges to B.S.T. "D" and "E" diameter

General Dimensions

Size	in. 2	in. 2½	in. 3	in. 3½	in. 4	in. 5	in. 6	in. 7	in. 8
A	5	5½	6	6½	7	8	9	10	11
C	2½	3	3	3½	4	4½	5	5½	5½
D	10	10	12	14	14	15	16	16	18
E	5	5½	6	8	8	9	10	11	11
J	3	3½	4	4	4	5	6	6	7
M	8	9½	10	11½	12	13½	14½	16½	17½
N	2½	2½	3	3	3	3½	3½	4	4½
O	5	5½	6	6½	7	8	9	10	11½
P	2½	2¾	3	3¼	3½	4	4½	5	5½

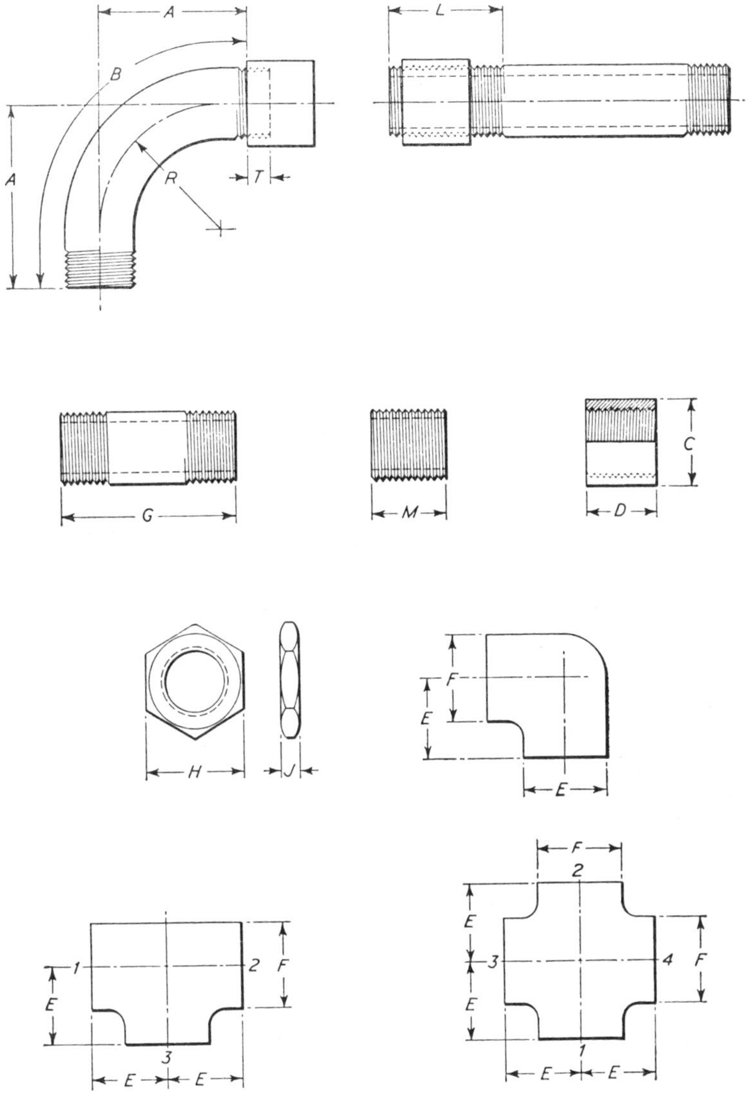

SPECIFY OUTLETS OF TEES AND CROSSES IN ORDER AS NUMBERED ABOVE

APPROXIMATE DIMENSIONS (in inches) OF STEEL TUBULAR FITTINGS TO B.S. 1387 : 1967 WROUGHT (Heavyweight) FITTINGS TO B.S. 1740 : 1971

(Subject to standard tolerances)

SIZE	in.	1/8	1/4	3/8	1/2	3/4	1	1 1/4	1 1/2	2	2 1/2	3	4	5	6
90° BENDS, No. 8	A	2	2 1/2	2 7/8	3 3/8	4	4 3/4	5 3/4	6 1/2	8	9 3/4	11 1/2	15	21	24 1/2
90° BENDS, No. 8 and SPRINGS, Nos. 9, 10 & 11	R	1 1/4	1 5/8	1 7/8	2 1/4	2 7/8	3 1/2	4 1/4	5	6 1/4	7 3/4	9 1/8	12 1/8	18	21
	B	3 3/4	4 1/4	4 7/8	5 1/2	6 3/4	8	10 1/8	11 3/8	13 1/4	16 1/4	19	24 3/4	34 1/4	40
SOCKETS, No. 18	C	19/32	23/32	7/8	1 1/16	1 9/32	1 9/16	1 15/16	2 3/16	2 11/16	3 5/16	3 7/8	4 7/8	5 15/16	7
	D	3/4	1 1/16	1 1/8	1 7/16	1 9/16	1 13/16	2	2	2 3/8	2 11/16	2 15/16	3 7/16	3 3/4	3 3/4
ELBOWS, No. 15 ... ... TEES, No. 16	E	5/8	7/8	15/16	1 1/4	1 3/8	1 11/16	2	2 1/16	2 1/2	3	3 1/2	4 1/2	5 1/2	6 1/4
CROSSES, No. 17	F	19/32	27/32	7/8	1 1/16	1 9/32	1 9/16	1 15/16	2 3/16	2 11/16	3 5/16	3 7/8	4 7/8	5 15/16	7
BARREL NIPPLES, No. 7 ...	G	1 1/4	1 1/2	1 1/2	2	2 1/8	2 3/8	2 3/4	2 3/4	3 1/8	3 1/2	4	4 1/2	4 7/8	5
BACKNUTS, No. 22	H	3/4	7/8	1 1/16	1 1/4	1 7/16	1 13/16	2 3/8	2 1/2	2 1/16	3 13/16	4 1/2	5 1/2	6 3/4	7 3/4
	J	1/4	1/4	9/32	5/16	11/32	3/8	7/16	15/32	17/32	11/16	13/16	7/8	13/16	1
RUNNING NIPPLES, No. 6 ...	M	1/2	6	13/16	1 1/16	1 3/16	1 5/16	15	15	1 7/8	2	2 1/4	2 3/4	3 1/16	3 1/16
CONNECTORS, Nos. 4 & 5 ...	L	1 1/8	1 15/32	1 9/16	2 23/32	2 1/8	2 7/16	2 3/4	2 25/32	3 7/32	3 13/16	4 3/16	4 3/4	5 1/8	5 13/16
APPROX. THREAD ENGAGEMENT WHEN WRENCH TIGHT	T	5/16	13/32	7/16	1/2	5/8	11/16	6	13/16	7/8	1 1/16	1 1/8	1 3/8	1 9/16	1 5/8

Section III

FUEL AND COMBUSTION

CLASSES OF FUELS

1. Solid. 2. Liquid. 3. Gaseous.

CLASSIFICATION OF COAL

Wood.
Peat.
Lignite.
Bituminous Coal.
Anthracite.

NAMES AND SIZES OF SOME TRADE SIZES OF COAL

Egg	$3\frac{1}{4}$ to $2\frac{3}{8}$ in.
Stove	$2\frac{3}{8}$ to $1\frac{5}{8}$ in.
Nut	$1\frac{5}{8}$ to $\frac{7}{8}$ in.
Pea	$\frac{7}{8}$ to $\frac{5}{8}$ in.
Buck-Wheat No. 1	$\frac{5}{8}$ to $\frac{3}{8}$ in.

BULK OF VARIOUS FUELS

Density and Specific Volume of stored fuel.

Fuel	*Density*		*Specific Volume*	
	lb. per cu. ft.	*kg. per m.*^{3}	*cu. ft. per ton*	*m.*^{3} *per 1,000 kg.*
Wood	22·5–24	360-385	90–100	2·5–2·8
Charcoal, hard wood ...	9·3	149	240	6·7
Charcoal, soft wood ...	13·5	216	165	4·6
Anthracite	45–53	720–850	42–50	1·2–1·4
Bituminous Coal ...	43–50	690–800	45–52	1·2–1·5
Peat	19·5–25	310–400	90–115	2·5–3·2
Coke	23·5–31	375–500	72–95	2·0–2·7
Gas Oil	52	835	43	1·2
Fuel Oil	58	930	39	1·1
Kerosene	49	790	47	1·3

HEAT OF COMBUSTION OF IMPORTANT CHEMICALS

Substance	*Products of Combustion*	*Chemical Formula*	*Heat of Combustion*	
			kJ./kg.	*B.t.u./lb.*
Carbon ...	Carbon Dioxide	$C+O_2 = CO_2$... ...	33,950	14,590
			Complete combustion	
Carbon ...	Carbon Monoxide ...	$2C+O_2=2CO$	9,210	3,960
			Incomplete combustion	
Carbon Monoxide ...	Carbon Dioxide	$2CO+O_2=2CO_2$ ...	10,150	4,367
Hydrogen ...	Water Vapour	$2H_2+O_2=2H_2O$ ...	144,200	62,000
Sulphur ...	Sulphur Dioxide Vapour ...	$S+O_2=SO_2$	9,080	3,900
Methane Marsh Gas ...	Carbon Dioxide & Water Vapour	$CH_4+2O_2=CO_2+2H_2O$	55,860	24,017

FUEL SIZE FOR CENTRAL HEATING BOILERS

Output of Boiler in B.t.u. per hr.	*Up to* 20,000	20,000 *to* 40,000	40,000 *to* 120,000	120,000 *to* 250,000	250,000 *up*
Size of Anthracite, in.	$\frac{3}{4}$	$\frac{3}{4}$ to $1\frac{1}{4}$	$\frac{3}{4}$ to 2	2—3	2—4

THEORETICAL EVAPORATIVE POWER OF FUELS

is the mass of steam theoretically evaporated (without losses) from and at 100°C. by burning unit mass of fuel.

In S.I. units

$$\text{Theoretical Evaporative Power (kg. per kg.)} = \frac{\text{calorific value of fuel in kJ/kg}}{2257}$$

In Imperial units

$$\text{Theoretical Evaporative Power (lb. per lb.)} = \frac{\text{calorific value of fuel in B.T.U./lb.}}{970}$$

Actual Evaporative Power of Fuels = Theoretical evaporative power × Boiler efficiency.

IGNITION (OR KINDLING) TEMPERATURES OF VARIOUS FUELS

Wood	300°C	570°F
Peat	227°C	440°F
Bituminous Coal	300°C	570°F
Semi Anthracite Coal	400°C	750°F
Coke	700°C	1290°F
Hydrogen ...	500°C	930°F
Carbon-Monoxide	300°C	570°F
Carbon	700°C	1290°F
Petroleum ...	400°C	750°F
Benzene	415°C	780°F
Coal-Tar Oil ...	580°C	1080°F
Producer Gas ...	750°C	1380°F
Light Hydrocarbons	650°C	1200°F
Heavy Hydrocarbons	750°C	1380°F
Light Gas	600°C	1110°F
Naphtha	550°C	1020°F

ATOMIC WEIGHTS OF ELEMENTS OCCURRING IN COMBUSTION CALCULATIONS

Element	*Symbol*	*Atomic No.*	*Atomic Weight*
Carbon	C	6	12·011
Hydrogen	H	1	1·008
Nitrogen	N	7	14·007
Oxygen	O	8	15·9994
Phosphorus	P	15	30·9738
Sulphur	S	16	32·06

The Heat of Combustion or Calorific Value of a fuel is the heat produced by the complete combustion of unit mass of the fuel. It is expressed in B.t.u. per lb., or J per kg.
The calorific value may be obtained experimentally (Calorimeter), or by calculation from chemical analysis (Dulong's formula).

The Higher or Gross Calorific Value of a fuel is the amount of heat given out in the complete combustion of unit weight of the fuel in oxygen, when the products of combustion are cooled down to the initial temperature (15°C. or 60°F.) at which the oxygen is supplied.

The Lower or Net Calorific Value is the heat obtained by the complete combustion of unit of weight of the fuel in oxygen, when the products are cooled down to 100°C. or 212°F. and the steam is not condensed to water.

Lower Calorific Value = Higher Calorific Value − Heat of steam formed by combustion =

Higher Calorific Value − $(9H \times 970)$ B.t.u. per lb.

= Higher Calorific Value − $(9H \times 2257)$ kJ. per kg.

Determination of the Calorific Value of a Fuel

1. *By Calorimeter Test*—Measuring the heat which is liberated by the combustion of a test sample in oxygen.
2. *By Analysis*—Dulong's formula for higher calorific value (approx.)

$$H_h = 14{,}500C. + 62{,}000\left(H - \frac{O}{8}\right) + 4{,}000S. \text{ B.t.u. per lb.}$$

$$H_h = 33{,}700C. + 144{,}200\left(H - \frac{O}{8}\right) + 9300S. \text{ kJ. per kg.}$$

C, H, O, S = Weight of carbon, hydrogen, oxygen, sulphur resp. in lb. per 1 lb. of fuel. = kg. per 1 kg. of fuel.

Weight of Dry Air theoretically required for the complete combustion:

$$A_{th} = 11{\cdot}47 + 34{\cdot}28\left(H - \frac{O}{8}\right) + 4{\cdot}31S. \text{ (kg. per kg. of fuel).}$$

Theoretical Weight of Dry Flue Gases:

$$F_{th} = 12{\cdot}47C. + 35{\cdot}28\left(H - \frac{O}{8}\right) + 5{\cdot}31S. + N \text{ (kg. per kg. of fuel).}$$

C, H, O = Weight in kg. per 1 kg. of fuel.

Actual Air supplied:

$A_a = X.A_{th}$ (kg. per 1 kg. of fuel)

X = Excess of air factor.

Actual Weight of Flue Gases:

$F_a\ F_{th} + (A_a - A_{th})$ (kg. per 1 kg. of fuel)

Excess of Air:

$$X = \frac{A_a}{A_{th}} = \frac{\text{Actual quantity of Air}}{\text{Theoretically required quantity of air}} =$$

$$\left(\frac{N}{N - 3{\cdot}782\,(O - \frac{1}{2}CO)} - 1\right) \times 100 \text{ (per cent).}$$

N, O, CO = Percentage of nitrogen, etc., by volume in flue gas.

$$X = \frac{\text{Maximum theoretical percentage of } CO_2 \text{ in flue gases}}{\text{Actual percentage of } CO_2 \text{ in flue gases}} \text{ (per cent).}$$

Excess of Air for good conditions:

For Anthracite and Coke	40%
For Semi-Anthracite, hand firing	70 to 100%
For Semi-Anthracite, with stoker ...	40 to 70%
For Semi-Anthracite, with travelling grate	30 to 60%
For Oil	10 to 20%
For Gas	10%

Temperature of Combustion, or Calorific Intensity of Dry Coal with Dry Air:

$$t = t_A + \frac{C_1 W_1}{S W_2}$$

t = Temperature produced in (°F or °C)

t_A = Temperature in boiler house in (°F or °C)

C_1 = Lower heat value in (B.t.u./lb. or kJ./kg.)

W_1 = Weight of fuel in (lb. or kg.)

W_2 = Weight of combustion products including surplus air (lb. or kg.)

S = Mean specific heat capacity of combustion products at constant pressure.

= 0·24 B.t.u. per lb. per deg.F.

= 1·0 kJ. per kg. per deg.C.

HEAT LOSSES IN A BOILER FURNACE

1. Sensible heat carried away by dry flue gases.
2. Heat lost by free moisture in fuel.
3. Heat lost by incomplete combustion.
4. Heat lost by unburned carbon in the ash.
5. Heat lost by radiation and unaccounted losses.

1. Sensible Heat carried away by Dry Flue Gases:

(in good practice about 15%)

$$L_1 = Wc_p\,(t_1 - t_A)\ \text{(kJ. per kg. of fuel)}$$

$$= Wc_p\,(t_1 - t_A)\frac{100}{CO_2}\ \text{(per cent).}$$

W = Weight of combustion products (kJ. per kg. of fuel)

t_1 = Temperature of flue gas in °C.

t_A = Temperature in boiler room in °C.

c_p = 1·0 = Mean specific capacity heat of flue gas in kJ. per kg. per deg.C.

2. Heat Lost by Free Moisture in Fuel:

$$L_2 = w\,(H - h)\ \text{(kJ. per kg. of fuel)}$$

$$= w\,(H - h)\ \frac{100}{C_1}\ \text{(per cent).}$$

w = Weight of water kg. per kg. of burned fuel.

H = Total heat of superheated steam at temperature t_1 and atmospheric pressure.

h = Sensible heat of water at temperature t_A.

3. Heat Lost by Combustible in the Flue Gas (incomplete combustion):

$$L_3 = 24{,}000\,\frac{CO}{CO_2 + CO} \times C\ \text{(kJ. per kg. of fuel)}$$

$$= 24{,}000\,\frac{CO}{CO_2 + CO} \times C \times \frac{100}{C_1}\ \text{(per cent)}$$

CO, CO_2 = Percentage of carbon-monoxide and carbon-dioxide respectively by volume in flue gas.

C = Weight of carbon in kg. of fuel.

C_1 = Lower calorific value of fuel, kJ. per kg.

The heat loss occasioned by CO escaping in the flue gas is approximately 4% to 5% for 1% CO in the flue gas.

4. Heat Lost by Unburned Carbon in the Ash:

$L_4 = W_1 \times$ calorific value of carbon (kJ. per kg. of fuel).

$= W_1 \times$ calorific value of carbon $\frac{100}{C_1}$ (per cent).

2·5 to 5 per cent for good practice.

W_1 = weight of carbon in ash (kg. per kg. of fuel).

5. Heat Lost by Radiation and Unaccounted Losses

L_5 = Lower calorific value. $-(\text{Utilized heat} + L_1 + L_2 + L_3)$.

HEAT BALANCE FOR BOILER FURNACE

$$C_1 = H_u + (L_1 + L_2 + L_3 + L_4 + L_5)$$

C_1 = Lower calorific value of fuel (kJ. per kg.)

H_u = Heat utilized from 1 kg. of fuel (kJ. per kg.).

$L_1 \ldots L_5$ = Heat losses kJ. per kg.

The Output of a Boiler is generally stated as the equivalent evaporation from and at 212°F. = 100°C.

Equivalent Evaporation: $W_1 = \frac{W_s\,(H-h)}{L}$ (kg. per hr.).

Efficiency of Boiler $= \frac{\text{Heat in steam supplied}}{\text{Heat in fuel supplied}} = \frac{W_s\,(H-h)}{B \times C_1}$.

In which:

W_1 = Equivalent evaporation (kg. per hr.).

Ws = Weight of steam supplied produced at any given temperature and pressure (kg. per hr.).

H = Total heat of steam raised (kJ. per kg.).

h = Sensible heat of feed water (kJ. per kg.).

L = Latent heat of water at atmospheric pressure (kJ. per kg.).

B = Weight of fuel burned per hour (kg. per hr.).

C = Lower calorific value of fuel (kJ. per kg.).

CONSTITUENTS AND HEATING VALUES OF FUELS

Fuel	*Composition per cent by Weight*						*Higher Calorific Value*	
	C	*H*	*O+N*	*S*	H_2O	*Ash*	*kJ./kg.*	*B.t.u./lb.*
Anthracite	83–87	3·5–4·0	3·0–4·7	0·9	1–3	4–6	32,500–34,000	14,000–14,500
Semi-Anthracite	63–76	3·5–4·8	8–10	0·5–1·8	5–15	4–14	26,700–32,500	11,500–14,000
Bituminous Coal	46–56	3·5–5·0	9–16	0·2–3·0	18–32	2–10	17,000–23,250	7,300–10,000
Lignite	37	7	13·5	0·5	37	5	16,300	7,000
Peat	38–49	3·0–4·5	19–28	0·2–1·0	16–29	1–9	13,800–20,500	5,500–8,800
Coke	80–90	0·5–1·5	1·5–5·0	0·5–1·5	1–5	5–12	28,000–31,000	12,000–13,500
Charcoal	84	1	—	—	12	3	29,600	12,800
Wood (Dry)	35–45	3·0–5·0	34–42	—	7–22	0·3–3·0	14,400–17,400	6,200–7,500
							kJ./m.³	*B.t.u./cu. ft.*
Town Gas	26	56	18	—	—	—	18,600	500
Natural Gas	75	25	—	—	—	—	37,200	1000
Propane C_3H_8	82	18	—	—	—	—	93,900	2520
Butane C_4H_{10}	83	17	—	—	—	—	130,000	3490
							kJ./l.	*B.t.u./gal.*
Gas Oil (35 secs.)	86·2	13·0	—	0·8	—	—	38,000	164,000
Heavy Fuel Oil (3500 secs.)	85·0	10·8	—	3·8	—	—	41,200	177,000

Fuel	*Theoretical Air for Combustion Volume at S.T.P.*		*Theoretical Flue Gas Produced Volume at S.T.P.*	
	m.³/kg.	*ft.³/lb.*	*m.³/kg.*	*ft.³/lb.*
Anthracite ...	9·4	150	9·5	152
Semi-Anthracite	8·4	135	8·6	137
Bituminous Coal	6·9	110	7·0	112
Lignite	5·7	92	5·8	93
Peat	5·7	92	5·9	94
Coke	8·4	134	8·4	135
Charcoal	8·4	134	8·4	135
Wood (Dry) ...	4·4	70	5·0	80
	m.³/m.³	*ft.³/ft.³*	*m.³/m.³*	*ft.³/ft.³*
Town Gas ...	4	4	3·8	3·8
Natural Gas ...	9·5	9·5	8·5	8·5
Propane C_3H_8 ...	24·0	24·0	22	22
Butane C_4H_{10} ...	31	31	27	27
	m.³/l.	*ft.³/gal.*	*m.³/l.*	*ft.³/gal.*
Gas Oil (35 secs.)	9·8	1,570	10·4	1,670
Heavy Fuel Oil ... (3,500 secs.)	10·8	1,730	11·6	1,860

SENSIBLE HEAT CARRIED AWAY BY FLUE GASES

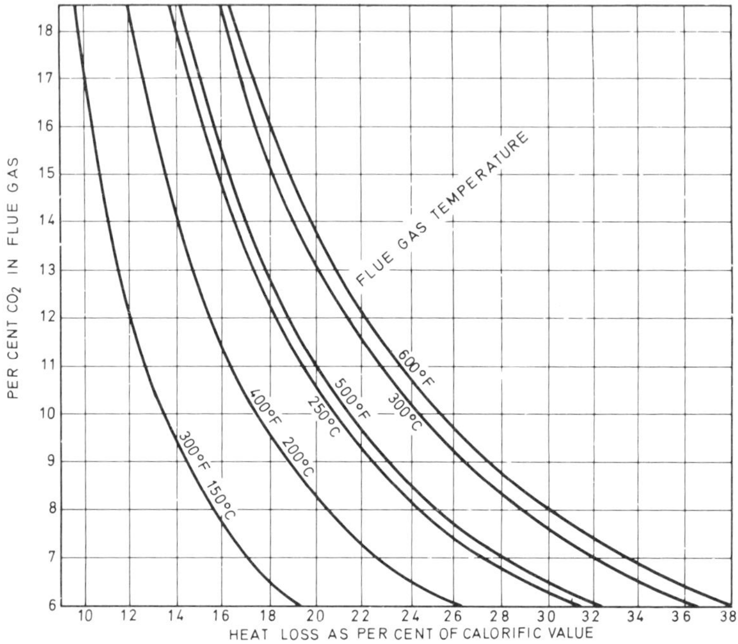

HEAT LOSS IN FLUE GASES FOR COAL

THE RINGELMANN SCALE FOR GRADING DENSITY OF SMOKE						
NUMBER	*0*	*1*	*2*	*3*	*4*	*5*
LINES, mm	-	*1*	*2·3*	*3·7*	*5·5*	*ALL BLACK*
SQUARES, mm	*ALL WHITE*	*9*	*7·7*	*6 3*	*4·5*	-

SENSIBLE HEAT CARRIED AWAY BY FLUE GASES

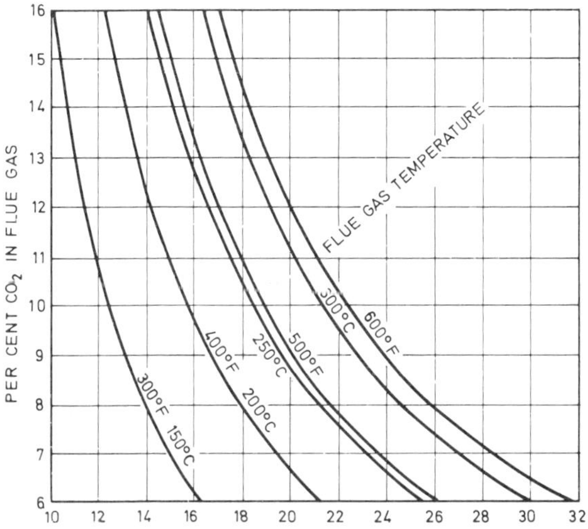

HEAT LOSS IN FLUE GAS FOR HEAVY OIL

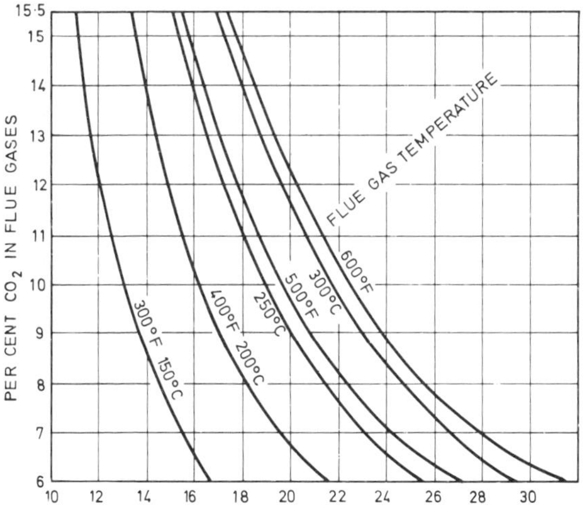

HEAT LOSS IN FLUE GAS FOR GAS OIL

ORSAT'S APPARATUS FOR ANALYSING FLUE GASES

For the Absorption of Carbon-dioxide, 50 g. of potassium hydrate or a little less sodium hydrate is dissolved in 150 g. of water.

For the Absorption of Oxygen, 17 g. of pyrogallol are dissolved in 35 g. of hot water. After cooling, this solution is mixed with 40 g. of potassium hydrate in 115 g. of water.

For the Absorption of Carbon Monoxide a solution is prepared by boiling a solution of cuprous chloride with metallic copper. When the solution has assumed a very dark colour it is poured into water and the Cu_2Cl_2 is precipitated as a white powder. This is thoroughly washed, out of contact with air if possible, and is dissolved in ammonia or hydrochloric acid to give the absorption solution.

The cuprous chloride and pyrogallol are liable to oxidation, for which reason precautions are taken to exclude the air as much as possible.

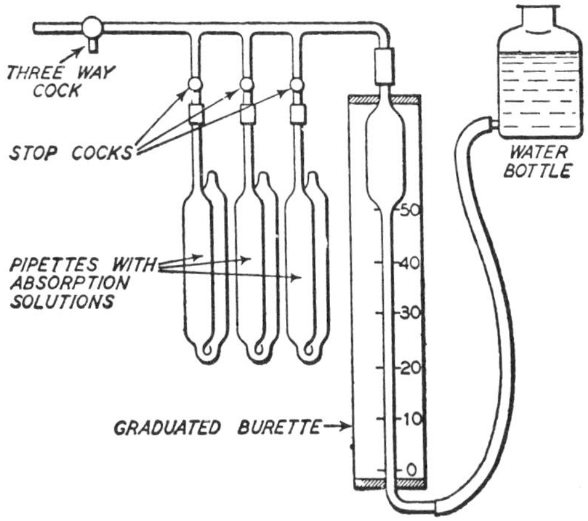

By manipulating the water bottle 100cm³ of fuel gas is drawn into the graduated burette. By again manipulating this 100 cm³ gas sample is passed into the pipette containing the solution absorbing carbon dioxide. After the carbon-dioxide has thus been removed the remainder of the sample is passed back into the burette and measured again.

The decrease in volume indicates the percentage, by volume, of CO_2 in the dry flue gas. In like manner the percentage of carbon-monoxide and oxygen is found.

Theoretical Chimney Draught

$$h = 354\,H\left(\frac{1}{T_1} - \frac{1}{T_2}\right)$$

where h = draught in mm. water
H = chimney height in m.
T_1 = absolute temperature outside chimney °K.
T_2 = absolute temperature inside chimney °K.

$$h = 7{\cdot}64H\left(\frac{1}{T_1} - \frac{1}{T_2}\right)$$

where h = draught in in. water.
H = chimney height in ft.
T_1 = absolute temperature outside chimney °R.
T_2 = absolute temperature inside chimney °R.

CHIMNEY AREA

The chimney should be designed to give a maximum velocity of 2 m./s. (7 ft./s.) for small furnaces, and 10 m./s. (30 ft./s.) for large furnaces.

$$A = \frac{Q}{V}$$

where A = cross-sectional area of chimney, $m.^2$
Q = volume of flue gases at chimney temperature, $m.^3/s.$
V = velocity, m./s.

An empirical rule is to provide 1100 $mm.^2$ chimney area per 1 kW boiler rating (1 $in.^2$ per 2000 B.t.u./hour boiler rating).

RECOMMENDED SIZES OF EXPLOSION DOORS OR DRAUGHT STABILIZERS FOR OIL FIRING INSTALLATIONS

Cross-sectional area of chimney, in.²	*Release opening of stabilizer, approx., in.*
40—80	6 × 9
80—200	8 × 13
200—300	13 × 18
300—600	16 × 24
600—1500	24 × 32

Cross-sectional area of chimney, m.²	*Release opening of stabilizer, approx., mm.*
0·025—0·050	150 × 230
0·050—0·125	200 × 330
0·125—0·200	330 × 450
0·200—0·400	400 × 600
0·400—1·000	600 × 800

COMBUSTION AIR

A boiler house must have openings to fresh air to allow combustion air to enter. An empirical rule is to allow 1600 $mm.^2$ free area per 1 kW boiler rating (1·5 $in.^2$ per 2000 B.t.u./hour boiler rating).

SPECIFICATIONS OF OIL FUELS

Fuel	*Gas Oil*	*Light Fuel Oil*	*Medium Fuel Oil*	*Heavy Fuel Oil*
Corresponding Class of B.S.2869	D	E	F	G
Viscosity ... ... (Redwood No. 1 at 100 secs.)	35	220	950	3,500
Density ... ... lb. per gal.	8·35	9·3	9·5	9·7
kg. per litre	0·835	0·93	0·95	0·97
Gross Calorific Value ... ... B.t.u./lb. ...	19,600	18,700	18,500	18,300
B.t.u./gal. ...	164,000	175,000	176,000	177,000
kJ./kg. ...	45,600	43,500	43,000	42,600
kJ./litre ...	38,000	40,400	40,800	41,200
Flash Point (Pensky-Martens closed) °F.	150	150	150	150
°C.	65·6	65·6	65·6	65·6
Pour Point ... °F. max.	20	35	70	70
°C. max.	–6·7	17	21	21
Temperature requirements: Storage Tank ... ... °F.	No heating required	45	65	75
°C.		7	18	24
Suction Line ... ... °F.		45	65	100
°C.		7	18	38
Max. Sulphur content per cent. by weight	0·8	3·2	3·5	3·5

FLOW OF OIL IN PIPES

Head Loss of Various Viscosities for Laminar Flow.

Visc. at Temp. in pipe, Redwood Sec. No. 1	35	100	200	1,000	2,000
i_1	$0{\cdot}49\times10^{-4}\frac{f_1}{d_1{}^4}$	$3{\cdot}25\times10^{-4}\frac{f_1}{d_1{}^4}$	$6{\cdot}4\times10^{-4}\frac{f_1}{d_1{}^4}$	$33{\cdot}3\times10^{-4}\frac{f_1}{d_1{}^4}$	$66{\cdot}1\times10^{-4}\frac{f_1}{d_1{}^4}$
i_2	$1{\cdot}6\times10^4\frac{f_2}{d_2{}^4}$	$10{\cdot}8\times10^4\frac{f_2}{d_2{}^4}$	$20{\cdot}5\times10^4\frac{f_2}{d_2{}^4}$	$109\times10^4\frac{f_2}{d_2{}^4}$	$220\times10^4\frac{f_2}{d_2{}^4}$

(For meaning of symbols see next page.)

$i_1 = i_2 =$ head loss in feet of oil per foot of pipe or metres of oil per metre of pipe. (Length of pipe to include allowances for bends, valves and fittings).

$f_1 =$ flow of oil in gal./hr.

$f_2 =$ flow of oil in litre/sec.

$d_1 =$ internal diameter of pipe in inches.

$d_2 =$ internal diameter of pipe in mm.

The above formulae are for laminar flow. Flow is laminar if Reynolds Number (Re) is less than 1500. Reynolds number can be checked from the following formulae. As Re is a dimensionless ratio it is the same in all consistent systems of units. The coefficients in the following formulae take into account the dimensions of f_1 d_1, f_2 d_2, respectively.

The viscosity to be taken is that at the temperature of the oil in the pipe. Heavier oils should be heated until the viscosity in the pipe is not more than 2000 secs. Redwood No. 1.

Viscosity Redwood No. 1	35	100	200	1,000	2,000
Re	$17{\cdot}5\frac{f_1}{d_1}$	$2{\cdot}6\frac{f_1}{d_1}$	$1{\cdot}44\frac{f_1}{d_1}$	$0{\cdot}26\frac{f_1}{d_1}$	$0{\cdot}13\frac{f_1}{d_1}$
	$35\times10^4\frac{f_2}{d_2}$	$5{\cdot}4\times10^4\frac{f_2}{d_2}$	$2{\cdot}8\times10^4\frac{f_2}{d_2}$	$0{\cdot}52\times10^4\frac{f_2}{d_2}$	$0{\cdot}26\times10^4\frac{f_2}{d_2}$

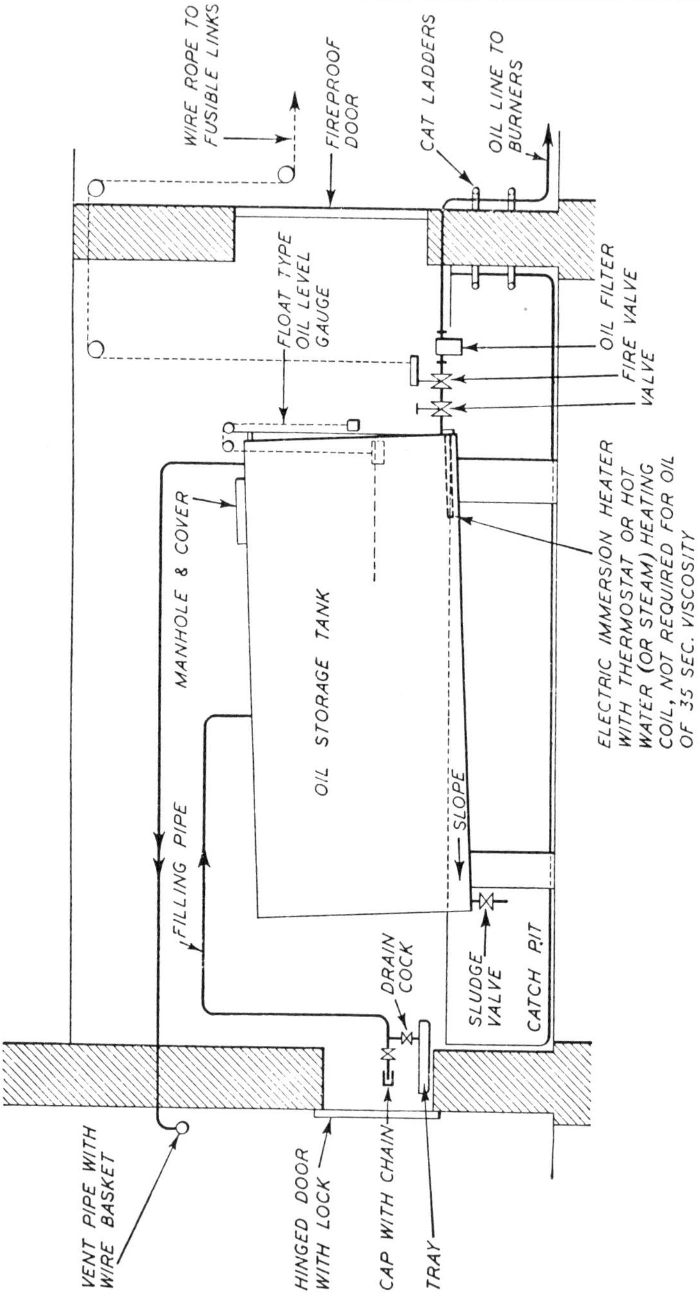

DIAGRAMMATIC ARRANGEMENT OF OIL STORAGE TANK

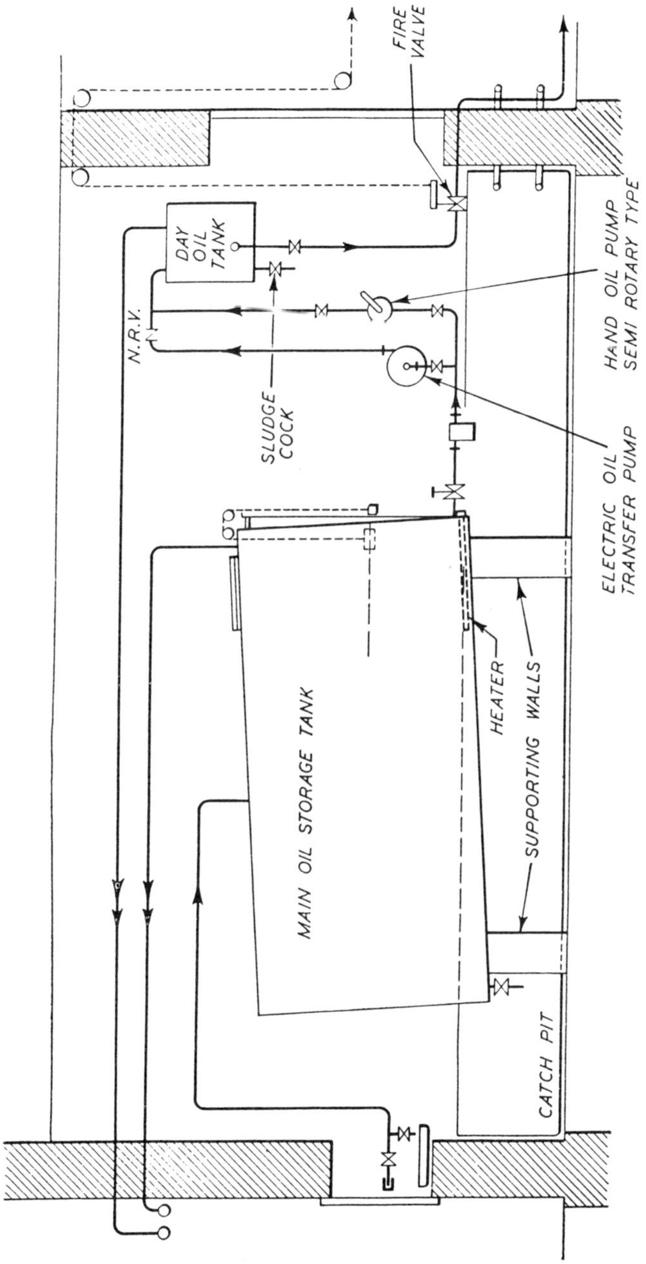

DIAGRAMMATIC ARRANGEMENT OF OIL STORAGE TANK AND DAY OIL TANK

APPROXIMATE CAPACITIES OF CYLINDRICAL TANKS

Approx. capacity in gallons	*Internal diameter ft.*	*in.*	*Overall length ft.*	*in.*	*Thickness of plate in.*	*Approx. weight in tons*	
						Fuel oil	*Tank*
500	4	6	6	0	$\frac{3}{16}$	2	0·4
650	4	6	7	0	$\frac{3}{16}$	2·6	0·45
750	4	6	8	6	$\frac{3}{16}$	3	0·5
1,000	4	6	11	0	$\frac{3}{16}$	4	0·6
1,250	4	10	11	8	$\frac{3}{16}$	5	0·7
1,500	5	0	13	6	$\frac{3}{16}$	6	0·9
1,500	6	0	9	6	$\frac{1}{4}$	6	1·0
2,000	6	0	12	6	$\frac{1}{4}$	8	1·25
2,000	6	6	10	9	$\frac{1}{4}$	8	1·2
2,500	6	6	13	3	$\frac{1}{4}$	10	1·5
3,000	7	0	13	6	$\frac{1}{4}$	12	1·5
4,000	7	0	17	9	$\frac{1}{4}$	16	1·9
4,000	7	6	15	6	$\frac{1}{4}$	16	1·9
5,000	7	6	19	6	$\frac{1}{4}$	20	2·3
5,000	8	0	15	3	$\frac{1}{4}$	20	2·0
6,000	8	6	18	3	$\frac{1}{4}$	24	2·5
6,000	9	0	16	6	$\frac{1}{4}$	24	2·9
7,000	8	6	21	0	$\frac{1}{4}$	28	2·6
7,000	9	0	19	0	$\frac{1}{4}$	28	2·8
8,000	8	6	24	0	$\frac{1}{4}$	32	3·2
8,000	9	0	21	6	$\frac{1}{4}$	32	3·1
9,000	8	6	27	0	$\frac{1}{4}$	36	3·6
9,000	9	0	24	0	$\frac{1}{4}$	36	3·4
10,000	8	6	30	0	$\frac{1}{4}$	40	3·9
10,000	9	0	27	0	$\frac{1}{4}$	40	3·8

RECTANGULAR TANKS

Approx. capacity in gallons	Length ft. in.	Width ft. in.	Height ft. in.	Thickness of plate (in.)	Approx. weight in tons: Fuel oil	Approx. weight in tons: Tank
500	5 0	4 0	4 0	3/16	2	0·4
625	6 3	4 0	4 0	3/16	2½	0·4
750	7 6	4 0	4 0	3/16	3	0·45
1,000	8 0	5 0	4 0	3/16	4	0·50
1,250	8 0	5 0	5 0	3/16	5	0·65
1,500	10 0	5 0	4 10	3/16	6	0·70
1,750	10 0	6 0	4 9	1/4	7	0·90
2,000	10 0	6 0	5 4	1/4	8	1·0
2,500	10 0	8 0	5 0	1/4	10	1·05
3,000	12 0	8 0	5 0	1/4	12	1·1

HEAT LOSS FROM OIL IN TANK AND PIPE LINES

	Position	Oil temperature		Heat Loss			
				Unlagged		Lagged	
		°F.	°C.	B.t.u./ft.²hr. °F.	W./m.² °C.	B.t.u./ft.²hr. °F.	W./m.² °C.
Tank ...	Sheltered	up to 50 50 – 80 80 – 100	up to 10 10 – 27 27 – 38	1·2 1·3 1·4	6·8 7·4 8·0	0·3 0·325 0·35	1·7 1·8 2·0
Tank ...	Exposed	up to 50 50 – 80 80 – 100	up to 10 10 – 27 27 – 38	1·4 1·5 1·6	8·0 8·5 9·0	0·35 0·375 0·4	2·0 2·1 2·25
Tank ...	In pit			NIL		NIL	
Pipe line ...	Sheltered	up to 80 80 – 260	up to 27 27 – 127				
Pipe line ...	Exposed	up to 80 80 – 260	up to 27 27 – 127				

For maximum heat loss of oil tanks in the open air the ambient temperature should be assumed as —4°C. (25°F.).

The specific heat capacity of heavy fuel oil is 1·88—2·00 kJ./kg. deg.C. (0·45—0·48 B.t.u./lb. deg. F.).

The heat transfer coefficients for coils are:

Steam to oil: approx. 11·3 W/m² deg.C. (20 B.t.u./ft.² hr. deg.F.).

Hot water to oil: approx. 5·7 W/m² deg.C. (10 B.t.u./ft.² hr. deg.F.).

APPROXIMATE CAPACITIES OF CYLINDRICAL TANKS

Approximate Capacity Litres	*Internal Diameter m.*	*Length m.*	*Thickness of Plate mm.*	*Approximate Weight 1,000 kg.*	
				Oil	*Tank*
2,000	1·25	1·7	5·0	1·9	0·36
2,500	1·25	2·0	5·0	2·4	0·40
3,000	1·25	2·5	5·0	2·8	0·50
3,500	1·25	2·85	5·0	3·3	0·55
4,000	1·25	3·25	5·0	3·8	0·60
5,000	1·5	2·8	5·0	4·7	0·67
6,000	1·5	3·4	5·0	5·7	0·80
7,000	1·75	3·0	6·5	6·6	1·1
8,000	1·75	3·3	6·5	7·5	1·2
9,000	1·75	3·75	6·5	8·5	1·3
10,000	2·0	3·25	6·5	9·4	1·4
15,000	2·25	3·75	6·5	14	1·75
20,000	2·5	4·1	6·5	19	2·1
25,000	2·5	5·1	6·5	24	2·6
30,000	2·5	6·1	6·5	28	3·0
30,000	2·75	5·1	6·5	28	2·9
35,000	2·75	5·9	6·5	33	3·2
40,000	2·75	6·75	6·5	38	3·6
45,000	2·75	7·6	6·5	43	4·0
45,000	3·0	6·4	6·5	43	3·8

RECTANGULAR TANKS

Approximate Capacity Litres	*Length m.*	*Width m.*	*Height m.*	*Thickness of Plate mm.*	*Approx. Weight 1,000 kg.*	
					Oil	*Tank*
2,000	1·5	1·2	1·2	5·0	2·0	0·5
2,500	1·8	1·2	1·2	5·0	2·4	0·5
3,000	2·0	1·25	1·2	5·0	2·8	0·5
3,500	2·0	1·5	1·2	5·0	3·4	0·6
4,500	2·0	1·8	1·25	5·0	4·3	0·7
4,500	2·5	1·5	1·2	5·0	4·3	0·7
5,000	2·5	1·5	1·4	5·0	4·7	0·7
6,000	3·0	1·5	1·4	6·5	5·7	1·1
7,500	3·0	1·8	1·4	6·5	7·1	1·2
9,000	3·0	1·8	1·7	6·5	8·5	1·4
12,000	3·0	2·75	1·5	6·5	11·3	1·7
15,000	3·5	2·75	1·6	6·5	14·1	1·8

Section IV

HEAT AND HEAT TRANSFER

Heat is a Form of Energy.

Amount or Quantity of Heat contained in any portion of matter is determined by:

1. The mass of the matter (w) in kg.
2. The specific heat capacity (c) in J/kg. deg. C.
3. The temperature (t) in °C.
4. The pressure (for gases only).

Amount of heat (sensible heat) (H) in J. $H = wct$.

One British Thermal Unit (B.t.u.) is the amount of heat required to increase the temperature of 1 lb. of water by 1°F. at atmospheric pressure.

One Kilogram Calorie (kcal) is the amount of heat required to increase the temperature of 1 kg. of water by 1°C.

One Joule is the energy produced by a force of one Newton moving through a distance of one metre. Since heat is a form of energy the Joule is also a unit of heat. In S.I. units mechanical energy and heat energy are measured in the same unit, namely the Joule.

Temperature is the intensity of heat.

Melting temperature of ice = 0°C. = 32°F. } at atm. pres.
Boiling temperature of water = 100°C. = 212°F. } at atm. pres.

Absolute temperature: $T = (t + 273°C.)$ (Celsius)
or $T = (t + 460°F.)$ (Fahrenheit)

Absolute zero temperature = −273°C.
= −460°F.

Conversion of degrees Fahrenheit into Celsius:

$$°F. = 32 + \frac{9}{5}C.; \qquad °C. = \frac{5}{9}(F - 32).$$

(Conversion table, see pages I. 6 to I. 8)

Expansion by Heat:

1. Linear Expansion is the increase in length
$$L_2 = L_1(1 + et)$$
2. Surface Expansion is the increase in area
$$A_2 = A_1(1 + 2et)$$

3. Volumetric expansion is the increase in volume

$$V_2 = V_1(1+3et)$$

where:

t	=	temperature difference in °C.
L_1, A_1 V_1	=	original length, area, volume respectively.
L_2, A_2 V_2	=	final length, area, volume respectively.
e	=	coefficient of linear expansion in metres per metre per °C.

The coefficient of linear expansion is the increase in length in metres per metre of original length per deg. C. temperature difference. (See IV. 15.)

Latent Heat is the heat energy which is required to produce changes in the physical state of a substance:

Latent heat of melting or fusion,

Latent heat of vaporization or evaporation,

Latent heat = Change of State Heat + External Work Heat.

Specific Heat Capacity of a Solid or Liquid is the amount of heat energy required to raise the temperature of 1 kg. of substance by 1 degree.

(a) Proper specific heat capacity

(b) Mean specific heat capacity

The specific heat capacity increases with the temperature.

Specific Heat Capacity of Gases:

1. Specific Heat Capacity at constant pressure C_p
2. Specific Heat Capacity at constant volume C_v

$$\frac{C_p}{C_v} = \gamma$$

(See Table page IV. 20)

Sensible Heat for Heating or Cooling:

$$H = cW\,(t_2 - t_1)$$

where			
H	=Heat	... ...	J.
W	=Mass	... ...	kg.
c	=specific heat capacity	...	J per kg. per deg. C.
t_1	=initial temperature	...	°C.
t_2	=final temperature	...	°C.

The Mechanical Equivalent of Heat is the number of foot-pounds in one British Thermal Unit.

1 B.t.u. =777·5 ft.-lb.
1 ft.-lb. =0·00128 B.t.u.
1 kcal =427 m.kg.
1 h.p. =42·42 B.t.u. per min.=2,545 B.t.u. per hr.

In S.I. units both mechanical energy and heat energy are measured in Joules and no conversion factor is needed.

Pressure is force per unit area:

$$P = \frac{W}{A}$$

P	=Pressure	... ...	N/m^2.
W	=Total load of force	...	N.
A	=Area	... ...	m^2.

Atmospheric Pressure is the pressure which is exerted by the earth's atmosphere.

The atmospheric pressure at sea-level at a temperature of 0°C. is 14·696 lb. per sq. in. = 1 atm.=29·921 ins. Hg.=$1{\cdot}013 \times 10^5$ N/m^2.

Absolute Pressure existing at any point in a fluid medium is the true total pressure.

(Absolute pressure)=(Gauge pressure) + (Atmospheric pressure).

Gauge Pressure is the difference between the true pressure and the pressure of the surrounding air.

(Gauge pressure) = (Absolute pressure) − (Atmospheric pressure).

Density is the weight of unit volume.

$$\rho = \frac{W}{V} = \frac{\text{Weight in kg.}}{\text{volume in m}^3\text{.}}$$

Specific Gravity is the ratio of the weight of a body to the weight of an equal volume of pure water at the temperature of maximum density (39°F.).

Specific Volume,is the volume of unit weight,

$$v = \frac{1}{\rho} = \frac{V}{W} = \frac{\text{volume in m}^3\text{.}}{\text{weight in kg.}}$$

Mass is the quantity of matter to which the unit of force will give unit acceleration.

$$\text{mass in slugs} = \frac{\text{weight in lb.}}{\text{gravitation of earth in ft. per sec.}^2}$$

$$\text{mass in kg.} = \frac{\text{weight in Newtons}}{\text{1 metre per sec.}^2}$$

First Law of Thermodynamics:

Heat and mechanical energy can be converted one to the other and when thus converted a definite relationship always exists.

Second Law of Thermodynamics:

1. Heat has never been known to flow of its own accord from a cold to a relatively hot body.
2. It is impossible to obtain by cooling any portion of matter below the temperature of the coolest surrounding objects.

THE LAWS OF PERFECT GASES

(These Laws do not apply to Vapours)

(Nearly true for Permanent Gases, such as Oxygen, Nitrogen, Hydrogen, and Atmospheric Air)

Relations between the Volume, Pressure and Temperature:

Normal Temperature and Pressure (N.T.P.): For convenience, the properties of gases are always given on a standard basis of temperature and pressure which is known as normal temperature and pressure.

Normal temperature = 0°C. = 32° F.

Normal pressure = 14·7 lb. per sq. in. = 1 atm. = $1{\cdot}013 \times 10^5$ N/m².

Notations:

W = Weight of gas

$V_1\ V_2$ = Initial and final volume resp.

$P_1\ P_2$ = Initial and final pressure, resp.

$T_1\ T_2$ = Initial and final absolute temperature.

R = Gas constant.

Rm = Gas constant for mixtures.

1. Boyle's Law—Temperature Constant:

If the temperature of a given weight of gas is kept constant, the absolute pressure of the gas will vary inversely as its volume.

$$V_2 = \frac{V_1 P_1}{P_2} \quad \text{Final volume.}$$

$$P_2 = \frac{V_1 P_1}{V_2} \quad \text{Final abs. pressure.}$$

$$\frac{P_1}{P_2} = \frac{V_2}{V_1} \quad \text{(Ratio).}$$

or: $P_1 V_1 = P_2 V_2$.

$PV = \text{Constant}$.

2. Gay-Lussac's Law—Volume Constant:

If the volume of a given weight of gas is kept constant, the absolute pressure will vary directly as the absolute temperature of the gas.

$$\frac{P_1}{P_2} = \frac{T_1}{T_2} \text{ (Ratio).}$$

$$P_2 = \frac{P_1 T_2}{T_1} \text{ (Final pressure).}$$

$$T_2 = \frac{T_1 P_2}{P_1} \text{ (Final absolute temperature).}$$

3. Charles' Law—Pressure Constant:

If the absolute pressure of a given weight of any gas is kept constant its volume will vary directly as the absolute temperature of the gas

$$\frac{V_1}{V_2} = \frac{T_1}{T_2} \text{ (Ratio).}$$

$$V_2 = \frac{V_1 T_2}{T_1} \text{ (Final volume).}$$

$$T_2 = T_1 \frac{V_2}{V_1} \text{ (Final absolute temperature).}$$

The volume of a mass of gas varies by $\frac{1}{492} = 0{\cdot}002032$ of its volume at 32°F. for every 1°F. change of temperature, the pressure being kept constant.

4. Combined Boyle's, Charles' and Gay-Lussac's Laws:

$$\frac{P_1 V_1}{T_1} = \frac{P_2 V_2}{T_2}.$$

$$P_2 = \frac{P_1 V_1 T_2}{V_2 T_1} \text{ (Final abs. pressure).}$$

$$V_2 = \frac{P_1 V_1 T_2}{P_2 T_1} \text{ (Final volume).}$$

$$T_2 = \frac{P_2 V_2 T_1}{P_1 V_1} \text{ (Final abs. temperature).}$$

5. General Gas Law (the Perfect-Gas Law) states the relation which exists between the pressure, volume, absolute temperature and weight of a perfect gas.

$$PV = wRT \qquad R = \frac{PV}{wT} = \text{Gas constant.}$$

6. Universal Gas Constant (applied for any gas):

$wR = 1{\cdot}985$ B.t.u./lb. mol. °F.
$= 1{\cdot}985$ kcal./kg. mol. °C.
$= 8{\cdot}3143$ kJ./kg. mol. °C.

(At N.T.P., 1 kg. mol. occupies 22·4 m.^{3}, 1 lb. mol. occupies 359 ft.^{3}.)

Mixture of Gases:

$$PV = wRmT.$$

$$Rm = \frac{R_1 w_1 + R_2 w_2 + R_n w_n}{w_1 + w_2 + w_n}.$$ Gas constant of mixture.

Dalton's Law of Gases:

Each separate gas, in a mixture of gases, responds to change of pressure, volume, and temperature exactly as though it were entirely isolated from the other gases.

The total pressure of a mixture of gases is equal to the sum of the pressures of all gases in the mixture.

P = Pressure of mixture.
P_1 P_2 = Partial gas or vapour pressure.
$P = P_1 + P_2$.

Avogadro's Law:

Equal volumes of all gases at the same temperature and pressure contain the same number of molecules.

METHODS OF HEATING OR EXPANDING GASES (NOT VAPOURS)

Type of Expansion.	*Remarks.*	*Work Done = W*	*Change of Intl. Energy = E*	*Heat Absorbed or rejected = H*	*Final Temp.*
Constant Pressure	Isobar	$\frac{P(V_2-V_1)}{J}$	$MCv(T_2-T_1)$	$MCp(T_2-T_1)$	$T_1 \frac{V_2}{V_1}$
Constant Temperature	Isotherm Boyle's Law PV = Constant	$\frac{P_1V_1 \log_e \frac{V_2}{V_1}}{J}$	O	W	T_1
Constant Heat	Adiabatic PV^γ=Constant	$\frac{P_1V_1-P_2V_2}{J(\gamma-1)}$	$MCv(T_2-T_1)$	O	$\left(\frac{V_1}{V_2}\right)^{\gamma-1} = \frac{T_2}{T_1}$
Int. Energy & Temp.Change	Polytrope PV^n=const.	$\frac{P_1V_1-P_2V_2}{J(n-1)}$	$MCv(T_2-T_1)$	W+E	$\left(\frac{V_1}{V_2}\right)^{n-1} = \frac{T_2}{T_1}$

in which: W = External work done by gas.
E = Increase of internal energy of the gas.
H = Total heat absorbed or rejected.
C_p = Specific heat of the gas at constant pressure.
C_v = Specific heat of the gas at constant volume.
$\gamma = \frac{C_p}{C_v}$.
J = mechanical equivalent of heat.
V_1 P_1 T_1 = Initial volume, pressure, temperature, resp.
V_2 P_2 T_2 = Final volume, pressure, temperature, resp.
$n = \frac{\log P_1 - \log P_2}{\log V_2 - \log V_1}$ (For indicator diagrams).
M = Mass.

ENTROPY OF GASES

Heat energy transfer may be expressed as the product of entropy and absolute temperature.

Definition:

The change in entropy of a substance, between two thermal conditions, is the heat energy transferred to the substance, as heat, per degree of average absolute temperature between the two conditions;

or: The change in entropy between two thermal conditions is a value such that when it is multiplied by the proper average of all the absolute temperatures which the substance experienced during the heat transfer between the two conditions, it will give as a product the total energy added or abstracted from the substance—as heat—during the transfer.

Usual symbol for Entropy $=\phi$.

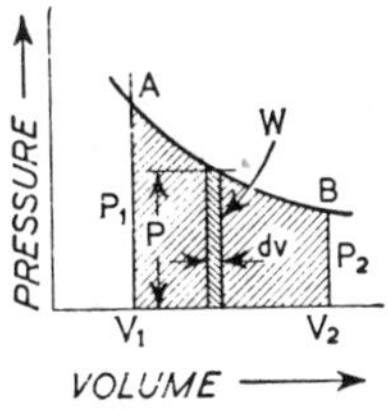

VOLUME - PRESSURE DIAGRAM

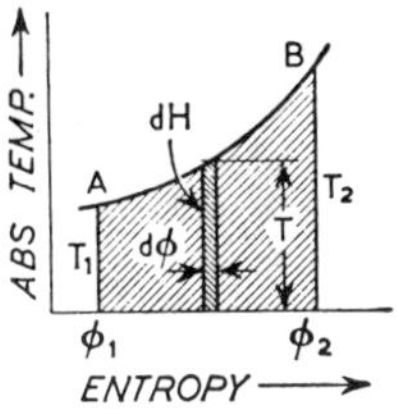

TEMP. - ENTROPY DIAGRAM

T = Absolute temperature.
dH = Small amount of heat.
W = Work done.
P = Pressure.
V = Volume.
ϕ = Entropy.
$d\phi$ = Change of Entropy.

General Formulæ:

$$dH = Td\phi.$$

$$d\phi = \frac{dH}{T}.$$

$$\phi_2 - \phi_1 = \int_{T_1}^{T_2} \frac{dH}{T}.$$

$$W = \int_{V_1}^{V_2} PdV.$$

Heat energy change = (Entropy change) × (Average absolute temperature).

The Critical Temperature of a substance is that temperature above which it cannot exist as a liquid.

The Critical Pressure is the pressure of a saturated vapour at its critical temperature.

CRITICAL TEMPERATURES AND PRESSURES OF VARIOUS SUBSTANCES

Substance.	*Critical Temperature* °F.	°C.	*Critical Pressure Absolute.* lb./sq. in.	atm.	*Boiling Temp. at Atm. Pres.* °F.	°C.
Air	−220	−140	573	39	...	...
Alcohol (C_2H_6O) ...	421	216	956	65	172·4	78
Ammonia (NH_3) ...	266	130	1691	115	− 27·4	−33
Benzol (C_6H_6) ...	554	292	735	50	176	80
Carbon-dioxide (CO_2)	88·2	31	1132	77	−110	− 79
Carbon-monoxide (CO)	−222	−141	528	35·9	−310	−190
Ether ($C_4H_{10}O$) ...	381·2	194	544	37	95	35
Hydrogen (H) ...	−402	−242	294	20	−423	−253
Nitrogen (N) ...	−236	−149	514	35	−321	−195
Oxygen (O_2) ...	−180	−118	735	50	−297	−183
Water (H_2O) ...	706–716	375–380	3200	217·8	212	100

(From Mark's Mech. Eng. Hand.)

ESTIMATIONS OF TEMPERATURES OF INCANDESCENT BODIES

COLOURS OF DIFFERENT TEMPERATURES

Faint red	960°F.	516°C.
Dull red	1290°F.	700°C.
Brilliant red ...	1470°F.	750°C.
Cherry red ...	1650°F.	900°C.
Bright cherry red...	1830°F.	1000°C.
Orange	2010°F.	1100°C.
Bright orange ...	2190°F.	1200°C.
White heat ...	2370°F.	1300°C.
Bright white heat...	2550°F.	1400°C.
Brilliant white heat	2750°F.	1500°C.

TRANSFER OF HEAT MAY OCCUR BY:

1. Conduction.
2. Convection.
3. Radiation.

1. Conduction is the transfer of heat through the molecules of a substance.

(a) *Internal Conduction* is transmission within a body.

(b) *External Conduction* is transmission from one body to another, when the two bodies are in contact.

Thermal Conductivity of a substance is the quantity of heat transferred through a body of one unit area and one unit thickness in one unit of time per one degree temperature difference. In S.I. units it is measured in watts/metre deg. C., and in Imperial units in B.t.u. in./hr. ft.^{2}. deg.F.

Heat Flow: $H = A\frac{K}{X}(t_2 - t_1)$ ... ... B.t.u./hr. or watts.

in which: K = Thermal conductivity ...

A = Area ... ... sq. ft. or m^2.

X = Thickness ... ... in. or m.

t_1 = Temperature at Cooler Section ... ... °F. or °C.

t_2 = Temperature at Hotter Section ... ... °F. or °C.

Thermal Resistance is numerically the reciprocal of the thermal conductance.

Heat Flow: $H = \frac{t_2 - t_1}{R}$ ... ... B.t.u./hr. or Watts.

Thermal resistance $= R = \frac{X}{K\ A}$

2. Convection is the transfer of heat by flow of currents within a fluid body. (Liquid or gas flowing over the surface of a hotter or cooler body.)

$H = aA\,(t_2 - t_1) = \frac{t_2 - t_1}{R_1}$ (B.t.u./hr. or watts.)

a = Thermal conductance (B.t.u./hr. sq. ft. °F. or W/m^2. °C.)

$R_1 = \frac{1}{aA}$ = Thermal resistance.

The amount of heat transferred per unit of time is affected by the velocity of moving medium, the area and form of surface and the temp. difference.

3. Radiation is the transfer of heat from one body to another by wave motion.

Stephan-Baltzmann Formula

$$E = C\left(\frac{T}{100}\right)^4$$

E = Heat emission of a body B.t.u./hr. or Watts.
T = Absolute temperature. °R. or °K.
C = Radiation constant.

Quantities of heat transferred between two surfaces:

$$Q_{Rad} = CA\left[\left(\frac{T_1}{100}\right)^4 - \left(\frac{T_2}{100}\right)^4\right]$$

A = Area.
T_1 T_2 = Absolute temperatures of hot and cold surfaces respectively.

For the absolute black body

C = 5·72 Watts per sq.m. per (deg.C.)^4
= 0·173 B.t.u. per hr. per sq. ft. per (deg.F.)^4

For other materials see table below.

RADIATION CONSTANT OF BUILDING MATERIAL (C)

	W./m^2 (°C)^4	B.t.u./hr. ft.^{2} (°F.)^4		W./m^2 (°C)^4	B.t.u./hr. ft.^{2} (°F.)^4		W./m^2 (°C)^4	B.t.u./hr. ft.^{2} (°F.)^4
Black body ...	5·72	0·173	Sand ...	4·20	0·127	Cast Iron, rough oxidized	5·09	0·154
Cotton ...	4·23	0·128	Shavings ...	4·10	0·124	Copper, polished	1·19	0·028
Glass ...	5·13	0·155	Silk ...	4·30	0·130	Brass, dull...	0·152	0·036
Wood ...	4·17	0·126	Water ...	3·70	0·112	Silver ...	1·19	0·0046
Brick ...	5·16	0·156	Wool ...	4·30	0·130	Zinc, dull ...	0·152	0·036
Oil Paint ...	4·30	0·130	Wrought iron, dull oxidized	5·16	0·156	Tin... ...	0·26	0·0077
Paper ...	4·43	0·134	Wrought iron, polished	1·55	0·047			
Plaster ...	5·16	0·156						
Lamp Black...	5·16	0·156						

CONDUCTION OF HEAT THROUGH PIPES OR PARTITIONS

Symbols

t_m = Logarithmic mean temperature difference.
t_{a1} = Initial temperature of heating medium.
t_{a2} = Final temperature of heating medium.
t_1 = Initial temperature of heated fluid.
t_2 = Final temperature of heated fluid.

The heat exchange can be classified as follows:

1. Parallel Flow, the fluids flow in the same directions over the separating wall.

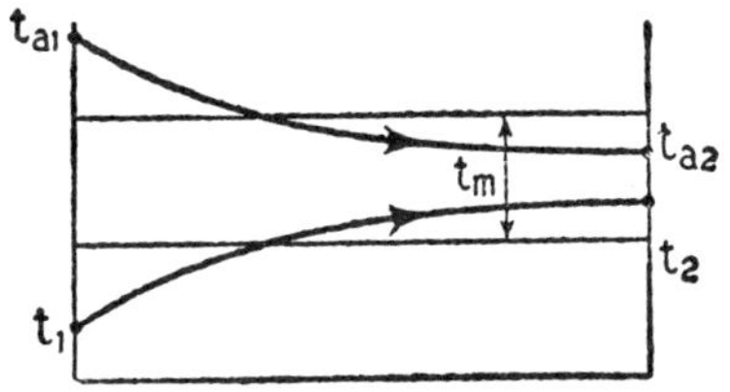

$$t_m = \frac{t_{a1} - t_{a2} + t_2 - t_1}{\log_e \dfrac{t_{a1} - t_1}{t_{a2} - t_2}} = \frac{\text{Initial temp. dif.} - \text{Final temp. dif.}}{2{\cdot}3 \log_{10} \dfrac{\text{Initial temp. dif.}}{\text{Final temp. dif.}}}$$

2. Counter Flow, the directions are opposite.

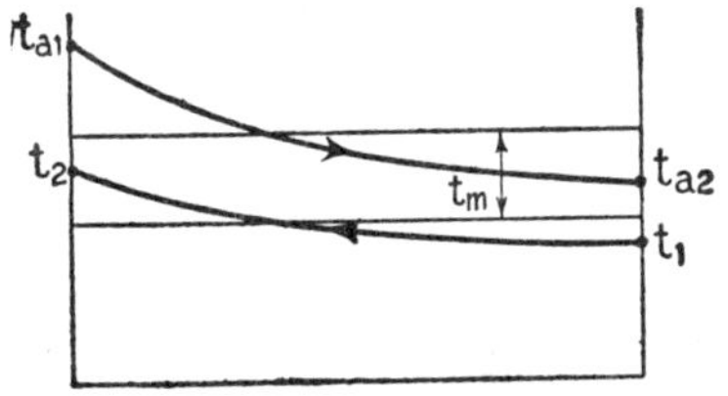

$$t_m = \text{(as before)} = \frac{\text{Initial temp. dif.} - \text{Final temp. dif.}}{2{\cdot}3 \log_{10} \dfrac{\text{Initial temp. dif.}}{\text{Final temp. dif.}}}$$

3. Evaporators or Condensers:

One fluid remains at a constant temperature while changing its state.

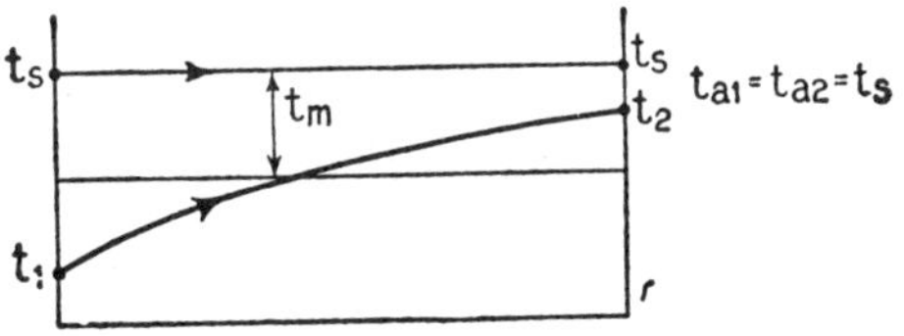

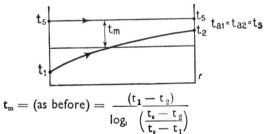

$$t_m = \text{(as before)} = \frac{(t_1 - t_2)}{\log_e \left(\dfrac{t_s - t_2}{t_s - t_1}\right)}$$

4. Mixed Flow:

One of the fluids takes an irregular direction with respect to the other.

$$t_m = \left(\frac{t_{a1} - t_{a2}}{2}\right) - \left(\frac{t_1 - t_2}{2}\right)$$

Newton's Law of Cooling: In the warming and cooling of bodies, the heat gain or loss respectively, is proportional to the difference between the temperatures of the body and the surroundings.

Let: t_s = Temperature of the cold surroundings.

t_1 = Initial Temperature of the Hot Body.

t_2 = Temperature of the Body.

C = Thermal Conductivity of the Body.

P = Density of the Body.

S = Specific Heat of the Body.

U = Coefficient of Heat Transfer between the Body and the surroundings.

R = Radius of a sphere or cylinder, or half thickness of a slab cooled or heated on both faces. Thickness of a slab cooled or heated on one face only.

$\theta = (\theta_1 - \theta_2)$ = Cooling time.

Then: $$\frac{t_2 - t_s}{t_1 - t_s} = e^{-K\theta}$$

and $\log_e (t_2 - t_s) - \log_e (t_1 - t_s) = - K\theta.$

where: K = Constant which can be found by measuring the temperatures of the body at different times θ_1 and θ_2 and which is:

$$K = \log_e \frac{t_2 - t_s}{t_1 - t_s} \div (\theta_1 - \theta_2).$$

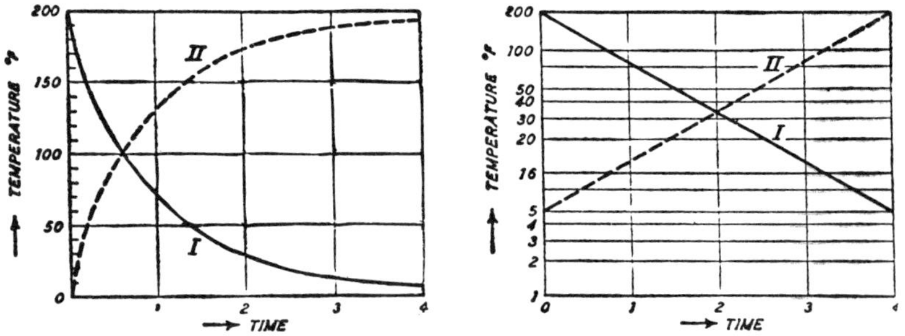

Cooling Curve (I) and Heating Curve (II) showing relation of Temperature and Time on Linear and Semi-logarithmic Paper.

Graphs showing how the Temperature of Cooling or Heating Up Bodies can be plotted on semi-logarithmic paper by introducing the following dimensionless ratios:

$$Y = \frac{t_s - t_1}{t_s - t_2} \qquad X = \frac{c\theta}{psR^2}, \qquad m = \frac{c}{UR}$$

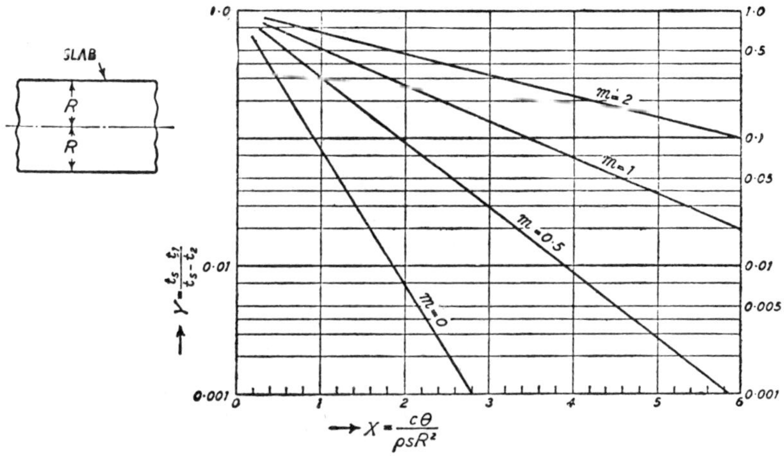

The Increased Heat Loss of Buildings during the heating up period causes a greater heat requirement than the steady state. This additional heat loss depends mainly on the type of building, length of heating interruption and heating up time, and type of heating installation. The allowance for covering the increased heat loss during heating up is usually expressed as a percentage of the heat loss in the steady state.

(See (VI. 4), Reitschel's Formula, and VI. 2, Allowances for intermittent Heating.)

The Temperatures during warming up of bodies are represented graphically by curves which are symmetrical to cooling down curves.

LOGARITHMIC MEAN TEMPERATURE DIFFERENCES

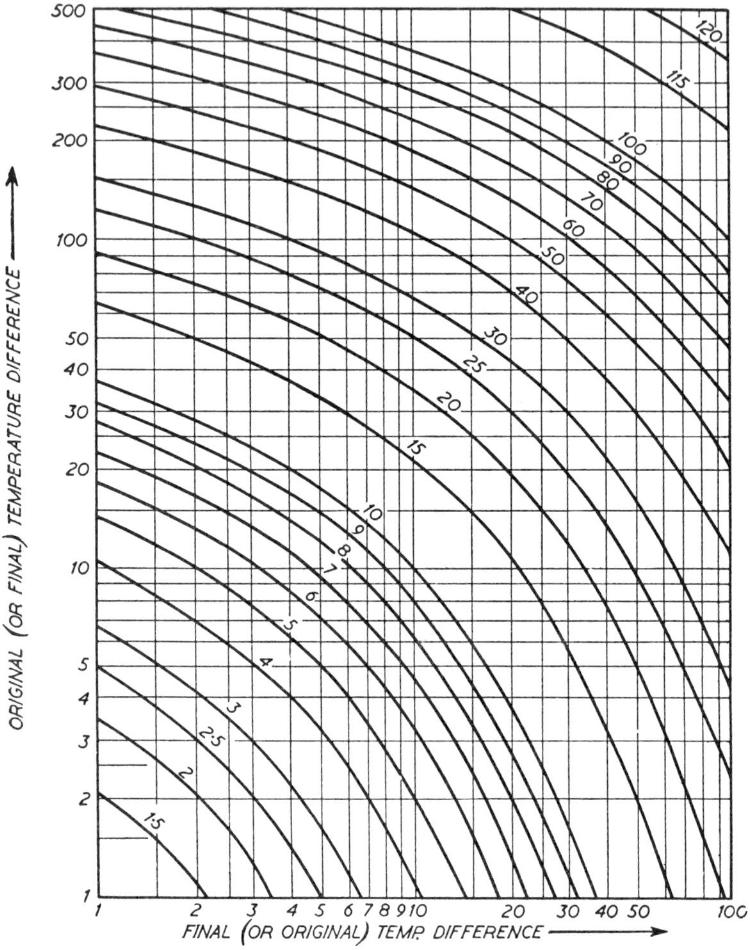

Example of Using the Chart

Water to Water Calorifier with Counter Flow.

Primary flow temperature 180°F. Secondary return temperature 50°F.

Primary return temperature 160°F. Secondary flow temperature 100°F.

Original temperature difference=180−50=130°F.

Final temperature difference=160−100=60°F.

From chart: Log. mean temp. difference=95°F.

The chart can be used equally well for °C. or F°.

HEAT TRANSMISSION COEFFICIENTS FOR METALS

			Watts / *m.2 deg. C.*	*B.t.u.* / *ft.2 hr. deg. F.*
Water	Cast Iron	Air or Gas	8·0	1·4
Water	Mild Steel	Air or Gas	11·0	2·0
Water	Copper	Air or Gas	11·0	2·25
Water	Cast Iron	Water	220 to 280	40 to 50
Water	Mild Steel	Water	340 to 400	50 to 70
Water	Copper	Water	350 to 450	62 to 80
Air	Cast Iron	Air	6·0	1·0
Air	Mild Steel	Air	8·0	1·4
Steam	Cast Iron	Air	11·0	2·0
Steam	Mild Steel	Air	11·0	2·5
Steam	Copper	Air	17·0	3·0
Steam	Cast Iron	Water	900	160
Steam	Mild Steel	Water	1050	185
Steam	Copper	Water	1170	205

The above values are average coefficients for practically still fluids. The coefficients are dependent on velocities of heating and heated media—on type of heating surface, temperature difference, and other circumstances. For special cases, see literature and manufacturers' data.

TABLE OF $n^{1\cdot3}$. For Radiator and Pipe Coefficients in relation to various Temperature Differences.

n	$n^{1\cdot3}$	n	$n^{1\cdot3}$	n	$n^{1\cdot3}$	n	$n^{1\cdot3}$	n	$n^{1\cdot3}$	n	$n^{1\cdot3}$
30	83	70	250	110	450	150	674	190	917	230	1176
35	102	75	273	115	477	155	704	195	948	235	1209
40	121	80	298	120	505	160	733	200	980	240	1242
45	141	85	322	125	533	165	763	205	1012	245	1219
50	162	90	347	130	560	170	793	210	1044	250	1310
55	183	95	372	135	589	175	824	215	1075		
60	205	100	398	140	617	180	855	220	1110		
65	226	105	424	145	645	185	887	225	1142		

TRANSMISSION TABLE, B.t.u./sq. ft. hr.

Coefficient B.t.u. /sq. ft. °F. hr.		*Heating System*	*Hot Water*						*Steam*		
		Mean Rad Temp. °F.	160			165			215		
Water.	*Steam.*	*Room Temp. °F.*	55	60	65	55	60	65	55	60	65
1·85	2·12	2-Col. Rad. ...	195	185	176	203	195	185	340	328	319
1·70	1·95	4-Col. Rad. ...	178	170	162	187	178	170	312	302	292
1·60	1·84	6-Col. Rad. ...	158	160	152	176	158	160	294	285	276
1·71	1·95	Wall	180	171	163	188	180	171	312	302	292
1·58	1·81	Window ...	166	158	150	174	166	158	290	282	270
1·85	2·12	3″ Hospital ...	195	185	176	203	195	185	340	328	319
1·58	1·80	5¾″ Hospital ...	166	158	150	174	166	158	290	278	270
1·51	1·72	7⅛″ Hospital ...	158	151	144	166	158	151	275	266	258
2·17	2·37	Pipe Coil, 1 row	228	217	206	239	228	217	380	367	356
1·76	1·90	Pipe Coil, 2 rows	185	176	167	191	185	176	304	284	285
1·03	1·13	Gilled Tube 1 row	107	103	98	118	107	103	181	175	170
0·93	1·01	„ 2 rows	98	93	88	108	98	93	160	155	150

Transmissions are based on:

30″ high Radiators.
2½″ between Radiator and Wall.

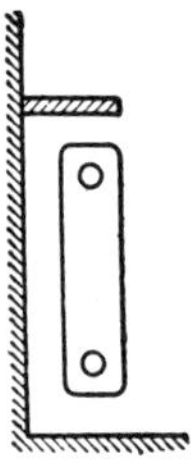

Fig. 1.

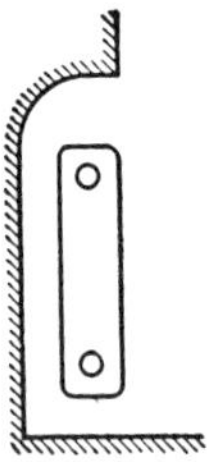

Fig. 2.

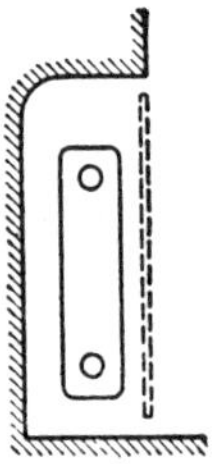

Fig. 3.

Transmissions are to be reduced by:

2– 5% for Radiators below Window-sill ... ... Fig. 1.
5–10% for Radiators in Wall Recess ... ... Fig. 2.
15–25% for Radiators in Wall Recess, with Grill ... ... Fig. 3.
25–30% for Radiators, with Aluminium Painting

TRANSMISSION TABLE W/m².

Coefficient W/m.² deg. C.		*Heating System*	*Hot Water*						*Steam*		
		Mean Rad. Temp. °C.	*70°C.*			*75°C.*			*102°C.*		
Water	*Steam*	*Room Temp. °C.*	12	16	18	12	16	18	12	16	18
10·5	12·0	2-Col. Rad. ...	608	566	545	661	619	597	1,080	1,030	1,010
9·6	11·1	4-Col. Rad. ...	559	520	501	607	569	549	996	952	930
9·1	10·4	6-Col. Rad. ...	526	490	472	572	536	517	940	898	877
9·7	11·1	Wall	562	524	501	611	572	553	996	952	930
9·0	10·3	Window	520	484	466	565	529	511	925	883	862
10·5	12·0	3″ Hospital ...	609	543	546	661	619	598	1,080	1,030	1,010
9·0	10·2	5½″ Hospital ...	520	484	466	565	529	511	919	878	858
8·6	9·7	8″ Hospital ...	496	462	445	540	506	488	875	836	817
12·3	13·5	Pipe Coil, 1 row	713	664	639	775	726	700	1,210	1,160	1,130
10·0	10·8	Pipe Coil, 2 rows	579	539	519	629	589	569	970	927	906
5·8	6·4	Gilled Tube, 1 row	338	315	303	368	344	333	578	552	539
5·3	5·7	,, ,, 2 rows	306	285	274	332	311	301	516	493	481

HEAT TRANSMISSION COEFFICIENTS FOR RADIATORS IN AIR DUCTS

Mean air temperatures 0°C. (32°F.) atmospheric pressure, for hot water heating:

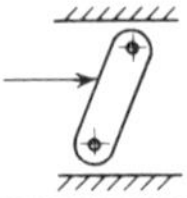

Mean water temperature 80°C., 2·0 m/s. water velocity, or steam heating of 10 to 30 N/m² working pressure. For various air velocities in m/s.

Air Velocity m./s.	*Heat Trans. Coefficient W/m.² deg. C.*	*Air Velocity m./s.*	*Heat Trans. Coefficient W/m.² deg. C.*
0·20	8·4	1·40	27·3
0·30	10·7	1·60	29·7
0·40	12·7	1·80	31·9
0·50	14·5	2·00	34·0
0·60	16·3	2·25	36·5
0·80	19·4	2·50	39·0
1·00	22·2	2·75	41·3
1·20	24·9	3·00	43·6

For Imperial units the following values apply for a mean water temperature of 170°F. and a water velocity of 6 ft./sec. or steam heating of 15 to 45 lb./sq. in. working pressure.

Air Velocity ft./min.	*Heat Trans. Coefficient B.t.u./ft.²°F. hr.*	*Air Velocity ft./min.*	*Heat Trans. Coefficient B.t.u./ft.²°F. hr.*
40	1·48	280	4·82
60	1·90	320	5·22
80	2·23	360	5·60
100	2·56	400	6·00
120	2·87	450	6·44
160	3·42	500	6·86
200	3·92	550	7·30
240	4·38	600	7·70

For water velocities lower than 2 m/s. or 6 ft./sec. the values in the above tables are to be reduced by 5 per cent.

For mean air temperatures other than 0°C (32°F.) the values in the above tables are to be multiplied by the following factors:

0·98	for mean air temperature of	10°C.	(50°F.)
0·96	,, ,, ,, ,, ,,	20°C.	(68°F.)
0·94	,, ,, ,, ,, ,,	30°C.	(86°F)
0·92	,, ,, ,, ,, ,,	40°C.	(104°F)
0·90	,, ,, ,, ,, ,,	50°C.	(122°F.)

Heat Loss of Steel Pipes:

For various Temperature Differences between pipe and air, in Imperial and S.I. units.

Nominal Bore		Surface Area		Heat Loss in W/m. length for Temperature Difference °C.						Heat Loss in B.t.u./hr. ft. for Temperature Difference °F.					
				Water					Steam	Water					Steam
in.	mm.	sq. ft. per ft.	sq. m. per m.	35°C.	45°C.	55°C.	85°C.	110°C	85°C.	60°F.	80°F.	100°F	150°F.	200°F.	150°F.
½	15	0·21	0·0640	38	40	54	91	132	99	29	42	56	95	138	103
¾	20	0·28	0·0854	35	50	67	113	165	124	36	52	70	118	172	129
1	25	0·33	0·101	39	58	77	129	190	141	41	60	80	135	198	147
1¼	32	0·43	0·131	48	70	94	158	232	173	50	73	98	165	241	180
1½	40	0·50	0·152	55	79	106	179	260	194	57	82	110	186	270	202
2	50	0·62	0·189	63	92	124	209	305	229	66	96	129	218	317	238
2½	65	0·75	0·229	76	110	148	250	365	272	79	115	154	260	380	282
3	80	0·92	0·280	93	135	181	305	445	332	97	141	188	318	462	345
4	100	1·19	0·363	115	167	223	376	548	409	120	174	232	392	570	426
5	125	1·46	0·445	140	212	271	458	668	502	146	221	282	477	695	522
6	150	1·75	0·533	164	334	334	538	783	586	171	348	331	560	815	610

The above Heat Transmissions are to be multiplied by:

Single Pipe	Along skirting or riser	1·0
More than 1 pipe	Along skirting or riser	0·91
Single pipe	Along ceiling	0·73
More than 1 pipe	Along ceiling	0·66
Single pipe	Freely exposed	1·1
More than 1 pipe	Freely exposed	1·0

Heat Loss of Copper Pipes:

For various Temperature Differences between pipe and air, in Imperial and S.I. units.

Nominal Bore		Heat Loss in W/m. for Temp. Diff.			Heat Loss in B.t.u. per hr. per ft. for Temp. Diff.			Nominal Bore		Heat Loss in W/m. for Temp. Diff.			Heat Loss in B.t.u. per hr. per ft. for Temp. Diff.		
in.	mm.	40°C.	55°C.	72°C.	70°F.	100°F.	130°F.	in.	mm.	40°C.	55°C.	72°C.	70°F.	100°F.	130°F
½	15	21	32	45	22	34	47	2	50	59	93	131	62	97	136
¾	20	28	43	60	29	45	53	2½	65	71	111	156	74	116	162
1	25	34	53	76	36	56	79	3	80	83	129	181	87	135	189
1¼	32	41	64	89	43	67	93	4	100	107	165	232	111	172	241
1½	40	47	74	104	49	77	108								

Heat Loss of Insulated Copper Pipes:

For 55°C. temperature difference (100°F.).

For 25mm. (1in.) thick insulation with K=0·043 W/m.deg.C. (0·3 B.t.u. in./ft.^{2} hr.deg.F.).

Nominal Bore		Heat Loss	
in.	mm.	W./m.	B.t.u. per hr. per ft.
¾	20	8	8
1	25	10	10
1½	40	11·5	12
2	50	14·5	15
2½	65	16	17
3	80	19	20

These Tables have been calculated from data published in the 1959 edition of the I.H.V.E. Guide to Current Practice.

SPECIFIC HEAT OF SUBSTANCES between 0°C. and 100°C.

	B.t.u./lb. °F.	kJ/kg. deg. C.		B.t.u./lb. °F.	kJ/kg. deg. C.		B.t.u./lb. °F.	kJ/kg. deg. C.		B.t.u./lb. °F.	kJ/kg. deg. C.
Metals			**Metal Alloys**						**Liquids**		
Alumin.	0·218	0·912	Ball Metal	0·086	0·360	Coal	0·314	1·31	Acetic Ac.	0·51	2·13
Antimony	0·051	0·214	Brass	0·090	0·377	Concrete	0·27	1·13	Alcohol	0·70	2·93
Copper	0·093	0·389	Bronze	0·104	0·435	Cork	0·485	2·03	Benzol	0·43	1·80
Gold	0·031	0·130	N'kel Steel	0·109	0·456	Glass	0·20	0·84	Ether	0·503	2·10
Iron	0·110	0·460	Solder	0·04	0·167	Ice	0·504	2·11	Glycerine	0·576	2·41
Lead	0·031	0·130				India Rub.	0·27	1·1	Mach. Oil	0·40	1·67
Nickel	0·108	0·452	**Miscellaneous**				to 0·98	to 4·1	Mercury	0·033	1·39
Platinum	0·032	0·134	Asbestos	0·20	0·84	Marble	0·21	0·88	Petroleum	0·50	2·09
Silver	0·056	0·234	Ashes	0·20	0·84	Paraffin	0·69	2·9	Sulph. Ac.	0·50	2·09
Tin	0·055	0·230	Brick	0·22	0·92	Porcelain	0·255	1·07	Turp'tine	0·472	1·98
Zinc	0·094	0·393	Coke	0·203	0·85	Sand	0·195	0·816	Water	1·000	4·187
						Wood	0·55	2·3			
							to 0·65	to 2·7			

SPECIFIC HEAT OF GASES

Name of Gas	Chemical Symbol	Specific Heat B.t.u./lb. deg. F.		Specific Heat Capac. kJ/kg. deg. C.		γ	Gas Constant $(C_p - C_v) \times J$	
		C_p	C_v	C_p	C_v	$=C_p/C_v$	ft. lb./lb. deg. F.	kJ/kg. deg.C
Sulphur Dioxide	SO_2	0·154	0·123	0·645	0·515	1·25	24·10	0·130
Carbon Dioxide	CO_2	0·210	0·160	0·827	0·632	1·31	38·86	0·189
Oxygen	O_2	0·217	0·155	0·917	0·656	1·40	48·25	0·260
Air	—	0·251	0·171	1·01	0·716	1·40	53·34	0·29
Nitrogen	N_2	0·247	0·176	1·034	0·737	1·40	54·99	0·297
Ethylene	$C2H_2$	0·400	0·330	1·67	1·38	1·20	55·08	0·29
Carbon Monoxide	CO	0·243	0·172	1·02	0·720	1·41	55·14	0·297
Acetylene	C_2H_2	0·350	0·270	1·47	1·13	1·28	59·34	0·34
Blast Furnace Gas	—	0·245	0·174	1·03	0·729	1·40	55·05	0·297
Amonia	NH_3	0·523	0·399	2·19	1·67	1·31	96·50	0·52
Methane	NH_3	0·523	0·399	2·19	1·67	1·31	96·50	0·52
Methane	CH_4	0·593	0·450	2·23	1·71	1·32	111·30	0·520
Hydrogen	H_2	3·42	2·44	14·24	10·08	1·40	765·90	4·16
Combustion Products	—	0·24	—	1·01	—	—	—	—

LATENT HEATS OF MELTING

	B.t.u./lb.	kJ./kg.		B.t.u./lb.	kJ./kg.		B.t.u./lb.	kJ./kg.
Aluminium ...	138·2	321	Iron Slag ...	90·0	209	Platinum ...	49·0	114
Bismuth ...	22·75	52·8	Ice	143·33	333	Silver ...	37·9	88·0
Copper ...	75·6	176	Lead ...	9·65	22·4	Sulphur ...	16·87	39·2
Iron, Grey Cast	41·4	96	Mercury ...	5·08	11·8	Tin... ...	25·2	58·5
Iron, White Cast	59·4	138	Nickel ...	8·35	19·4	Zinc ...	50·63	118

LATENT HEATS OF VAPORIZATION

	B.t.u./lb.	kJ./kg.		B.t.u./lb.	kJ./kg.		B.t.u./lb.	kJ./kg.
Acetone ...	233	542	Chloroform	110	258	Nitrogen ...	81·5	190
Alcohol ...	385	896	Ether ...	162	377	Oxygen ...	92	214
Aniline ...	198	461	Hydrogen ...	222	517	Sulphur ...	650	1,510
Benzol ...	169	393	Methyl			Turpentine...	126	293
Chlorine ...	112	261	Chlorine...	175	407	Water ...	970·4	2,257
			Mercury ...	122	284			

MELTING AND FREEZING POINTS at atmospheric pressure.

	°C.	°F.
Alcohol	–97	–143
Aluminium	658	1,218
Antimony	631	1,166
Barium	850	1,562
Bismuth	271	520
Borax	741	1,366
Calcium	810	1,490
Sulphur	106–119	234–247
Ice (Fresh Water)	0	
Ice (Sea Water)	–2·5	
Calcium Chloride	762	1,404
Carbon	3,480	6,300
Carbon Dioxide	–410	–706
Carbon Disulphide	–110	–166
Copper	1,083	1,981
Glycerine	–16	4
Gold	1,063	1,945
Tin	232	449
Iron (Pure)	1,530	2,786
Lead	327	621
Mercury	–39	–38
Nickel	1,455	2,646
Oxygen	–218	–360
Platinum	1,755	3,191
Silver	960	1,761
Zinc	419	787
Wax	64	149

BOILING TEMPERATURES at atmospheric pressure.

	°C.	°F.
Zinc	916	1,680
Sulphur	440	823
Mercury	357	675
Paraffin	300	572
Glycerine	290	554
Phosphorus	290	554
Linseed Oil	281	538
Naphthalene	218	424
Anilene	184	363
Calcium Chloride (Sat. Sol.)	180	356
Turpentine	160	320
Toluene	111	230
Sodium Chloride (Sat. Sol.)	108	226·4
Water	100	212
Alcohol	78	172·4
Helium	–269	–450

COEFFICIENTS OF LINEAR EXPANSION. Average Values between 0°C. and 100°C.

	in. per in. °F. × 10^6	m. per m. °C. × 10^6
Aluminium	12·3	22·2
Antimony	5·8	10·4
Brass	10·4	18·7
Bronze	10·0	18·0
Copper	9·3	16·5
Gold	8·2	14·2
Iron Pure	6·7	12·0
Iron Cast	5·9	10·4
Iron Forged	6·3	11·3
Lead	15·1	28·0
Silver	10·7	19·5
Solder	13·4	24·0
Nickel Steel	7·3	13·0
Type Metal	10·8	19·0
Zinc	16·5	29·7
Brick	3·1	55
Cement	6·0	10·0
Concrete	8·0	14·5
Glass Hard	3·3	5·9
Glass Plate	5·0	9·0
Marble	6·5	12
Rubber	42·8	77
Oak parallel to grain	2·7	4·9
Oak across grain	3·0	5·4
Porcelain	1·7	3·0
Masonry	2·5 to 5·0	4·5 to 9·0
Graphite	4·4	7·9

DENSITY OF GASES

Name of Gas	*Chemical Symbol*	*Mol. Weight*	*Density at 0°C and atm. press.*		*Name of Gas*	*Chemical Symbol*	*Mol. Weight*	*Density at 0°C and atm. press.*	
			kg./m³	*lb./ft.³*				*kg./m³*	*lb./ft³*
Sulphur Dioxide	SO_2	64	2·926	0·1828	Carbon Monoxide	CO	28	1·250	0·0780
Carbon Dioxide	CO_2	44	1·977	0·1234	Acetylene	C_2H_2	26	1·170	0·0729
Oxygen	O_2	32	1·429	0·0892	Blast Furnace Gas	—	—	1·250	0·0780
Air	—	28	1·293	0·0806	Ammonia	NH_3	17	0·769	0·0480
Nitrogen	N_2	28	1·250	0·0780	Methane	CH_4	16	0·717	0·0447
Ethylene	C_2H_4	28	1·260	0·0786	Hydrogen	H_2	2	0·00899	0·0056

DENSITIES OF VARIOUS SUBSTANCES

Metals

	kg. per m.^{3}	*lb. per cu. ft.*		*kg. per m.*^{3}	*lb. per cu. ft.*
Aluminium ...	2,690	168	Iron, wrought ...	7,850	486
Antimony ...	6,690	417	Lead	11,340	705
Brass, cast ...	8,100	505	Mercury ...	13,450	840
Brass, yellow ...	8,350	518	Nickel	8,830	551
Bronze, gunmetal	8,450	529	Platinum ...	21,450	1,340
Copper... ...	8,650	551	Silver	10,500	655
Gold, pure cast	19,290	1,200	Steel	7,900	493
Iron, cast ...	7,480	467	Zinc	7,200	444

Liquids

	kg. per m.^{3}	*lb. per cu. ft.*		*kg. per m.*^{3}	*lb. per cu. ft.*
Alcohol, pure 60°F.	790	49	Oil, olive ...	920	57
Alcohol, 80°F. ...	850	53	Oil, turpentine	870	54
Alcohol, 50°F. ...	910	57	Water, distilled	1,000	62
Ether	870	54	Water, sea 39°F.	1,030	64
Glycerine, acetic	1,270	79	Milk	1,030	64
Oil, naphtha ...	850	53	Beer	1,030	64

Miscellaneous

	kg. per m.^{3}	*lb. per cu. ft.*		*kg. per m.*^{3}	*lb. per cu. ft.*
Asbestos ...	3,060	191	Limestone ...	3,170	198
Atm. air ...	1·20	0·075	Marble ...	2,650	165
Asphalt... ...	1,650	103	Mortar	1,400-1,750	86-109
Cement, Port ...	3,000	187	Peat	600 –1,330	37–83
Cement, Roman	1,550	97			
Chalk	1,500–2,800	95–175	Plaster	1,180	73
Coal	1,500–1,650	95–103	Porcelain ...	2,300	143
Coke	1,000	62	Rubber ...	920	67
Concrete, mean	2,240	140	Salt, common ...	2,130	133
Glass, window	2,640	164	Soap	1,070	67
Granitstone ...	2,130	133	Starch	945	59
Gypsum ...	2,165	135	Sulphur ...	2,020	126
Ice at 32°F. ...	910	57	Wax, paraffin	930	58
Lime, hydraulic	2,740	171			

Material	Conductivity k		Resistivity 1/k	
	B.t.u. in. / ft.² hr. °F.	W. / m. deg. C.	ft.² hr. °F. / B.t.u. in.	m. deg. C. / W.
Air	0·18	0·026	5·56	38·6
Aluminium	1050	150		
Asbetolux	0·8	0·12	1·25	8·67
Asbestos:				
flues and pipes	1·9	0·27	0·53	3·68
insulating board	1·0	0·14	1·0	6·93
lightweight slab	0·37	0·053	2·70	18·7
Asphalt: light ...	4·0	0·58	0·25	1·73
heavy ...	8·5	1·23	0·12	0·83
Brass...	550	150		
Bricks: common ...	9·9	1·43	0·10	0·69
engineering	5·5	0·79	0·18	1·25
Brine	3·3	0·48	0·30	2·10
Building board ...	0·55	0·079	1·82	12·62
paper ...	0·45	0·065	2·22	15·39
Caposite	0·36	0·052	2·78	19·28
Cardboard	1·0 to 2·0	0·144 to 0·288	1·0 to 0·5	6·9 to 3·5
Celotex	0·33	0·048	3·0	21·0
Concrete: 1:2:4 ...	10·0	1·4	0·10	0·69
lightweight	2·8	0·40	0·36	2·5
Copper	2100	300		
Cork...	0·30	0·043	3·33	23·1
Densotape	1·7	0·25	0·58	4·0
Diatomaceous Earth	0·60	0·087	1·66	11·5
Econite	0·68	0·098	1·47	10·19
Felt	0·27	0·039	3·70	25·7
Fibreglass	0·25	0·036	4·0	27·7
Firebrick	9·0	1·30	0·11	0·76
Fosalsil	1·0	0·14	0 10	0·69
Glass...	7·3	1·05	0·14	0·97
Glasswool	0·28	0·04	3·6	24·8
Gold	2150	310		

Material	Conductivity k		Resistivity 1/k	
	B.t.u. in. / ft.² hr. °F.	W. / m. deg. C.	ft.² hr. °F. / B.t.u. in.	m. deg. C. / W.
Granwood floor blocks ...	2·20	0·32	0·45	3·1
Gyproc plasterboard ...	1·1	0·16	0·91	6·3
Gypsum plaster board ...	1·1	0·16	0·91	6·3
Hardboard	0·65	0·094	1·54	10·68
Holoplast: 1″ panel ...	0·95	0·14	1·05	7·3
Ice...	16·0	2·31	0·0625	0·43
Insulating board	0·41	0·059	2·45	16·99
Iron: cast	450	65		
wrought	400	58		
Jute	0·25	0·036	4·0	27·7
Kapok	0·25	0·036	4·0	27·7
Lead	240	35		
Linoleum: cork	0·5	0·072	2·0	13·9
p.v.c.	1·5	0·22	0·67	4·65
rubber ...	2·1	0·30	0·48	3·33
Marinite	0·74	0·11	1·35	9·36
Mercury	48	7		
Mica sheet	4·5	0·65	0·22	1·53
Mineral wool	0·39	0·056	3·33	23·1
Nickel	400	58		
On ozote	0·20	0·029	5·0	34·7
Paper	0·90	0·13		
Perspex	1·45	0·21	0·69	4·8
Plaster	3·3	0·48	0·30	2·1
Platinum	480	69		
Polystyrene: cellular ...	0·23	0·033	4·3	29·8
Polyurethane: cellular	0·29	0·042	3·45	23·9
Polyzote	0·22	0·032	4·55	31·5
Porcelain	7·2	1·04	0·14	0·96

Material	Conductivity k		Resistivity 1/k	
	B.t.u. in. / ft.² hr. °F.	W. / m. deg. C.	ft.² hr. °F. / B.t.u. in.	m. deg. C. / W.
Refractory brick alumina	2·2	0·32	0·45	3·1
diatomaceous ...	0·9	0·13	1·11	7·70
silica	10·0	1·44	0·10	0·69
vermiculite insulating	1·35	0·19	0·74	5·13
Refractory concrete:				
diatomaceous ...	1·8	0·26	0·56	3·9
aluminous cement...	3·2	0·46	0·31	2·15
Rubber: natural ...	1·1	0·16	0·91	6·3
silicone ...	1·6	0·23	0·63	4·4
Sand	2·9	0·42	0·35	2·4
Scale, boiler	16·0	2·3	0·0625	0·43
Silver	2,900	420		
Sisalkraft building paper	0·46	0·066	2·17	15·0
Slate	14·0	2·0	0·071	0·5
Snow	1·5	0·22	0·67	4·65
Steel, soft	320	46		
Steel wool	0·75	0·108	1·33	9·22
Stillite	0·25	0·036	4·0	27·7
Stone: granite	20·3	2·9	0·05	0·35
limestone ...	10·6	1·5	0·09	0·62
marble	17·4	2·5	0·06	0·42
sandstone ...	13·0	1·9	0·08	0·55
Sundeala: insulating board	0·36	0·052	2·78	19·3
medium hardboard	0·51	0·074	2·0	13·9
Tentest	0·35	0·05	2·86	19·8
Thermalite	1·4	0·20	0·71	4·9
Tiles: asphalt and asbestos	3·8	0·55	0·26	1·8
burnt clay... ...	5·8	0·84	0·17	1·2
concrete ...	8·0	1·2	0·13	0·90
cork	0·58	0·084	1·72	11·9
plaster	2·6	0·37	0·38	2·63

Material	*Conductivity k*		*Resistivity 1/k*	
	B.t.u. in. / ft.² hr. °F	*W. / m. deg. C.*	*ft.² hr. °F. / B.t.u. in.*	*m. deg. C. / W.*
Timber: balsa	0·33	0·048	3·0	20·8
beech	1·16	0·17	0·86	5·97
cypress ...	0·67	0·097	1·49	10·3
deal	0·87	0·13	1·15	7·97
fir	0·76	0·11	1·3	9·1
oak	1·11	0·16	0·90	6·24
plywood ...	0·96	0·14	1·04	7·21
teak	0·96	0·14	1·04	7·21
Treetex	0·39	0·056	2·56	17·8
Water	4·15	0·60	0·24	1·7
Weyboard	0·63	0·091	1·60	11·1
Weyroc	1·0	0·14	1·0	6·9
Woodwool	0·28	0·040	3·58	24·8
Wool	0·30	0·043	3·33	23·1
Zinc	440	64		
Sawdust	0·49	0·071	2·04	14·1
Cotton waste	0·41	0·059	2·4	16·9

MULTIPLYING FACTORS

For determining the capabilities of Unit Heaters for various conditions. Based on inlet air temperature 15°C., barometric pressure $101{\cdot}3 \times 10^3 N/m^2$.

HOT WATER HEATING — based on Mean Water Temperature 75°C.

		Air Inlet Temperature °C.					
		1	5	10	12·5	15	20
Mean Water Temperature °C.	45	0·78	0·67	0·58	0·54	0·50	0·42
	50	0·86	0·76	0·67	0·62	0·58	0·50
	55	0·94	0·84	0·75	0·70	0·67	0·58
	60	1·02	0·92	0·84	0·80	0·76	0·67
	65	1·10	0·96	0·92	0·88	0·89	0·75
	70	1·18	1·09	1·00	0·96	0·92	0·83
	75	1·27	1·16	1·09	1·00	1·00	0·91
	80	1·35	1·25	1·17	1·08	1·08	0·99
	85	1·43	1·34	1·25	1·16	1·16	1·07

STEAM HEATING Based on Steam Pressure $115 \times 10^3 N/m^2$ abs.

Absolute Steam Pressure kN/m.²	101·3	1·20	1·02	1·05	1·00	0·98	0·91
	115	1·26	1·14	1·07	1·02	1·00	0·93
	150	1·33	1·25	1·18	1·13	1·11	1·04
	180	1·40	1·31	1·24	1·20	1·18	1·10
	250	1·51	1·42	1·36	1·32	1·30	1·22
	300	1·59	1·50	1·44	1·39	1·37	1·30
	350	1·66	1·57	1·50	1·46	1·44	1·36
	400	1·71	1·62	1·56	1·51	1·49	1·42
	450	1·76	1·68	1·60	1·56	1·54	1·46
	500	1·81	1·72	1·65	1·61	1·58	1·51
	550	1·85	1·76	1·69	1·65	1·62	1·54
	600	1·89	1·80	1·73	1·68	1·66	1·59
	700	1·92	1·87	1·80	1·76	1·74	1·66
	800	2·03	1·93	1·86	1·82	1·79	1·71

MULTIPLYING FACTORS

For determining the capacities of Unit Heaters for various conditions. Based on inlet air temperature 60°F., barometric pressure 30 in. Hg.

HOT WATER HEATING — based on Mean Water Temp. 170°F.

		Air Inlet Temperature °F.					
		30	40	50	55	60	70
Mean Water Temperature °F.	110	0·73	0·64	0·53	0·50	0·45	0·36
	120	0·82	0·73	0·64	0·59	0·55	0·45
	130	0·91	0·82	0·73	0·68	0·64	0·55
	140	1·00	0·91	0·82	0·77	0·73	0·64
	150	1·10	1·00	0·91	0·86	0·82	0·73
	160	1·18	1·09	1·00	0·95	0·91	0·82
	170	1·27	1·18	1·09	1·05	1·00	0·91
	180	1·36	1·27	1·18	1·14	1·09	1·00
	190	1·46	1·36	1·27	1·23	1·18	1·09

STEAM HEATING — based on Steam Pressure 2 lb./in.^{2} gauge

Steam Pressure lb./in.^{2} gauge	0	1·190	1·115	1·041	0·994	0·967	0·896
	2	1·226	1·148	1·074	1·037	1·000	0·927
	5	1·286	1·211	1·136	1·098	1·062	0·989
	10	1·368	1·291	1·214	1·177	1·140	1·067
	20	1·496	1·417	1·340	1·302	1·265	1·191
	30	1·596	1·517	1·438	1·401	1·361	1·286
	40	1·677	1·598	1·519	1·480	1·442	1·366
	50	1·751	1·670	1·592	1·552	1·513	1·436
	60	1·812	1·732	1·656	1·612	1·573	1·496
	70	1·871	1·788	1·708	1·669	1·629	1·552
	80	1·923	1·841	1·760	1·720	1·681	1·603
	90	1·969	1·886	1·805	1·770	1·725	1·646
	100	2·015	1·932	1·850	1·809	1·764	1·686

Temp. °F.	Abs. pressure lb./sq. in.	Density lb./cu. ft.	Specific Gravity	Specific Volume cu. ft./lb.	Specific Heat B.t.u./lb. °F.	Entropy B.t.u./lb. °F.	Absolute Viscosity in poises	Total Heat B.t.u./lb.
32	0·088	62·42	1·000	0·0160	1·0093	0·0000	0·0179	0
40	0·122	62·42	1·000	0·0160	1·0048	0·01615	0·0155	8
50	0·178	62·42	1·000	0·0160	1·0015	0·03595	0·0131	18
60	0·256	62·38	1·000	0·0160	0·9995	0·05765	0·0113	28
62	0·275	62·35	1·000	0·0160	0·9992	0·05919	0·0110	30
70	0·363	62·30	0·999	0·0160	0·9982	0·0754	0·0098	38
80	0·507	62·22	0·998	0·0160	0·9975	0·0929	0·0086	48
90	0·698	62·11	0·996	0·0161	0·9971	0·1112	0·0076	58
100	0·949	61·99	0·994	0·0161	0·9970	0·1292	0·0088	68
110	1·27	61·86	0·992	0·0161	0·9971	0·1469	0·0062	78
120	1·69	61·71	0·990	0·0162	0·9974	0·1641	0·0056	88
130	2·22	61·55	0·987	0·0162	0·9978	0·1816	0·0051	98
140	2·89	61·38	0·984	0·0163	0·9984	0·1981	0·0047	108
150	3·72	61·20	0·982	0·0163	0·9990	0·2147	0·0043	118
160	4·74	61·00	0·979	0·0164	0·9988	0·2309	0·0040	128
170	5·99	60·80	0·975	0·0164	1·0007	0·2472	0·0037	138
180	7·51	60·58	0·971	0·0165	1·0017	0·2629	0·00345	148
190	9·33	60·36	0·969	0·0166	1·0028	0·2787	0·00323	158
200	11·53	60·12	0·965	0·0166	1·0039	0·2938	0·00302	168
210	14·13	59·92	0·958	0·0167	1·0052	0·3089	0·00287	178
212	14·70	59·88	0·957	0·0167	1·0055	0·3118	0·00285	180
220	17·19	59·66	0·955	0·0168	1·0068	0·3237	0·00272	188·1
230	20·77	59·37	0·950	0·0168	1·0087	0·3385	0·00257	198·2
240	24·97	59·17	0·946	0·0169	1·0104	0·3531	0·00254	208·3
250	29·81	58·84	0·941	0·0170	1·0126	0·3676	0·00230	218·4
260	35·42	58·62	0·940	0·0171	1·0148	0·3818	0·00217	228·6
270	41·85	58·25	0·933	0·0172	1·0174	0·3962	0·00208	238·7
280	49·18	58·04	0·929	0·0172	1·0200	0·4097	0·00200	248·9
290	57·55	57·65	0·923	0·0173	1·0230	0·4236	0·00193	259·2
300	67·00	57·41	0·920	0·0174	1·0260	0·4272	0·00186	262·5
310	77·67	57·00	0·913	0·0175	1·0296	0·4507	0·00179	279·8
320	89·63	56·65	0·906	0·0177	1·0332	0·4643	0·00173	290·2
330	103·00	56·31	0·900	0·0178	1·0368	0·4777	0·00188	300·6
340	118·0	55·95	0·897	0·0179	1·0404	0·4908	0·00163	311·1
350	134·6	55·65	0·890	0·0180	1·0440	0·5040	0·00158	321·7
360	153·0	55·19	0·883	0·0181	1·0486	0·5158	0·00153	332·3
370	173·3	54·78	0·876	0·0182	1·0532	0·5292	0·00149	342·9
380	195·6	54·36	0·870	0·0184	1·0578	0·5420	0·00145	353·5
390	220·2	53·96	0·865	0·0187	1·0624	0·5548	0·00141	364·3
400	247·1	53·62	0·834	0·0186	1·0670	0·5677	0·00137	375·3
450	422	51·3	0·821	0·0195	1·0950	0·6298	—	430·2
500	679	48·8	0·781	0·0205	1·1300	0·6907	—	489·1
550	1043	45·7	0·730	0·0219	1·2000	0·7550	—	553·5
600	1540	41·5	0·666	0·0241	1·3620	0·8199	—	623·2
706·1	3226	19·2	0·307	0·0522	—	1·0785	—	925·0

Temp. °C.	Abs. Pressure kN/m.2	Density kg./m.3	Specific Gravity	Specific Volume m.3/kg.	Specific Heat Capacity kJ./kg. °C.	Specific Entropy kJ./kg. K	Dynamic Viscosity g./m^2	Specific Enthalpy kJ./kg.
0·01	0·6	1,000	1·00	0·00100	4·217	0	1·755	0
10	1·2	1,000	1·00	0·00100	4·193	0·150	1·301	41·9
20	2·3	1,000	1·00	0·00100	4·182	0·296	1·002	83·8
30	4·25	1,000	1·00	0·00100	4·179	0·438	0·797	125·7
40	7·67	991	0·991	0·00101	4·179	0·581	0·651	170·4
50	12·5	991	0·991	0·00101	4·181	0·707	0·544	210·3
60	20	980	0·980	0·00102	4·185	0·832	0·462	251·5
70	31·25	980	0·980	0·00102	4·190	0·966	0·400	293·1
80	47·5	971	0·971	0·00103	4·197	1·076	0·350	336·1
90	70·0	962	0·962	0·00104	4·205	1·192	0·311	376·8
100	101·325	962	0·962	0·00104	4·216	1·307	0·278	419·1
125	228	935	0·935	0·00107	4·254	1·565	0·219	534·3
150	477	918	0·918	0·00109	4·310	1·842	0·180	632·2
175	890	893	0·893	0·00112	4·389	2·090	0·153	740·9
200	1,550	863	0·863	0·00116	4·497	2·329	0·133	851·7
225	2,550	834	0·834	0·00120	4·648	2·569	0·1182	966·8
250	4,000	800	0·800	0·00125	4·867	2·797	0·1065	1,087·4
275	5,950	756	0·756	0·00132	5·202	3·022	0·0972	1,210·7
300	8,600	714	0·714	0·00140	5·762	3·256	0·0897	1,345·4
325	12,125	654	0·654	0·00153	6·861	3·501	0·0790	1,494·3
350	16,540	575	0·575	0·00174	10·10	3·781	0·0648	1,672·2
360	18,680	526	0·526	0·00190	14·6	3·921	0·0582	1,764·6

Note: For convenience of tabulation the dynamic viscosity is given in g./m.^{2} instead of kg./m.^{2}.

Tabulated by permission from Haywood: *Thermodynamic Tables in S.I. (metric) Units*. Cambridge University Press.

GENERAL DATA, WATER

Maximum density at a temperature of: 39·2°F. or 4°C.

1 cu. ft.	=	6·23 gal.	=	62·355 lb.	at 62°F.
1 lb.	=	0·1 gal.	=	27·72 cu. in.	at 62°F.
1 gal.	=	0·16 cu. ft.	=	10 lb.	at 62°F.
1 ton	=	35·9674cu.ft.	=	224 gal.	at 62°F.
1 litre	=	0·22 gal.	=	1 kg.=0·45 lb. at 39·2°F. (4°C.).	

Freezing temperature ...	32°F.	0°C.	at atm. pres.
Boiling temperature ...	212°F.	100°C.	at atm. pres.
Latent heat of melting...	144 B.t.u./lb.	334 kJ/kg.	at atm. pres.
Latent heat of evaporation	977 B.t.u./lb.	2,270 kJ/kg.	at atm. pres.
Critical temperature ...	706–716°F.	380–386°C.	at atm. pres.
Critical pressure ...	3,200 lb./sq. in.	23·52 MN/m.^{2}	at atm. pres.

Pressure of water:

1 lb. per sq. in. = 2·31 ft. of water.
= 27·71 in. of water.
= 703·03 kg. per m^2.
= 703·03 mm. of water.

(See also Conversions Table, page I. 2)

Specific heat of water = 1·0000 B.t.u./lb. °F. = 4·187 kJ/kg. deg. C.
Specific heat of ice = 0·504 B.t.u./lb. °F. = 2·108 kJ/kg. deg. C.
Specific heat of water vapour = 0·477 B.t.u./lb. °F. = 1·996 kJ/kg. deg. C.

Thermal expansion: Water expands in bulk from 40°F. to 212°F. $\frac{1}{23}$ of its original volume.

Bulk elastic modulus of water = 300,000 lb. per sq. in.

Section V

PROPERTIES OF STEAM AND AIR

PROPERTIES OF STEAM AND OTHER VAPOURS

A Vapour is any substance in the gaseous state which does not even approximately follow the general gas laws.
Highly superheated vapours are gases, if the superheat is sufficiently great, and they then approximately follow the general gas law.

Conditions of Vapours:

1. **Dry Saturated Vapour** is free from unvaporized liquid particles.
2. **Wet Saturated Vapour** carries liquid globules in suspension.
3. **Superheated Vapour** is vapour the temperature of which is higher than that of the boiling point corresponding to the pressure.

Dryness Fraction or Quality of Saturated Vapour (X) is the percentage of dry vapour present in the given amount of the wet saturated vapour.

$$X = \frac{Ws}{Ws + Ww} \times 100 \text{ per cent.}$$

Ws = Weight of dry steam in steam considered.
Ww = Weight of water in steam.

The Heat of the Liquid "h" is the heat in Joules per kg. required to raise the temperature of the liquid from 0°C. to the temperature at which the liquid begins to boil at the given pressure.

$$h = ct.$$

c = Mean specific heat of water.
t = Temperature of formation of steam at pressure considered °C.

The Latent Heat of Evaporation "L" is the heat required to change a liquid at a given temperature and pressure into a vapour at the same temperature and pressure. It is divided into two parts:

1. External Latent Heat of Vapour = External Work Heat.
2. Internal Latent Heat of Vapour = Heat due to change of state.

The Total Heat of a Vapour (or Enthalpy) is the amount of heat which must be supplied to 1 kg. of the liquid which is at 0°C. to convert it at constant pressure into vapour at the temperature and pressure considered.

Total Heat of Dry Saturated Vapour:

$$H = h + L \text{ (Joules per kg.).}$$

h = Heat of liquid at the temperature of the wet vapour, Joules per kg.
L = Latent heat, Joules per kg.

Total Heat of Wet Saturated Vapour:

$$H_w = h + xL \text{ (Joules per kg.).}$$

x = Dryness Factor.

Total Heat of Superheated Vapour:

$$H_s = h + L + c(t_s—t\ ). \text{ (Joules per kg.).}$$

c = Mean specific heat capacity of superheated vapour at the pressure and degree of superheat considered.
t_s = Temperature of superheat °C.
t_1 = Temperature of formation of steam °C.

Specific Volumes of Wet Vapour:

$$V_W = (1-x)\ V + xV_D.$$

when x = very small.

$$V_W = xV_D, \quad x = \frac{V_W}{V_D}.$$

V_W = Specific volume of the wet vapour, m.³ per kg.
V_D = Specific volume of dry saturated vapour of the same pressure, m.³ per kg. (Can be found from the Vapour Tables.)

Specific Volume of Superheated Vapour:

Approximate method by using Charles' Law:

$$V = \frac{V_s T_s}{T_1}.$$

ENTROPY OF STEAM

1. Entropy of Water:

$$\text{Change of Entropy} = \log_e \frac{T_1}{T}.$$

T_1, T = Absolute temperature.

$$\text{Entropy of water above freezing-point} = \phi_w = \log_e \frac{T_1}{273}.$$

2. Entropy of Evaporation:

$$\text{Change of Entropy during evaporation} = \frac{dL}{T}.$$

Entropy of 1 kg. of wet steam above freezing point:

$$\phi_s = \phi_w + \frac{xL^1}{T_1}.$$

3. Entropy of Superheated Steam:

Change of Entropy per kg. of steam during superheating.

$$= C_p \log_e \frac{T}{T_1}.$$

Total Entropy of 1 kg. of superheated steam above freezing point

$$= \phi_w + \frac{L_1}{T_1} + C_p \log_e \frac{T_s}{T_1}.$$

L_1 = Latent heat of evaporation at T_1 °C. absolute.
T_1 = Absolute temperature of evaporation.
T_s = Absolute temperature of superheat.

TEMPERATURE—ENTROPY DIAGRAM FOR STEAM

shows the relationship between Pressure, Temperature, Dryness Fraction and Entropy.

When two of these factors are given the two others can be found on the chart.

The ordinates represent the Absolute Temperature and the Entropy.

The chart consists of the following lines:

1. Isothermal Lines.
2. Pressure Lines.
3. Lines of Dryness Fraction.
4. Water Line between Water and Steam.
5. Dry Steam Lines.
6. Constant Volume Lines.

The total heat is given by the area, enclosed by absolute zero base water line and horizontal and vertical line from the respective points.

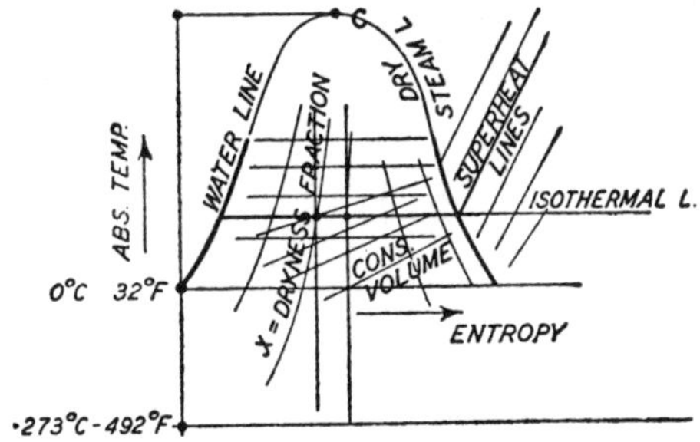

An adiabatic expansion is a vertical line (expansion at Constant Entropy, no transfer of heat).

C = Critical temperature of steam
= 706°F. to 716°F.
= 375°C. to 380°C.

Critical Pressure: 3,200 lb. per sq. in. = 217·8 atm. = 23·5 MN/m^2.

MOLLIER OR TOTAL HEAT—ENTROPY CHART

Contains the same lines as the Temperature-Entropy Diagram, but with ordinates representing the Total Heat and Entropy of Steam. This diagram is used to find the drop in the total heat of steam during an adiabatic expansion.

TOTAL HEAT OF SUPERHEATED STEAM (B.t.u. per lb.)

Abs.Pres. lb. per sq.in.	*Sat.Temp. °F.*	*Degrees of Superheat °F.* 0	40	80	120	160	200	280
20	228	1157·1	1177·2	1197·2	1216·9	1236·6	1256·1	1294·9
30	250·3	1165·5	1185·9	1206·1	1226·1	1245·9	1265·6	1304·7
40	267·2	1171·6	1192·3	1212·9	1233·0	1253·0	1272·8	1312·2
50	280·9	1176·3	1197·3	1218·1	1238·5	1258·6	1278·5	1317·9
60	292·6	1180·1	1201·4	1222·2	1242·8	1263·1	1283·2	1322·9
70	302·8	1183·3	1204·7	1225·8	1246·6	1266·9	1287·2	1327·1
80	311·9	1186·1	1207·9	1229·1	1250·0	1270·5	1290·9	1330·8
90	320·2	1188·5	1210·5	1232·1	1253·0	1273·7	1294·0	1334·2
100	327·9	1190·7	1212·9	1234·6	1255·7	1276·5	1297·0	1337·3
120	341·3	1194·3	1216·9	1239·0	1260·4	1281·3	1302·0	1342·6
140	353·0	1197·2	1220·2	1242·5	1264·2	1285·5	1306·3	1347·1
160	363·6	1199·7	1222·9	1245·6	1267·6	1289·1	1310·0	1351·1
180	373·1	1201·7	1225·5	1248·3	1270·7	1292·2	1313·2	1354·6
200	381·8	1203·5	1227·6	1250·7	1273·1	1295·0	1316·2	1358·0
250	401·0	1207·0	1231·7	1255·7	1278·9	1301·2	1322·6	1364·9
300	417·4	1209·4	1235·0	1259·5	1283·2	1305·8	1327·6	1370·3
400	444·7	1212·1	1239·6	1265·4	1289·9	1313·3	1335·8	1379·6
500	467·1	1213·2	1242·2	1269·1	1294·7	1318·8	1341·9	1386·6

ENTROPY OF SUPERHEATED STEAM (B.t.u. per °F.)

Abs.Pres. lb. per sq.in.	*Sat.Temp. °F.*	*Degrees of Superheat °F.* 0	40	80	120	160	200	280
20	228	1·7333	1·7617	1·7883	1·8134	1·8372	1·8596	1·9017
30	250·3	1·7017	1·7298	1·7560	1·7807	1·8041	1·8261	1·8472
40	267·2	1·6793	1 7071	1·7331	1·7575	1·7806	1·8025	1·8233
50	280·9	1·6619	1·6895	1·7153	1·7397	1·7626	1·7843	1·8049
60	292·6	1·6477	1·6752	1·7010	1·7253	1·7480	1·7694	1·7899
70	302·8	1·6357	1·6632	1·6889	1·7130	1·7357	1·7570	1·7774
80	311·9	1·6254	1·6527	1·6784	1·7024	1·7251	1·7463	1·7665
90	320·2	1·6161	1·6436	1·6692	1·6931	1·7157	1·7367	1·7569
100	327·9	1·6079	1·6353	1·6608	1·6847	1·7073	1·7283	1·7484
120	341·3	1·5935	1·6210	1·6467	1·6705	1·6928	1·7138	1·7337
140	353·0	1 5813	1·6088	1·6345	1·6583	1·6805	1·7014	1·7212
160	363·6	1·5706	1·5983	1·6240	1·6479	1·6701	1·6909	1·7107
180	373·1	1·5610	1·5890	1·6148	1·6386	1·6607	1·6815	1·7013
200	381·8	1·5525	1·5806	1·6063	1·6301	1·6523	1·6730	1·6929
250	401·0	1·5342	1·5628	1·5886	1·6125	1·6347	1·6554	1·6751
300	417·4	1·5190	1·5479	1·5740	1·5980	1·6203	1·6410	1·6607
400	444·7	1·4941	1·5240	1·5506	1·5749	1·5973	1·6181	1·6379
500	467·1	1·4740	1·5049	1·5322	1·5568	1·5795	1·6004	1·6201

SPECIFIC ENTHALPY OF SUPERHEATED STEAM (kJ./kg.)

Abs. Press. MN/m.2	*Sat. Temp. °C.*	*Steam Temperature, °C.*							
		100	125	150	175	200	225	250	275
0·1	99·6	2,676	2,726	2,776	2,826	2,875	2,925	2,975	3,024
0·5	151·8	—	—	—	2,800	2,855	2,909	2,961	3,013
1·0	179·9				—	2,827	2,886	2,943	2,998
2·0	212·4					—	2,834	2,902	2,965
4·0	250·3						—	—	2,886

SPECIFIC ENTROPY OF SUPERHEATED STEAM

Abs. Press. MN/m.2	*Sat. Temp. °C.*	*Steam Temperature, °C.*							
		100	125	150	175	200	225	280	275
0·1	99·6	7·362	7·492	7·614	7·728	7·835	7·937	8·034	8·127
0·5	151·8				6·940	7·059	7·169	7·272	7·369
1·0	179·9					6·692	6·815	6·926	7·029
2·0	212·4						6·412	6·545	6·663
4·0	250·3								6·229

These Tables are taken by permission from Haywood: *Thermodynamic Tables in S.I. (metric) Units*: Cambridge University Press.

(Based on Callendar's Values)

Abs. Pres. p. lb./sq. in.	Temp. t. °F.	Specific Volume. v. cu. ft./lb.	Density. w. lb./cu. ft.	Heat of Liquid. h. B.t.u./lb.	Heat of Evap. L. B.t.u./lb.	Heat of Sat. Vap. H. B.t.u./lb.	Entropy S. B.t.u./lb.°F.
0·5	79·5	640·5	0·00156	47·4	1045	1092	2·0299
1	101·7	333·1	0·0030	69·5	1033	1102	1·9724
2	126·1	173·5	0·0058	93·9	1020	1114	1·9159
3	141·5	118·6	0·0085	109·3	1012	1121	1·8833
4	153·0	90·5	0·0111	120·8	1005	1126	1·8600
5	162·3	73·4	0·0136	130·1	1000	1130	1·8422
6	170·1	61·9	0·0162	137·9	995	1133	1·8277
7	176·9	53·6	0·0187	144·8	991	1136	1·8156
8	182·9	47·3	0·0212	150·8	988	1139	1·8049
9	188·3	42·4	0·0236	156·3	985	1141	1·7956
10	193·2	38·4	0·0261	161·1	982	1143	1·7874
12	202·0	32·4	0·0309	169·9	977	1147	1·7731
14	209·6	28·0	0·0357	177·6	972	1150	1·7611
14·7	212·0	26·8	0·0373	180·0	970	1151	1·7573
16	216·3	24·7	0·0404	184·4	968	1152·5	1·7506
18	222·4	22·2	0·0451	190·5	964	1155	1·7414
20	228·0	20·1	0·0498	196·1	961	1157	1·7333
22	233·1	18·37	0·0545	201·3	958	1159	1·7258
24	237·8	16·93	0·0591	206·1	955	1161	1·7189
26	242·2	15·71	0·0636	210·5	952	1162·5	1·7126
28	246·4	14·66	0·0682	214·8	949	1164	1·7069
30	250·3	13·72	0·0728	218·8	947	1165·5	1·7017
35	259·3	11·86	0·8425	228	941	1169	1·6898
40	267·2	10·48	0·0953	236	936	1172	1·6793
45	274·4	9·37	0·1067	243	931	1174	1·6701
50	281·0	8·50	0·1175	250	926	1176	1·6619
55	287·0	7·74	0·1292	256	922	1178	1·6547
60	292·6	7·16	0·1397	262	919	1180	1·6479
65	297·9	6·64	0·1506	267	914	1182	1·6415
70	303·0	6·20	0·1613	272	911	1183	1·6357
75	307·5	5·81	0·1721	277	907	1185	1·6304
80	312·0	5·47	0·1828	282	904	1186	1·6254
85	316·2	5·16	0·1938	286	901	1187	1·6206
90	320·2	4·89	0·2045	290	898	1189	1·6161
95	324·1	4·65	0·2150	295	895	1190	1·6120
100	327·9	4·43	0·2257	298	893	1191	1·6079
105	331·4	4·23	0·2364	302	890	1192	1·6041
110	334·8	4·04	0·2475	306	887	1193	1·6004
115	338·1	3·88	0·2577	309	884	1194	1·5969

Abs. Pres.	Temp.	Spec. Volume.	Density	Heat of			Entropy.
				Liquid.	Evap.	Sat. Vap.	
p. lb./sq. in.	t. °F.	v. cu. ft./lb.	w. lb./cu. ft.	h. B.t.u./lb.	L. B.t.u./lb.	H. B.t.u./lb.	S. B.t.u./lb.°F.
120	341·3	3·73	0·2681	312	882	1194	1·5935
125	344·4	3·59	0·2786	316	879	1195	1·5903
130	347·3	3·46	0·2890	319	877	1196	1·5872
135	350·2	3·33	0·3003	322	875	1197	1·5842
140	353·0	3·22	0·3106	325	872	1197	1·5813
145	355·8	3·12	0·3205	328	870	1198	1·5785
150	358·4	3·02	0·3311	331	868	1199	1·5758
160	363·6	2·84	0·3521	336	864	1200	1·5706
170	368·4	2·68	0·3731	341	860	1201	1·5657
180	373·1	2·54	0·3937	346	856	1202	1·5610
190	377·5	2·41	0·4149	351	852	1203	1·5567
200	382	2·29	0·4347	356	848	1203	1·5525
220	390	2·09	0·4785	364	841	1205	1·5448
240	387	1·93	0·5181	372	834	1206	1·5376
260	404·5	1·78	0·5618	380	827	1207·5	1·5310
280	411·1	1·66	0·6024	387	821	1208·5	1·5241
300	417·4	1·55	0·6452	394	815	1209·4	1·5190
350	431·8	1·34	0·7463	410	801	1211·1	1·5058
400	444·7	1·17	0·8547	425	787·5	1212·1	1·4941
450	456·4	1·04	0·9615	437·8	775	1212·8	1·4836
500	467·1	0·94	1·0638	450·1	763·1	1213·2	1·4740

Abs. Press. kN/m.^{2}	Temp. °C.	Specific Volume m.^{3}/kg.	Density kg./m.^{3}	Specific Enthalpy of			Specific Entropy of Steam kJ./kg. K.
				Liquid kJ./kg.	Evaporation kJ./kg.	Steam kJ./kg.	
0·8	3·8	160	0·00626	15·8	2,493	2,509	9·058
2·0	17·5	67·0	0·0149	73·5	2,460	2,534	8·725
5·0	32·9	28·2	0·0354	137·8	2,424	2,562	8·396
10·0	45·8	14·7	0·0682	191·8	2,393	2,585	8·151
20·0	60·1	7·65	0·131	251·5	2,358	2 610	7·909
28	67·5	5·58	0·179	282·7	2,340	2,623	7·793
35	72·7	4·53	0·221	304·3	2,327	2,632	7·717
45	78·7	3·58	0·279	329·6	2,312	2,642	7·631
55	83·7	2·96	0·338	350·6	2,299	2,650	7·562
65	88·0	2·53	0·395	368·6	2,288	2,657	7·506
75	91·8	2·22	0·450	384·5	2,279	2,663	7·457
85	95·2	1·97	0·507	398·6	2,270	2,668	7·415
95	98·2	1·78	0·563	411·5	2,262	2,673	7·377
100	99·6	1·69	0·590	417·5	2,258	2,675	7·360
101·33	100	1·67	0·598	419·1	2,257	2,676	7·355
110	102·3	1·55	0·646	428·8	2,251	2,680	7·328
130	107·1	1·33	0·755	449·2	2,238	2,687	7·271
150	111·4	1·16	0·863	467·1	2,226	2,693	7·223
170	115·2	1·03	0·970	483·2	2,216	2,699	7·181
190	118·6	0·929	1·08	497·8	2,206	2,704	7·144
220	123·3	0·810	1·23	517·6	2,193	2,711	7·095
260	128·7	0·693	1·44	540·9	2,177	2,718	7·039
280	131·2	0·646	1·55	551·4	2,170	2,722	7·014
320	135·8	0·570	1·75	570·9	2,157	2,728	6·969
360	139·9	0·510	1·96	588·5	2,144	2,733	6·930
400	143·6	0·462	2·16	604·7	2,133	2,738	6·894
440	147·1	0·423	2·36	619·6	2,122	2,742	6·862
480	150·3	0·389	2·57	633·5	2,112	2,746	6·833
500	151·8	0·375	2·67	640·1	2,107	2,748	6·819
550	155·5	0·342	2·92	655·8	2,096	2,752	6·787
600	158·8	0·315	3·175	670·4	2,085	2,756	6·758
650	162·0	0·292	3·425	684·1	2,075	2,759	6·730
700	165·0	0·273	3·66	697·1	2,065	2,762	6·705
750	167·8	0·255	3·915	709·3	2,056	2,765	6·682
800	170·4	0·240	4·16	720·9	2,047	2,768	6·660
850	172·9	0·229	4·41	732·0	2,038	2,770	6·639
900	175·4	0·215	4·65	742·6	2,030	2,772	6·619
950	177·7	0·204	4·90	752·8	2,021	2,774	6·601
1,000	179·9	0·194	5·15	762·6	2,014	2,776	6·583

Abs. Pressure kN/m.²	Temp. °C.	Specific Volume m.³/kg.	Density kg./m.³	Specific Enthalpy of			Specific Entropy kJ./kg. K.
				Liquid kJ./kg.	Evaporation kJ./kg.	Steam kJ./kg.	
1,050	182.0	0.186	5·39	772	2,006	2,778	6·566
1,150	186·0	0·170	5·89	790	1,991	2,781	6·534
1,250	189·8	0·157	6·38	807	1,977	2,784	6·505
1,300	191·6	0·151	6·62	815	1,971	2,785	6·491
1,500	198·3	0·132	7·59	845	1,945	2,790	6·441
1,600	201·4	0·124	8·03	859	1,933	2,792	6·418
1,800	207·1	0·110	9·07	885	1,910	2,795	6·375
2,000	212·4	0·0995	10·01	909	1,889	2,797	6·337
2,100	214·9	0·0949	10·54	920	1,878	2,798	6·319
2,300	219·6	0·0868	11·52	942	1,858	2,800	6·285
2,400	221·8	0·0832	12·02	952	1,849	2,800	6·269
2,600	226·0	0·0769	13·01	972	1,830	2,801	6·239
2,700	228·1	0·0740	13·52	981	1,821	2,802	6·224
2,900	232·0	0·0689	14·52	1,000	1,803	2,802	6·197
3,000	233·8	0·0666	15·00	1,008	1,794	2,802	6·184
3,200	237·4	0·0624	16·02	1,025	1,779	2,802	6·158
3,400	240·9	0·0587	17·04	1,042	1,760	2,802	6·134
3,600	244·2	0·0554	18·06	1,058	1,744	2,802	6·112
3,800	247·3	0·0524	19·08	1,073	1,728	2,801	6·090
4,000	250·3	0·0497	21·0	1,087	1,713	2,800	6·069

Taken by permission of Cambridge University Press from *Thermodynamic Tables in S.I. (metric) Units* by Haywood.

PROPERTIES OF AIR

Symbols:

V = Volume of dry air-vapour mixture in cu. ft. or m.³

W = Weight of air-vapour mixture in lb. or kg.

P_2 = Partial pressure of vapour
P_{wa} = Actual partial pressure of water vapour
P_{ws} = Saturation pressure of water vapour
} inches of mercury (or mm.).

P = Total pressure of mixture.

t = Dry bulb temperature.

T = $t + 460$ = absolute temperature °F. or $t + 273$°C.

ϕ = Relative humidity. per cent.

X = Actual specific humidity, in gr. per lb. or g. per kg.

Xs = Specific humidity of saturated air gr. per lb. or g. per kg.

d_a = Density of dry air in lb. per cu. ft. or kg./m.³

d_w = Density of water vapour in lb. per cu. ft. or kg./m.³

d = Density of air-water vapour mixture lb. per cu. ft. or kg./m.³

R = Gas constant, dry air = 53·34; water vapour = 86·00.

1 mm. of mercury = $1{\cdot}33 \times 10^2$ N per m.²

Atmospheric Air is a mixture of dry air and water vapour. It can be considered as an ideal gas without great discrepancies and the gas laws can be applied. (Boyle, Charles, Gay-Lussac, General Gas Law, Dalton, Avogadro, see pages IV. 4–6).

General Gas Law: $PV = WRT$. $W = \frac{PV}{T} \cdot \frac{1}{R}$.

Density of Dry Air:

$$d_a = \frac{1{\cdot}323\,P_a}{T}$$

$$d_a = \frac{0{\cdot}470\,P_a}{T}$$

Density of Water Vapour:

$$d_w = \frac{0{\cdot}821\,P_w}{T}$$

$$d_w = \frac{0{\cdot}293\,P_w}{T}$$

Density of Air/Water Vapour Mixture:

$$d = \frac{1{\cdot}323 P_t}{T} - \frac{0{\cdot}503 P_{ws}\phi}{T}$$

$$d = \frac{0{\cdot}470 P_t}{T} - \frac{0{\cdot}177 P_{ws}\phi}{T}$$

d_a, d_w, d, in lb. per cu. ft.
P_a, P_w, P_t, P_{ws}, in in. Hg

d_a, d_w, d, in kg. per m.³
P_a, P_w, P_t, P_{ws}, in mm. Hg.

Air-Water Vapour mixture is always lighter than dry air.

Humidity is the term applied to the quantity of water vapour present in the air.

Absolute Humidity is the actual weight of water vapour in grains or lb. per cu. ft. of a mixture, or grammes per kilogramme.

Specific Humidity is the actual weight of water vapour in gr. or lb. per 1 lb. of dry air, or gramme per 1 kg.

$$X = 4354 \frac{\phi P_{ws}}{P - P_{ws}} \text{ grains/lb.} = 622 \frac{\phi P_{ws}}{P - P_{ws}} \text{ grammes/kg.}$$

Specific Humidity of Saturated Air:

$$X_s = 4354 \frac{P_{ws}}{P - P_{ws}} \text{ grains/lb.} = 622 \frac{P_{ws}}{P - P_{ws}} \text{ grammes/kg.}$$

Relative Humidity is either the ratio of the actual partial pressure of water vapour to the vapour pressure at saturation at the existent dry bulb temperature;

or the ratio of the actual vapour density to the density at saturation at the dry bulb temperature;

or the ratio of actual amount of moisture in given air volume to amount of moisture required for saturation of that volume.

Usually expressed in per cent.

$$\phi = \frac{P_{wa}}{P_{ws}} = \frac{d_w}{d_{ws}} = \frac{X}{X_s} \text{ (per cent).}$$

Saturated Air holds the maximum amount of water at the given temperature; any lowering of the air temperature will cause condensation of water vapour.

Dry Bulb Temperature is the air temperature as indicated by a thermometer which is not affected by the moisture of the air.

Wet Bulb Temperature is the temperature of adiabatic saturation. It is the lowest temperature indicated by a moistened thermometer bulb when it is exposed to a current of air.

Dew Point Temperature is the temperature to which air with any given moisture content must be cooled to produce saturation of the air and begin condensation of its vapour.

Total Heat of Dry Air:

$$H_1 = 0{\cdot}24t \text{ (B.t.u./lb.)} \quad t \text{ in } ^\circ\text{F.}$$
$$= 1{\cdot}01t \text{ (kJ/kg.)} \quad t \text{ in } ^\circ\text{C.}$$

Total Heat of Air-Water Vapour Mixture is composed of the sensible heat of air and the latent heat of vaporization of the moisture of vapour in the air.

The total heat is constant for any certain wet-bulb temperature irrespective of any change in the dry-bulb temperature.

$H = 1{\cdot}01t + X\ (2463 + 1{\cdot}88t)$ kJ/kg.

where t = dry bulb temperature °C.

1·01 = specific heat capacity of dry air kJ/kg. deg. C.

2463 = latent heat of vaporization of water at 0°F. kJ/kg. deg. C.

1·88 = specific heat capacity of water vapour at constant pressure, kJ/kg. deg. C.

$H = 0{\cdot}24t + X\ (1059{\cdot}2 + 0{\cdot}45t)$ B.t.u./lb.

where t = dry bulb temperature °F.

Thermal Expansion of Air:

Dry air expands or contracts uniformly $\frac{1}{886}$ of its volume per deg. C. difference, or $\frac{1}{492}$ per deg. F. difference, under constant pressure.

THE HUMIDITY CHART FOR AIR (Mollier)
(See Chart No. 5)

The Chart shows the relationship between:

1. Dry bulb temperature.
2. Wet bulb temperature.
3. Dew point.
4. Relative humidity.
5. Specific humidity.
6. Vapour pressure.
7. Total heat.

When two of these factors are given the others can be found on the chart.

The Chart consists of the following lines:

1. Temperature lines.
2. Constant heat lines (constant wet bulb temperature).
3. Relative humidity lines.
4. Dew point lines (constant moisture).

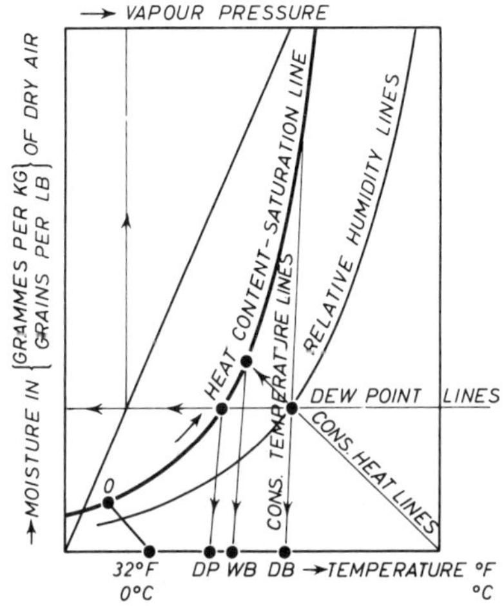

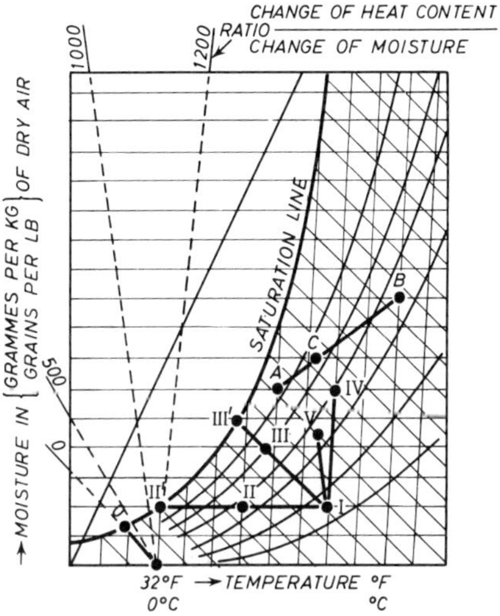

D.P. = Dew Point Temperature.
W.B. = Wet Bulb Temperature.
D.B. = Dry Bulb Temperature.

Change of Condition of Air.	*Indicated in above sketch.*	*Remarks.*
Cooling with constant moisture.	From I to II.	Dew point temperature in intersection II', with saturation line.
Adiabatic change.	From I to III, saturation at III'.	No heat is added or extracted.
Temperature constant.	From I to IV.	

Mixing of air, air volume V_A of condition "A" is to be mixed with air volume V_B of condition "B". Condition of mixture is "C"

$$\frac{\text{Distance "AC"}}{\text{Distance "BC"}} = \frac{\text{Air Volume } V_B}{\text{Air Volume } V_A}.$$

MAN AND AIR

(a) Respiration

An adult breathes at rest at 16 respirations per minute, about 50 m.^{3}/hr. = about 17·5 cu. ft. of air per hour. When working 3 to 6 times more.

Average condition of expired air:

Oxygen	16·5 per cent
Carbon dioxide	4·0 per cent
Nitrogen and Argon	79·5 per cent

Quantity of expired carbon dioxide in 24 hours—approx. 2·2 lb.
—approx. 1 kg.

(b) Equilibrium of Heat

Heat is generated within the human body by combustion of food.

Heat is lost from the human body by:

1. Conduction and Convection	about	25 per cent
2. Radiation	about	43 per cent
3. Evaporation		30 per cent
4. Expired Air		2 per cent

Evaporation prevailing in high temperature.

Conduction and convection prevailing in low temperatures.

Heat is liberated at such a rate that the internal body temperature is maintained at about 98·6°F. = 37°C.

PROPORTION OF SENSIBLE AND LATENT HEAT DISSIPATED BY MAN AT FAIRLY HARD WORK

Dry Bulb Temperature °C. ...	13	15	18	21	24	27	30	32
Dry Bulb Temperature °F. ...	55	60	65	70	75	80	85	90
Sensible Heat, per cent	75	68	60	51	42	31	20	10
Latent Heat, per cent	25	32	40	49	58	69	80	90

(c) Heat Loss of Human Body

The total heat loss of an adult (sensible and latent) is approximately 400 B.t.u. per hr. in room temperatures from 65° to 85°F. = 117 W (see Table VI. 15).

(d) Comfort Air Conditions

Optimum Winter Temperature	18°C. to 20°C.
	65°F. to 68°F.
Optimum Summer Temperature	20°C. to 22°C.
	68°F. to 71°F.
Optimum Relative Humidity	40% to 65%

Temperature of Heated Rooms (see Tables VI. 1 *et seq.*).

DESIRABLE INDOOR AIR CONDITIONS IN SUMMER APPLICABLE TO EXPOSURES LESS THAN 3 HOURS

Outside Temperature Dry bulb		*Inside Air Conditions with Dew Point Constanet at 14°C. or 57°F.*							
		Dry bulb		*Wet bulb*		*Relative Humidity*	*Eff. Temperature*	
°F.	°C.	°F.	°C.	°F.	°C.	%	°F.	°C.
95	35	80	27	65	18·5	44	73	23·0
90	32	78	26	64·5	18·0	46	72	22·0
85	29	76·5	25	64	17·8	52	71	21·5
80	27	75	24	63·5	17·5	51	70	21·0
75	24	73·5	23	63	17·2	57	69	20·5
70	21	72	22	62·5	17·0	57	68	20·0

Thermo-Equivalent Conditions are combinations of temperature, humidity and air movement which produce the same feeling of warmth (effective temperature, equivalent temperature).

British Equivalent Temperature of an environment (B.E.T.) is the temperature of a uniform enclosure in which, in still air, a sizeable black body maintained at a temperature of 75°F. would lose heat at the same rate as in the environment. (Measured by the Eupatheoscope.)

The Effective Temperature (used in America) is an arbitrary index of the degree of warmth or cold felt by the human body in response to temperature, humidity and air movement. It combines the reading of these three values into a single one.

The Numerical Value is that of the temperature of still, saturated air which would induce an identical sensation of warmth.

The Kata-Thermometer (Hill, 1926) is an alcohol thermometer graduated from 95°F. to 100°F. for measuring the effective temperature and especially the velocity of air. It is warmed above 100°F. and the time required for cooling from 100°F. to 95°F. is measured. This time is a scale for the cooling effect of the air.

The Hygrometer (Psychrometer) is an instrument for the determination of relative humidity and dew point by measuring the dry-bulb temperature by an ordinary thermometer and the wet bulb temperature by a thermometer the bulb of which is kept moist.

AIR CONDITIONING

CHART SHOWING RELATION OF EFFECTIVE TEMPERATURE WET-BULB, DRY-BULB TEMPERATURE AND RELATIVE HUMIDITY

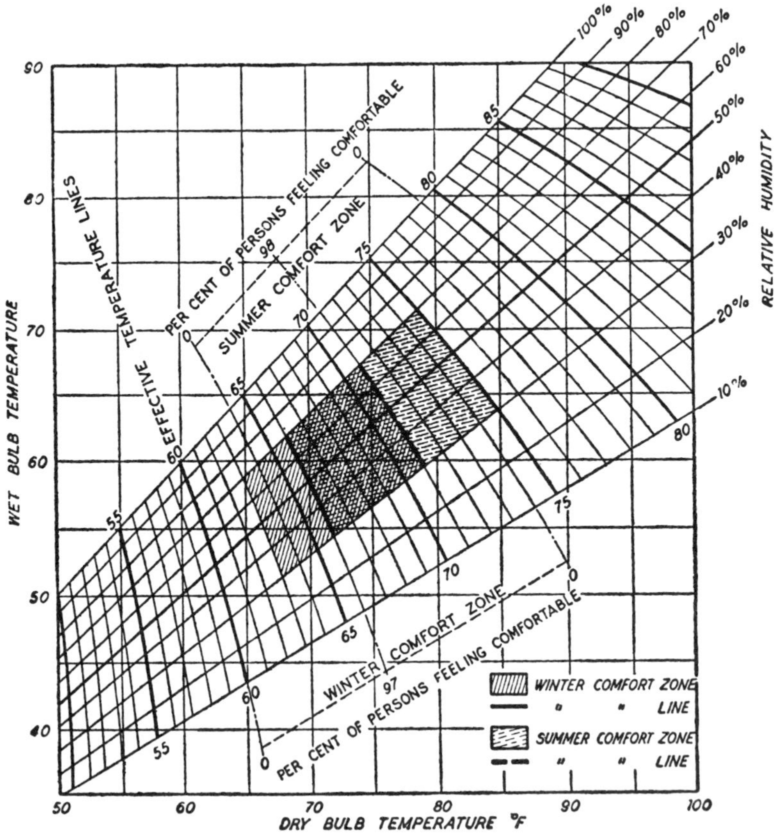

The Chart is for air velocities up to 25 ft/min (i.e. practically still air). For an air velocity of 100 ft/min the effective temperature decreases by 2°F.

RELATIVE HUMIDITY IN PER CENT

For various Room Temperatures and various Differences between Wet- and Dry-Bulb Temperatures.

(Accurate for Atm. Pressure of 14·25 lb. per sq. in. = 29 in. of Mercury).

Dry-Bulb Temp. °F.	Difference between Dry Bulb and Wet Bulb Temperature. °F.										
	0	2	4	6	8	10	12	14	16	18	20
50	100	87	74	62	50	39	28	17	7	0	0
52	100	88	75	63	52	41	30	20	10	0	0
54	100	88	76	65	54	43	33	23	14	5	0
56	100	88	77	66	55	45	35	26	17	8	0
58	100	88	77	67	57	47	38	28	20	11	3
60	100	89	78	68	58	49	40	31	22	14	6
62	100	89	79	69	60	50	41	33	25	17	9
64	100	90	79	70	61	52	43	35	27	20	12
66	100	90	80	71	62	53	45	37	29	22	15
68	100	90	81	72	63	55	47	39	31	24	17
70	100	90	81	72	64	56	48	40	33	26	20
72	100	91	82	73	65	57	49	42	35	28	22
74	100	91	82	74	66	58	51	44	37	30	24
76	100	91	83	74	67	59	52	45	38	32	26
78	100	91	83	75	67	60	53	46	40	34	28
80	100	91	83	76	68	61	54	47	41	35	29
82	100	92	84	76	69	62	55	49	43	37	31
84	100	92	84	77	70	63	56	50	44	38	32
86	100	92	85	77	70	63	57	51	45	39	34
88	100	92	85	78	71	64	58	52	46	41	35
90	100	92	85	78	71	65	59	53	47	42	37
92	100	92	85	78	72	65	59	54	48	43	38
94	100	93	86	79	72	66	60	54	49	44	39
96	100	93	86	79	73	67	61	55	50	45	40
98	100	93	86	79	73	67	61	56	51	46	41
100	100	93	86	80	74	68	62	57	52	47	42

RELATIVE HUMIDITY IN PER CENT

For various Room Temperatures and various Differences between Wet- and Dry-Bulb Temperatures.

Dry Bulb Temp. °C.	*Difference between Dry Bulb and Wet Bulb °C.*											
	0	1	2	3	4	5	6	7	8	9	10	11
10	100	88	77	66	55	44	34	25	15	6	0	0
11	100	89	78	67	56	46	36	27	18	9	2	0
12	100	89	78	68	58	48	39	29	21	12	3	0
13	100	90	79	69	60	50	41	32	34	15	6	1
14	100	90	80	70	61	52	43	34	26	17	8	2
15	100	90	81	71	62	53	44	46	28	20	11	5
16	100	90	81	71	63	54	46	37	30	22	14	8
17	100	91	82	72	64	56	48	39	32	25	17	11
18	100	91	82	73	65	57	49	41	34	27	20	13
19	100	91	83	74	66	59	51	43	36	29	23	16
20	100	91	83	74	67	59	52	44	38	31	25	18
21	100	92	83	75	68	61	53	46	40	33	27	20
22	100	92	83	75	68	61	54	47	41	34	28	22
23	100	92	84	76	69	62	55	48	42	36	30	24
24	100	92	84	76	70	63	56	49	43	37	31	26
25	100	92	85	77	71	64	57	51	45	39	33	28
26	100	92	85	77	71	64	58	52	46	40	34	29
27	100	93	85	78	72	65	59	53	47	42	36	31
28	100	93	85	78	72	66	59	54	48	43	37	32
29	100	93	86	79	73	67	60	55	49	44	39	34
30	100	93	86	79	73	67	61	56	50	45	40	35
31	100	93	86	80	74	68	62	57	51	46	41	36
32	100	93	86	80	74	68	63	57	52	47	42	37
33	100	93	87	81	75	69	64	58	53	48	43	38
34	100	93	87	81	75	69	64	59	53	49	44	39
35	100	94	87	82	76	70	65	60	54	50	45	40
36	100	94	87	82	76	70	65	60	55	50	46	41
37	100	94	88	82	76	71	66	61	56	51	47	42
38	100	94	88	82	76	71	66	61	56	52	47	43

Temp.	Press. of Sat. Vapour.	Weight of Sat. Vapour.		Volume in cu. ft.		Heat Content of Mixture.
		Grains. per cu. ft.	per lb. of Dry Air. Grains. per lb.	of 1 lb. of Dry Air.	of 1 lb. of Dry Air. & Vapour to saturate.	
°F.	in. of Hg.					B.t.u./lb.
0	0·0375	0·472	5·47	11·58	11·59	0·852
10	0·00628	0·772	9·16	11·83	11·58	3·831
20	0·01027	1·238	15·01	12·09	12·13	7·137
30	0·1646	1·943	24·11	12·34	12·41	10·933
32	0·1806	2·124	26·47	12·39	12·47	11·83
33	0·1880	2·206	27·57	12·41	12·49	12·18
34	0·1957	2·292	28·70	12·44	12·52	12·60
35	0·2036	2·380	29·88	12·47	12·55	13·02
36	0·2119	2·471	31·09	12·49	12·58	13·44
37	0·2204	2·566	32·35	12·52	12·61	13·87
38	0·2292	2·663	33·66	12·54	12·64	14·31
39	0·2384	2·764	35·01	12·57	12·67	14·76
40	0·2478	2·868	36·41	12·59	12·70	15·21
41	0·2576	2·976	37·87	12·62	12·73	15·67
42	0·2678	3·087	39·38	12·64	12·76	16·14
43	0·2783	3·201	40·93	12·67	12·79	16·62
44	0·2897	3·319	42·55	12·69	12·82	17·10
45	0·3003	3·442	44·21	12·72	12·85	17·59
46	0·3120	3·568	45·94	12·74	12·88	18·09
47	0·3240	3·698	47·73	12·77	12·91	18·60
48	0·3364	3·832	49·58	12·79	12·94	19·12
49	0·3492	3·970	51·49	12·82	12·97	19·65
50	0·3624	4·113	53·47	12·84	13·00	20·19
51	0·3761	4·260	55·52	12·87	13·03	20·74
52	0·3903	4·411	57·64	12·89	13·07	21·30
53	0·4049	4·568	59·83	12·92	13·10	21·87
54	0·4200	4·729	62·09	12·95	13·13	22·45
55	0·4356	4·895	64·43	12·97	13·16	23·04
56	0·4517	5·066	66·85	13·00	13·20	23·64
57	0·4684	5·242	69·35	13·02	13·23	24·25
58	0·4855	5·424	71·93	13·05	13·26	24·88
59	0·5032	5·611	74·60	13·07	13·30	25·52
60	0·5214	5·804	77·30	13·10	13·33	26·18

Temp. °F.	Press. of Sat. Vap. in. of Hg.	Weight of Sat. Vap. Grains. per cu. ft.	Weight of Sat. Vap. Grains. per lb.	Volume in cu. ft. of 1 lb. of Dry Air.	Volume in cu. ft. of 1 lb. of Dry Air & Vapour to saturate.	Heat Content of Mixture. B.t.u./lb.
61	0·5403	6·003	80·2	13·12	13·36	26·84
62	0·5597	6·208	83·2	13·15	13·40	27·52
63	0·5798	6·418	86·2	13·17	13·43	28·22
64	0·6005	6·633	89·3	13·20	13·47	28·93
65	0·6218	6·855	92·6	13·22	13·50	29·65
66	0·6438	7·084	95·9	13·25	13·54	30·39
67	0·6664	7·320	99·4	13·27	13·58	31·15
68	0·6898	7·563	103·0	13·30	13·61	31·92
69	0·7139	7·813	106·6	13·32	13·65	32·71
70	0·7386	8·069	110·5	13·35	13·69	33·51
71	0·7642	8·332	114·4	13·38	13·73	34·33
72	0·7906	8·603	118·4	13·40	13·76	35·17
73	0·8177	8·882	122·6	13·43	13·80	36·03
74	0·8456	9·168	126·9	13·45	13·84	36·91
75	0·8744	9·46	131·4	13·48	13·88	37·81
76	0·9040	9·76	135·9	13·50	13·92	38·73
77	0·9345	10·07	140·7	13·53	13·96	39·67
78	0·9658	10·39	145·6	13·55	14·00	40·64
79	0·9981	10·72	150·6	13·58	14·05	41·63
80	1·0314	11·06	155·8	13·60	14·09	42·64
85	1·212	12·89	184·4	13·73	14·31	48·04
90	1·421	14·96	217·6	13·86	14·55	54·13
95	1·659	17·32	256·3	13·98	14·80	61·01
100	1·931	19·98	301·3	14·11	15·08	68·79
105	2·241	22·99	354	14·24	15·39	77·63
110	2·594	26·38	415	14·36	15·73	87·69
115	2·993	31·8	486	14·49	16·10	99·10
120	3·444	34·44	569	14·62	16·52	112·37
125	3·952	39·19	667	14·75	16·99	127·54
130	4·523	44·49	780	14·88	17·53	145·06
135	5·163	50·38	913	15·00	18·13	165·34
140	5·878	56·91	1,072	15·13	18·84	189·22
150	7·566	72·10	1,485	15·39	20·60	250·30

Temp. °C.	Pressure of Sat. Vapour mm. Hg.	Weight of Sat. Vap.		Volume in m.³		Specific Entropy of mixture kJ./kg.
		per m.³ of mixture g./m.³	per kg. dry air g./kg.	of dry air m.³/kg.	of mixture m.³/kg.	
–15	1·2	1·6	1·0	0·731	0·732	–12·6
–10	2·0	2·3	1·6	0·745	0·746	–6·1
–5	3·0	3·4	2·5	0·759	0·761	+1·09
0	4·6	4·9	3·8	0·773	0·775	9·4
1	4·9	5·2	4·1	0·776	0·778	11·3
2	5·3	5·6	4·4	0·779	0·781	12·9
3	5·7	6·0	4·7	0·782	0·784	14·7
4	6·1	6·4	5·0	0·784	0·787	16·6
5	6·5	6·8	5·4	0·787	0·790	18·5
6	7·0	7·3	5·8	0·791	0·793	20·5
7	7·5	7·7	6·2	0·793	0·796	22·6
8	8·0	8·3	6·7	0·796	0·800	24·7
9	8·6	8·8	7·1	0·799	0·802	26·9
10	9·2	9·4	7·6	0·801	0·805	29·2
11	9·8	10	8·2	0·805	0·808	31·5
12	10·5	11	8·8	0·807	0·812	34·1
13	11·2	11	9·4	0·810	0·814	36·6
14	12·0	12	10·0	0·813	0·818	39·2
15	12·8	13	10·6	0·816	0·821	41·8
16	13·6	14	11·4	0·818	0·824	44·8
17	14·5	14	12·1	0·822	0·828	47·7
18	15·5	15	12·9	0·824	0·831	50·7
19	16·5	16	13·8	0·827	0·833	54·0
20	17·5	17	14·7	0·830	0·837	57·8
21	18·7	18	15·6	0·833	0·840	61·1
22	19·8	19	16·6	0·835	0·844	64·1
23	21·1	20	17·7	0·838	0·847	67·8
24	22·4	22	18·8	0·841	0·850	72·0
25	23·8	23	20·0	0·844	0·854	75·8
26	25·2	24	21·4	0·847	0·858	80·4
27	26·7	26	22·6	0·850	0·861	84·6
28	28·4	27	24·0	0·853	0·865	89·2
29	30·0	29	25·6	0·855	0·869	94·3
30	31·8	30	27·2	0·858	0·873	99·6
35	42·2	39	36·6	0·873	0·892	129
40	55·3	51	48·8	0·887	0·912	166
45	71·9	65	65·0	0·901	0·935	213
50	92·5	82	86·2	0·915	0·959	273
55	118·0	104	114	0·929	0·987	352
60	149·4	130	152	0·943	1·020	456
65	187·5	161	204	0·958	1·057	599

COMPOSITION OF AIR

Dry air is a mechanical mixture of gases.

	Dry Air per cent		*Atmospheric at Sea Level*	
	By Volume	*By Weight*	*By Volume*	
Oxygen	21·00	23·2	20·75	
Nitrogen	78·03	75·5	77·08	
Carbon Dioxide ...	0·03	0·046	0·03	
Hydrogen	0·01	0·007	0·01	
Rare Gases	0·93	1·247	0·93	
Water Vapour	—	—	1·20	

The composition of air is unchanged to a height of approximately 10,000 metres. The average air temperature diminishes at the rate of about 0·6°C. for each 100 metres of vertical height.

ALTITUDE-DENSITY TABLES FOR AIR

Altitude m.	*Barometer mm. Hg.*	*Altitude m.*	*Barometer mm. Hg.*	*Altitude m.*	*Barometer mm. Hg.*
0	749	600	695	1,350	632
75	743	750	681	1,500	620
150	735	900	668	1,800	598
250	726	1,000	658	2,100	577
300	723	1,200	643	2,400	555
450	709				

Altitude ft.	*Barometer in. Hg.*	*Altitude ft.*	*Barometer in. Hg.*	*Altitude ft.*	*Barometer in. Hg.*
0	29·92	2,000	27·72	4,500	25·20
250	29·64	2,500	27·20	5,000	24·72
500	29·36	3,000	26·68	6,000	23·79
750	29·08	3,500	26·18	7,000	22·90
1,000	28·80	4,000	25·58	8,000	22·04
1,500	28·31				

Normal Temperature and Pressure (N.T.P.) is 0°C. and 101·325 kN/m.^{2} Standard Temperature and Pressure (S.T.P.) used for determination of fan capacities is 20°C. and 101·6 kN/m.^{2} or 60°F. and 30 in. Hg. (These two sets of conditions do not convert directly, but the density of dry air is 1·22 kg./m.^{3} = 0·0764 lb./ft.^{3} at both conditions.)

Section VI

HEAT LOSSES

1. HEAT LOSS BY CONDUCTION AND CONVECTION THROUGH WALLS, WINDOWS, ETC.

2. HEAT LOSS DUE TO INFILTRATION.

1. Heat Loss through Walls, Windows, Doors, Ceilings, Floor, etc.

$$H = AU(t_i - t_o).$$

$$U = \frac{1}{\frac{1}{f_i} + \frac{X_1}{K_1} + \frac{X_2}{K_2} + \frac{X_3}{K_3} + \frac{1}{f_0}}$$

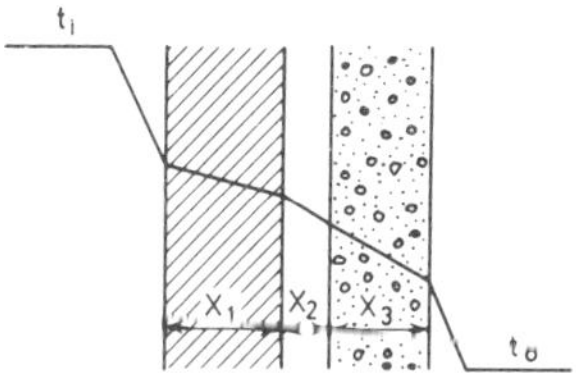

H = Heat transmitted — B.t.u. per hr. or Watts.
A = Area of exposed surface — sq. ft. or m.^{2}
t_i = Inside air temperature — °F. or °C.
t_o = Outside air temperature — °F. or °C.
U = Overall coefficient of heat transmission B.t.u. per sq. ft. per hr. per °F. or W/m.^{2} deg. C.
X = Thickness of material, in. or m.
K = Thermal conductivity in B.t.u. per in. thickness per sq. ft. per hr. per °F. or W/m. deg. C.
f_i = Surface conductance for inside wall (B.t.u. per sq. ft. per hr. per °F. or W/m.^{2} deg. C.); ranges from 1·4 to 2·1 for still air conditions with different materials, surface conditions, and mean temperatures; commonly taken—$f_i = 1{\cdot}65$.
f_o = Surface conductance for outside wall (B.t.u. per sq. ft. per hr. per °F. or W/m.^{2} deg. C.). Commonly taken—$f_0 = 6{\cdot}00$.
$C = \frac{K}{X}$ = Conductance is the amount of heat in B.t.u. per hr. passing through 1 sq. ft. of area of any material of the thickness X per °F. difference of the material surface temperatures B.t.u. per sq. ft. per hr. per °F. or W/m.^{2} deg. C.
$R = \frac{X}{K} = \frac{1}{C}$ = Thermal resistance or resistivity.

SAFETY ADDITIONS TO HEAT LOSS CALCULATIONS

1. For Aspect: North, East ... ... 10 per cent
   West ... ... 5 per cent
   To the loss of the respective exposure.
2. For Exposure: 5 to 10 per cent to surface exposed to wind.
3. For Intermittent Heating:
   Buildings heated during day only 10–15%
   Buildings not in use daily 25–30%
   Buildings without heat for long period (Churches) ... up to 50%
4. For Height (rooms over 15 ft. high):

Height (in ft.)	15	18	21	24	27	30	33	36 and above
Percentage ...	2½	5	7½	10	12½	15	17½	20

Air Movement: Air movement makes any conditions of temperature and humidity feel colder; it lowers the effective temperature. An air velocity of 25 ft. per min. may be considered as practically still air. A slight air movement is desirable for comfort to remove layers of humid and warm air from the surface of the human body. A higher air velocity is required in air of high temperature and high relative humidity than in air of low temperature and low humidity.

The Entering Air Temperature in plenum heating systems must not be too much below the room temperature.

Recommended air entering temperatures:

for heating, normally, 80–90°F., 26–32°C.

for heating, when good mixing, 100–120°F., 38–49°C.

for cooling, inlets near to occupied zone,	10–15°F.	below room temp.	5–9°C.
for cooling, high velocity jets, diffusion nozzles,	30°F.	below room temp.	17°C.

2. Heat Loss of a Room due to Infiltration:

$$H_2 = s.d.n.V\ (t_o - t_i)$$

in which:

H_2 = Heat loss (B.t.u./hr.; W)

s = Specific heat of air (B.t.u./lb.; kJ/kg.)

d = Density of air (lb./ft.^{3}; kg./m.^{3})

$s \times d$ = 0·019 B.t.u./ft.^{3} °F.
= 1·27 kJ/m.^{3} °C.

n = Number of air changes per hr.

V = Volume of room (ft.^{3}; m.^{3})

t_o = Outside temperature (°F.; °C.)

t_i = Inside temperature (°F.; °C.)

Number of Air Changes per hour usually assumed, due to Infiltration (not for ventilation—Ventilation, see X. 3–6).

Residences	1–2	Halls	3
Offices	1–2	Sitting-rooms	2
Factories	1–1½	Bedrooms	1
Large Rooms with small exposure	½–1.	Stores	2–3

Winter Inside Temperatures:

Usually assumed in Building Heating Calculations

Heated Rooms	°F.	°C.		°F.	°C.	Unheated Rooms	°F.	°C.
Residences	70	21	Churches	50	10	Cellars and Closed Rooms	32	0
Lecture Rooms	65	18	Museums	50	10	Vestibules frequently opened to outside	32	0
Schoolroom	65	18	Lavatories	50	10	Ditto, not opened	40	4
Offices	68	20	Cloakroom	55	13	Attic under a Roof	32	0
Stores	62	17	Prisons	50	10	Ditto, with composition covering	40	4
Bedrooms	62	17	Hot-houses	78	26			
Bathrooms	70	21	Warm air baths	120	49			
Wards	70	21	Steam Baths	110	43			
Operating-Th.	80	27	Factories	62	17			
Halls	55	13						
Corridors	55	13						
Gymnasiums	55	13						

Outside Design Temperature in England is {−4°C. to 0°C. / 25°F. to 32°F.}

HEAT LOSS CALCULATIONS FOR HIGH BUILDINGS

Designation	*Floor*	*No. of Air Changes*		*U—values*
		Living-room	*Bedroom*	
Sheltered ...	Ground, 1st	1	1½	Normal
Normal ...	2nd to 4th	1	1½	Normal
Severe ...	5th to 11th	1½	2	Normal
Very Severe ...	Above 11th	2	2½	Severe

1 air change per hour = 0·018 B.t.u./hr. ft.^{3} °F.
= 0·34 W/m.^{3} °C.

Air Temperatures at various Levels:

Assumption: Increase of air temperature per foot of height above the 5 ft. level is at the rate of 2 per cent of the breathing-line temperature up to ceiling heights of 20 ft. No further increase above 20 ft.

$$t^1 = t + 0{\cdot}02\,(h - 5)t$$

t^1 = air temperature °F. at a level h ft. above the floor.
t = air temperature °F. at breathing line 5 ft. above the floor.

Temperature of Unheated Spaces:

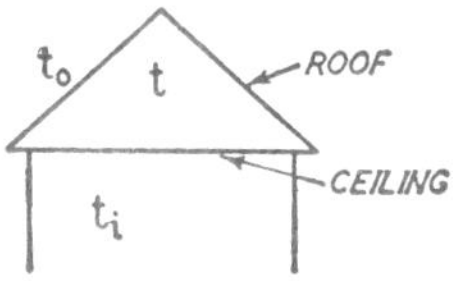

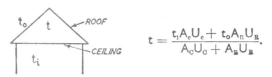

$$t = \frac{t_i A_c U_c + t_o A_R U_R}{A_C U_C + A_R U_R}.$$

$A_C A_R$ = Areas of ceilings and roof resp. sq. ft. or m.²
$U_C U_R$ = Coefficient of heat transmission of ceiling and roof, resp. B.t.u. per sq. ft. per hr. per °F. W/m.² deg. C.
t = Temperature of unheated space °F. °C.
t_i = Temperature of adjacent room °F. °C.
t_o = Outside temperature °F. °C.

Combined Coefficient for Ceiling and Roof:

$$U_{CR} = \frac{U_R\,U_C}{U_R + \dfrac{U_C}{r}} \text{ (B.t.u. per sq. ft. per °F. per hr.) (or W/m.}^2\text{ deg. C.).}$$

r = Ratio of roof area to ceiling area.

Allowance for Warming-up (Rietschel):

(a) Rooms heated daily (not by night):

$$H = \frac{0{\cdot}063\,(n-1)\,H_o}{z} \text{ B.t.u./hr. or W.}$$

(b) Rooms not heated daily:

$$H = \frac{0{\cdot}1\,H_t\,(8+z)}{z} \text{ B.t.u./hr. or W.}$$

H = Heat required for warming-up in B.t.u./hr. or W.
H_t = Total heat loss, B.t.u. per hr. or W.
H_o = Heat loss through outside surface, B.t.u. per hr. or W.
n = Interruption of heating in hr.
z = Warming-up time in hr.

CONTRACT TEMPERATURES AND THEIR EQUIVALENTS

Inside temperatures obtained with a certain system with outside temperatures other than for which the system is designed.
(Empirical formula by J. Roger Preston).

$$t_4 = (t_1^{12} - t_2^{12} + t_3^{12})^{\frac{1}{12}}$$

t_1 = Contract inside temperature °C. absolute.
t_2 = Contract outside temperature °C. absolute.
t_3 = Existing outside temperature °C. absolute.
t_4 = Estimated inside temperature °C. absolute.

(Formula remains unchanged if all temperatures are in °F. absolute.)

Table for 30°F. contract outside and 60°F. contract inside:

Existing Outside temp. °F.	20	22	24	26	28	**30**	32	34	36	38	40
Inside temp. °F.	55	56	57	58	59	**60**	61	62	63	64	65

Table for 0°C. contract outside and 20°C. contract inside:

Existing Outside temp. °C.	—5	—4	—3	—2	—1	**0**	+1	2	3	4	5
Inside temp. °C. ...	17·8	18·3	18·8	19·0	19·6	**20**	20·5	21·0	21·4	21·7	22·5

The data and tables on pages VI. 6 to VI. 8 are reprinted from **A Guide to Current Practice, 1970,** by courtesy of the publishers, the Institution of Heating and Ventilating Engineers. The S.I. figures have been calculated by the author.

SURFACE RESISTANCES

Surface resistances: Internal, R_{S_1}, and External, R_{S_2}, for walls, floors and roofs; sq. ft. h. deg. F./B.t.u.

Note: The following data are applicable to plain surfaces but not to bright metallic surfaces. The resistance of a corrugated surface is less than that of a plain one, generally by about 20 per cent.

INTERNAL RESISTANCES, R_{S_1}

	Sq. ft. hr.² °F. / B.t.u.	*m.² deg. C./W.*
Walls	0·70	0·123
Floors or Ceilings		
heat flow upwards	0·60	0·106
heat flow downwards	0·85	0·150
Roofs, flat or sloping	0·60	0·106

EXTERNAL RESISTANCES, R_{S_2}

Orientation	*Sheltered*		*Normal*		*Severe*	
	ft.³ hr. °F. / B.t.u.	*m.² °C. / W*	*ft.² hr. °F. / B.t.u.*	*m.² °C. / W*	*ft.² hr. °F. / B.t.u.*	*m.² °C. / W*
S.	0·73	0·128	0·57	0·100	0·43	0·076
W., SW., SE.	0·57	0·100	0·43	0·076	0·30	0·053
N., W.	0·43	0·076	0·30	0·053	0·18	0·032
N., NE., E.	0·43	0·076	0·30	0·053	0·07	0·012
Roof	0·40	0·070	0·25	0·044	0·10	0·018

These figures, in conjunction with the following values, were used in computing the overall coefficients in the tables on pages VI. 13 to VI. 20.

	Conductivity k		*Resistivity 1/k*	
	B.t.u. in. / ft.2 hr. °F.	*W. / m. deg. C.*	*ft.2 hr. °F. / B.t.u. in.*	*m. deg. C. / W.*
Brickwork	8·0	1·5	0·125	0·87
Plaster	4·0	0·58	0·25	1·73
Concrete	10·0	1·4	0·10	0·69
Stone	12·0	1·7	0·083	0·58
Wood	1·0	0·14	1·0	6·90

The Calculation of Overall Coefficients

Conductivities and resistivities of various materials given on the following pages enable overall transmittance factors to be calculated for composite walls, floors and roofs, or by the following formulae:

$$U = \frac{1}{R_{S_1} + R_{S_2} + r_1L_1 + r_2L_2, \text{ etc., } + R_a + R_h}$$

or

$$U = \frac{1}{R_{S_1} + R_{S_2} + \frac{L_1}{k_1} + \frac{L_2}{k_2}, \text{ etc., } + R_a + R_h}$$

COEFFICIENTS FOR WALLS

Thermal Transmittance, U_1 B.t.u./sq. ft. hr. deg. F.
U_2 W/m.^{2} deg. C.

		Exposure to Wind											
	S	*Sheltered*		*Normal*		*Severe*		—		—		—	
Orientation	W, SW, SE	—		*Sheltered*		*Normal*		*Severe*		—		—	
	NW ...	—		—		*Sheltered*		*Normal*		*Severe*		—	
	N, NE, E	—		—		*Sheltered*		*Normal*		—		*Severe*	
		A		B		C		D		E		F	
		U_1	U_2	U_1	U_2	U_1	U_2	U_1	U_2	U_1	U_2	U_1	U_2
Brickwork													
Solid, unplastered	4½ in.	0·50	2·9	0·55	3·1	0·59	3·4	0·64	3·6	0·69	3·9	0·75	4·3
	9 in.	0·39	2·2	0·42	2·4	0·44	2·5	0·47	2·7	0·50	2·9	0·53	3·0
	13½ in.	0·32	1·8	0·34	1·9	0·35	2·0	0·37	2·1	0·39	2·2	0·41	2·3
Solid, plastered	4½ in.	0·46	2·6	0·49	2·8	0·53	3·0	0·57	3·2	0·61	3·5	0·65	3·7
	9 in.	0·36	2·1	0·38	2·2	0·41	2·3	0·43	2·4	0·45	2·6	0·48	2·7
	13½ in.	0·30	1·7	0·32	1·8	0·33	1·9	0·35	2·0	0·36	2·1	0·38	2·2
	18 in.	0·26	1·5	0·27	1·5	0·28	1·6	0·29	1·6	0·30	1·7	0·31	1·8
	22 in.	0·23	1·3	0·23	1·3	0·24	1·4	0·25	1·4	0·26	1·5	0·26	1·5
Cavity, plastered (unventilated) ...	11 in.	0·27	1·5	0·28	1·6	0·29	1·6	0·30	1·7	0·31	1·8	0·32	1·8
	15½ in.	0·23	1·3	0·24	1·4	0·25	1·4	0·26	1·5	0·27	1·5	0·27	1·5
	20 in.	0·21	1·2	0·21	1·2	0·22	1·2	0·22	1·2	0·23	1·3	0·24	1·4
(ventilated)	11 in.	0·30	1·7	0·31	1·8	0·33	1·9	0·34	1·9	0·36	2·0	0·37	2·1
	15½ in.	0·26	1·5	0·27	1·5	0·28	1·6	0·29	1·6	0·30	1·7	0·31	1·8
	20 in.	0·22	1·2	0·23	1·3	0·24	1·4	0·25	1·4	0·25	1·4	0·26	1·5
Concrete ...	4 in.	0·55	3·1	0·60	3·4	0·66	3·8	0·71	4·0	0·78	4·4	0·85	4·8
	6 in.	0·49	2·8	0·53	3·0	0·58	3·3	0·63	3·6	0·68	3·9	0·73	4·1
	8 in.	0·45	2·5	0·48	2·7	0·52	3·0	0·56	3·2	0·60	3·4	0·64	3·6
	10 in.	0·41	2·3	0·44	2·5	0·47	2·7	0·50	2·8	0·53	3·0	0·57	3·2
Glass													
Single Windows	...	0·70	4·0	0·79	4·5	0·88	5·0	1·00	5·7	1·14	6·5	1·30	7·4
Double Windows	...	0·41	2·3	0·44	2·5	0·47	2·7	0·50	2·8	0·53	3·0	0·56	3·2
Stone													
	12 in.	0·41	2·3	0·44	2·5	0·47	2·7	0·50	2·8	0·53	3·0	0·56	3·2
	18 in.	0·34	1·9	0·36	2·0	0·38	2·2	0·40	2·3	0·42	2·4	0·44	2·5
	24 in.	0·29	1·6	0·31	1·8	0·32	1·8	0·33	1·9	0·35	2·0	0·36	2·0
Wood													
Tongued and grooved	1 in.	0·41	2·3	0·44	2·5	0·47	2·7	0·50	2·8	0·53	3·0	0·56	3·2
	1½ in.	0·34	1·9	0·36	2·0	0·38	2·2	0·40	2·3	0·42	2·4	0·44	2·5
Sheets													
Asbestos ...	¼ in.	0·64	3·1	0·72	4·1	0·80	4·6	0·89	5·1	1·00	5·7	1·12	6·4

COEFFICIENTS FOR WALLS

Thermal Transmittance, U_1 B.t.u./sq. ft. hr. deg. F.
U_2 W/m.² deg. C.

		Exposure to Wind											
Orientation	*S*	*Sheltered*		*Normal*		*Severe*		—		—		—	
	W, SW, SE	—		*Sheltered*		*Normal*		*Severe*		—		—	
	NW	—		—		*Sheltered*		*Normal*		*Severe*		—	
	N, NE, E	—		—		*Sheltered*		*Normal*		—		*Severe*	
		A		B		C		D		E		F	
		U_1	U_2	U_1	U_2	U_1	U_2	U_1	U_2	U_1	U_2	U_1	U_2
Sheets													
Corrugated Asbestos ...		0·77	4·4	0·88	5·0	1·00	5·7	1·15	6·6	1·33	7·6	1·56	8·9
Corrugated Iron $\frac{1}{16}$ in.		0·79	4·5	0·91	5·2	1·04	5·9	1·20	6·8	1·40	8·0	1·67	9·5
Corrugated Iron on 1 in. t.and g. board		0·34	1·9	0·36	2·0	0·38	2·2	0·40	2·3	0·42	2·4	0·44	2·5

VALUES OF $\frac{X}{K}$ FOR AIR SPACES

Material bounding space	*Resistivity in m.² deg. C. per W. for thickness of air space in mm.*									
	15	20	25	35	50	65	75	90	100	115
Glass	0·141	0·145	0·148	0·155	0·165	0·172	0·176	0·183	0·186	0·188
Brick	0·150	0·153	0·158	0·165	0·175	0·185	0·190	0·197	0·200	0·203

Material bounding space	*Resistivity in ft.² hr. °F. per B.t.u. for thickness of air space in inches*										
	$\frac{1}{2}$	$\frac{3}{4}$	1	$1\frac{1}{2}$	2	$2\frac{1}{2}$	3	$3\frac{1}{2}$	4	$4\frac{1}{2}$	5
Glass	0·79	0·82	0·85	0·89	0·94	0·97	1·00	1·04	1·06	1·07	1·08
Brick	0·84	0·87	0·90	0·95	1·04	1·04	1·08	1·11	1·14	1·16	1·17

COEFFICIENTS FOR ROOFS

Thermal Transmittance, U_1 B.t.u./sq. ft. h. deg. F.
U_2 W/m.^{2} deg. C.

Construction of Roof	*Sheltered*		*Normal*		*Severe*	
	U_1	U_2	U_1	U_2	U_1	U_2
Flat Roofs—						
Asphalt on 6 in. concrete	0·58	3·3	0·64	3·6	0·70	4·0
Asphalt on 6 in. concrete, plastered	0·51	2·9	0·55	3·1	0·61	3·5
Asphalt on 6 in. concrete with 1 in. cork ...	0·21	1·2	0·22	1·2	0·22	1·2
Asphalt on 6 in. concrete with 1 in. cork, plastered	0·20	1·1	0·21	1·2	0·21	1·2
Asphalt on 6 in. concrete with 2 in. cork ...	0·13	0·74	0·13	0·74	0·13	0·74
Asphalt on 6 in. concrete with 2 in. cork, plastered	0·12	0·68	0·12	0·68	0·13	0·74
Asphalt on 6 in. thick hollow tile	0·45	2·55	0·48	2·7	0·52	3·0
Asphalt on 6 in. thick hollow tile, plastered ...	0·41	2·3	0·44	2·5	0·47	2·7
Asphalt on 6 in. hollow tile with 1 in. cork ...	0·19	1·1	0·20	1·1	0·21	1·2
Asphalt on 6 in. hollow tile with 1 in. cork, plastered	0·18	1·0	0·19	1·0	0·20	1·1
Asphalt on 6 in. hollow tile with 2 in. cork ...	0·12	0·68	0·12	0·68	0·12	0·68
Asphalt on 6 in. hollow tile with 2 in. cork, plastered	0·12	0·68	0·12	0·68	0·12	0·68
Asphalt, 1 in. cork, 1¼ in. boards, joists and plaster ceiling	0·16	0·91	0·16	0·91	0·16	0·91
Pitched Roofs—						
Corrugated aluminium (bright)	0·90	5·1	1·15	6·6	1·45	8·3
Corrugated asbestos	1·20	6·8	1·40	8·0	1·70	9·7
Corrugated asbestos lined ½ in. boards	0·47	2·7	0·50	2·9	0·53	3·0
Corrugated iron	1·25	7·1	1·50	8·6	1·80	10·3
Corrugated iron lined 1 in. boards and felt ...	0·33	1·9	0·35	2·0	0·37	2·1
Tiles on boards and felt	0·33	1·9	0·35	2·0	0·37	2·1
Tiles on battens	1·22	6·9	1·50	8·6	2·00	11·4
Tiles on battens, felted	0·63	3·6	0·70	4·0	0·78	4·4
Tiles on feather-edged boarding	0·87	5·0	1·00	5·7	1·17	6·7
Plaster ceiling with roof space above—						
(*a*) with tiles and battens	0·50	2·8	0·56	3·2	0·64	3·6
(*b*) with tiles or slates on boards and felt ...	0·28	1·6	0·30	1·7	0·32	1·8
Roof Glazing—						
Skylight	1·00	5·7	1·20	6·8	1·40	8·0
Laylight, with lantern over...	0·57	3·2	0·60	3·4	0·63	3·6

For Air to Air.

	W. / m.2 deg. C.	B.t.u. / Sq. ft. hr. deg. F.
Partitions—		
4-in. stud lath and plaster, one side	3·0	0·53
4-in. stud lath and plaster, both sides ...	1·9	0·33
½ in. thick, plain wood	3·0	0·52
¾ in. thick, plain wood	2·6	0·45
1 in thick, plain wood	2·3	0·41
2 in. thick, solid plaster	3·4	0·60
3 in. thick, solid plaster	3·0	0·52
4 in. thick, solid plaster	2·7	0·48
4 in. hollow partition, plastered one side ...	2·4	0·43
4 in. hollow partition, plastered both sides	2·0	0·35
Windows—		
Single window	6·3	1·10
Double window	3·4	0·60
Single skylight	6·8	1·20
Double skylight	4·0	0·70
Doors—		
All wood, 1 in. thick	4·0	0·70
All wood, 2 in. thick	2·6	0·45
All wood, upper portion glass	4·3	0·75
All glass, wood framing	5·4	0·95
All steel	6·8	1·20
All steel, upper portion glass	6·5	1·15
All glass, steel framing	6·0	1·05
Inner vestibule door, all wood	1·7	0·30
Inner vestibule door, half glass	2·6	0·45
Inner vestibule door, all glass	3·4	0·60

For Air to Air.

	Coefficient for Cold Air			
	B.t.u./sq. ft. hr. °F.		*W./m.² deg. C.*	
	Below	*Above*	*Below*	*Above*
FLAT ROOFS				
Flat roof, reinforced concrete with asphalt, thickness				
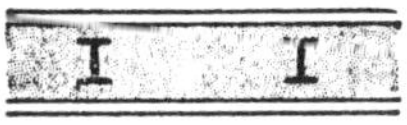 of concrete ... 5″	—	0·30		1·7
Ditto 6″	—	0·28		1·6
Ditto 7″	—	0·27		1·5
Ditto 9″	—	0·25		1·4
Ditto 11″	—	0·23		1·3
Ditto 13″	—	0·21		1·2
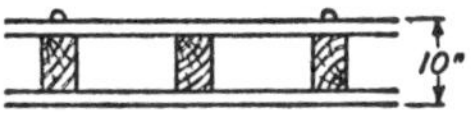 Flat roof covered with metal with lath and plaster ceiling	—	0·17		1·0
CLOSED ROOF SPACE (Ceiling & Roofs)				
 Slates or tiles on battens and rafters, lath and plaster ceiling ...	—	0·35		2.0
Ditto, with 1″ tongued and grooved boards under tiles	—	0·25		1·4
Ditto, with roofing felt between tiles and boards	—	0·19		1·1
OPEN ROOF:				
Slates or tiles on batten and rafters only	—	0·80		4·6
 Ditto, with 1″ tongued and grooved boards... ...	—	0·35		2·0
Ditto, with roofing felt between tiles and boards	—	0·25		1·4
 Slates or tiles on 1″ tongued and grooved boards, and lath and plaster ceiling...	—	0·30		1·7
Corrugated iron, unlined...	—	1·80		10·3
Corrugated asbestos, cement sheeting, unlined ...	—	1·45		8·3
Protected sheet metal ...	—	0·90		5·1
Single glass, skylights ...	—	1·20		6·8

	Coefficient for Cold Air			
	B.t.u./sq. ft. hr. °F.		W./m.2 deg. C.	
	Below	Above	Below	Above
ROOFS—continued				
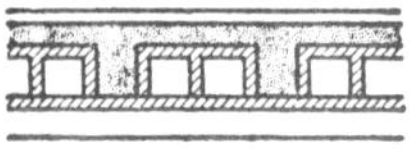 Asphalt ¼″ thick on cement screen and 6″ hollow tile fire-proof roof with plaster under ...	—	0·22		1·2
Tile roof	—	1·00		5·7
Zinc or copper on 1″ boards	—	0·44		2·5
Flat asbestos sheets	—	1·00		5·7
Corrugated asbestos sheets, ¼″ asbestos boards	—	0·44		2·5
Corrugated asbestos sheets, ½″ asbestos boards	—	0·37		2·1
Corrugated steel with felt and 1″ boards	—	0·37		2·1
FLOORS AND CEILINGS:				
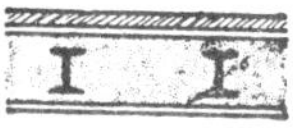 Concrete and steel, laid with 1¼″ wood blocks and plaster ceiling, concrete = 5″ thick	0·20	—	1·1	—
Ditto 7″ ,,	0·19	—	1·0	—
Ditto 9″ ,,	0·18	—	1·0	—
 Ditto 11″ ,,	0·17	—	0·97	—
Ditto 13″ ,,	0·16	—	0·91	—
 Board floor 1″ thick on 7″ joists with lath and plastered ceiling	0·08	0·18	0·45	1·02
Board floor 1″ thick on 7″ joists	0·20	0·44	1·1	2·5
FLOORS:				
 4″ concrete on earth ...	0·55	—	3·1	—
6″ ditto	0·49	—	2·8	—
 1¼″ wood blocks, ¼″ screed and 5″ concrete on earth	0·25	—	1·4	—
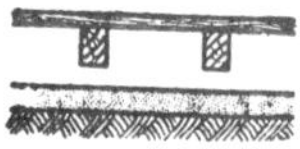 Board floor 1″ thick on joists, 9″ air space and 4½″ concrete	0·07	—	0·40	—

The Cooling Load for summer air conditions is made up from the following components:

1. Normal Heat Transfer through Walls, Windows, Ceilings, etc.
2. Transfer of Solar Radiation through Walls, etc.
3. Heat Emission of Occupants.
4. Heat introduced by Infiltration of Outside Air or Ventilation.
5. Heat Emission of Appliances (Mechanical, Electrical, Gas, etc.).

1. Normal Heat Transfer through Walls, etc. (B.t.u. per hr. or W)

$$H_1 = AU\,(t_1 - t_0)$$

A = Net area of wall, etc. (ft^2 or m^2).
U = Coefficient of transmission (B.t.u./ft^2 per hr. per °F. or W/m^2 deg. C.).
t_o = Outside temperature (°F or °C).
t_i = Inside temperature (°F or °C).

2. Transfer of Solar Radiation: (B.t.u. per hr. or W)

$$H_2 = AFaJ$$

A = Net area of wall, etc. (ft^2 or m^2).
F = Radiation factor—percentage of absorbed solar radiation which is transmitted to the inside (for Glass : F=1 ; for walls, see Table).
a = Absorption coefficient—percentage of the incident solar radiation which is absorbed by the surface (see Table below).
J = Intensity of solar radiation striking the surface in B.t.u. per hr./ft.^{2} (W/m^2). (see Table below).

INTENSITY OF SOLAR RADIATION J
(B.t.u. per hr./ft.^{2}) For Latitude 45°

Sun Time	*North East*	*East*	*South East*	*South*	*South West*	*West*	*North West*	*Hori-zontal*
5	25	24	9					2
6	89	99	52					26
7	149	194	125					90
8	140	219	171	22				156
9	92	194	183	65				210
10	33	144	171	98				251
11		75	139	121	32			274
12			91	128	91			282
1			32	121	139	75		274
2				98	171	144	33	251
3				65	183	194	92	210
4				22	171	219	140	156
5					125	194	144	90
6					52	99	89	26

SOLAR ABSORPTION COEFFICIENT a

Surface Material.	a
Very Light Coloured Surface White Stone Very Light Cement	0·4
Medium Dark Surface Unpainted Wood Brown Stone, Brick, Red Tile	0·7
Very Dark Surface Slate Roofing Very Dark Paints	0·9

RADIATION FACTOR F FOR DIFFERENT WALL TRANSMISSION COEFFICIENTS U

U =	0·1	0·2	0·3	0·4
F =	0·025	0·0475	0·07	0·095
U =	0·5	0·6	0·7	0·8
F =	0·115	0·14	0·16	0·18

ALTERNATIVE CALCULATION OF COOLING LOAD

(For England)

Skylights	200 B.t.u./sq. ft. hr.	630 W./m.^{2}
Windows, west, east	125 ,,	400 ,,
Windows, south	75 ,,	240 ,,

Additional Temperature Difference for Walls and Flat Roofs:

Walls, west, east, black finish	25°F.	14°C.
Walls, west, east, red finish ...	18°F.	10°C.
Walls, west, east, white finish	10°F.	5·5°C.
Walls, south, black finish ...	12·5°F.	7°C.
Walls, south, red finish ...	9°F.	5°C.
Walls, south, white finish ...	5°F.	2·8°C.
Horizontal roofs, black finish...	40°F.	22°C.
Horizontal roofs, red finish ...	28°F.	15·5°C.
Horizontal roofs, white finish	13°F.	7°C.

The Latent Heat Load consists of the moisture given up by people, from infiltration of outside air, and from appliances.

The Total Refrigeration Load consists of the total heat gain of the room or building plus the sensible and latent heat removed from the outside air introduced into the room.

INTENSITY OF SOLAR RADIATION

W/m.^{2} for latitude 45°.

Solar Time	*North-East*	*East*	*South-East*	*South*	*South-West*	*West*	*North-West*	*Horizontal*
5	79	75	28					6
6	281	312	164					82
7	470	612	394					284
8	441	691	539	69				492
9	290	612	577	205				663
10	104	455	539	309				791
11		237	438	382	101			864
12			287	404	287			890
1			101	382	438	237		864
2				309	539	455	104	791
3				205	577	612	290	663
4				69	539	691	441	492
5					394	612	470	284
6					164	312	281	82

TIME LAG IN TRANSMISSION OF SOLAR RADIATION THROUGH WALLS

Type of Wall	*Time Lag in Hours*
6 in. Concrete	3
4 in Gypsum	2½
22 in. Brick	10
3 in. Concrete + 1 in. Cork	2
2 in. Pine	1½

SOLAR RADIATION TRANSMITTED THROUGH SHADED WINDOWS

Type	*Per cent. Transmitted*
Canvas Awning, Plain...	28
Canvas Awning, Aluminium	22
Inside Shade, full drawn	45
Inside Shade, half drawn, Buff	68
Inside Venetian Blind, Slats at 45°, Aluminium	58
Outside Venetian Blind, Slats at 45°, Aluminium	22

3. Heat Emission by Occupants:

3. Heat Emission by Occupants

Air Velocity	*Air Temp. °C.*	10	12	14	16	18	20	22	24	26	28	30	32
Still Air	Sensible Heat W	136	126	115	106	98	92	85	77	69	58	47	33
	Latent Heat W	21	21	21	21	23	27	33	41	49	60	69	81
	Total W	157	147	136	127	121	119	118	118	118	118	116	114
	Moisture grammes/hr.	31	31	31	31	34	40	48	60	73	88	102	120
1 m/s	Sensible Heat W	152	142	131	122	112	104	97	88	81	69	55	38
	Latent Heat W	19	19	19	19	19	20	25	32	38	49	61	77
	Total W	171	161	150	143	131	124	122	120	119	118	116	115
	Moisture grammes/hr.	28	28	28	28	28	29	36	47	57	73	89	114
	Air Temp. °F.	50	54	57	61	65	68	72	75	79	82	86	90
Still Air	Sensible Heat B.t.u./hr.	468	432	396	364	336	316	292	264	236	200	160	112
	Latent Heat B.t.u./hr.	72	72	72	72	80	92	112	140	168	204	236	280
	Total B.t.u./hr.	540	504	468	438	416	408	404	404	404	404	396	392
	Moisture grains/hr.	480	480	480	480	520	620	740	930	1120	1350	1570	1850
3ft./sec.	Sensible Heat B.t.u./hr.	520	484	448	416	384	356	332	300	276	236	188	128
	Latent Heat B.t.u./hr.	64	64	64	64	64	68	84	108	132	168	208	264
	Total B.t.u./hr.	584	548	512	490	448	424	416	408	408	404	396	392
	Moisture grains/hr.	430	430	430	430	430	450	560	720	880	1120	1380	1760

4. Heat introduced by Infiltration and Ventilation:

Total Heat Gain = Sensible Heat Gain + Latent Heat Gain

$$H_T = d.\ Q.\ (h_o = h_i).$$

H_T =	Total heat gain	in Watts	(or B.t.u./hr.).
Q =	Volume of outside air entering	in m.³/s.	(or cu. ft./hr.).
d =	Density of air at outside temperature	in kg./m.³	(or lb./cu. ft.).
h_o =	Enthalpy of mixture of outside dry air and water vapour	in J./kg.	(or B.t.u./lb.).
h_i =	Enthalpy of mixture of inside dry air and water vapour	in J./kg.	(or B.t.u./lb.).

5. Heat Emission of Appliances

In S.I. units both appliance rating and heat emission are measured in watts and no conversion is necessary.

In other units:

Electric Appliances:	3,415 B.t.u. per kW.
Motors:	2,546 B.t.u. per h.p.
Gas Lights:	500–1,000 B.t.u./cu. ft. of gas used.
Machinery driven from outside:	2,546 B.t.u. per b.h.p. supplied to the machine.

SOLAR RADIATION THROUGH HEAT ABSORBING GLASS

Solar Heat Gain can be reduced by using "Heat Absorbing Glass." This glass ("Anti-sun"—Pilkington Bros. Ltd.) is bluish-green with relatively high light transmission but restricting the passage of Solar Radiant Heat.

Type and Thickness of Glass	*Transmission in per cent.*	
	Visible Light	*Solar Radiation*
"Calorex"—Pilkington ... ⅛ in.	65	32
3/16 in.	55	25
¼ in.	48	20
"Anti-sun"—Pilkington ... 3/16 in.	78	53
¼ in.	74	46
Ordinary Window Glass	85	82

VERTICAL GLAZING — THERMAL TRANSMITTANCE ("U"-value) B.t.u./ft.² hr. °F.

Orientation	*Exposure to Wind*																	
S	*Sheltered*			*Normal*			*Severe*											
W, SW, SE				*Sheltered*			*Normal*			*Severe*								
NW							*Sheltered*			*Normal*			*Severe*					
N, NE, E							*Sheltered*			*Normal*						*Severe*		
	A			B			C			D			E			F		
Single Glazing	·70			·79			·88			1·00			1·14			1·30		
D'ble or Multiple Glazing No. of Air Spaces	1	2	3	1	2	3	1	2	3	1	2	3	1	2	3	1	2	3
Air Space ¾ in.	·41	·29	·22	·44	·31	·23	·47	·32	·24	·50	·33	·25	·53	·35	·26	·56	·36	·26
½ in.	·42	·30	·24	·46	·32	·25	·48	·33	·25	·52	·35	·26	·55	·37	·27	·59	·38	·28
⅜ in.	·44	·32	·25	·47	·34	·26	·51	·35	·27	·54	·37	·28	·58	·39	·29	·62	·41	·30
¼ in.	·47	·35	·28	·51	·37	·29	·54	·39	·31	·58	·41	·32	·63	·43	·33	·67	·46	·34
⅛ in.	·52	·41	·34	·57	·44	·37	·61	·47	·38	·68	·50	·40	·73	·54	·43	·79	·57	·45

HORIZONTAL AND SLOPING GLAZING THERMAL TRANSMITTANCE ("U"-value), B.t.u. per sq. ft. per hour °F.

Exposure	*Sheltered*			*Normal*			*Severe*		
Single Glazing	0·97			1·13			1·36		
Double and Multiple Glazing No. of Air Spaces	1	2	3	1	2	3	1	2	3
Air Space ¾ in.	·52	·36	·27	·57	·38	·28	·62	·40	·30
½ in.	·53	·36	·28	·57	·38	·29	·63	·41	·30
⅜ in.	·53	·37	·28	·58	·39	·29	·64	·41	·31
¼ in.	·56	·40	·30	·61	·42	·32	·67	·45	·33
⅛ in.	·65	·49	·39	·72	·52	·41	·80	·57	·44

MULTIPLYING FACTORS FOR "U"-VALUES OF WINDOWS TO ALLOW FOR HEAT TRANSFER OF WINDOW FRAME

Type of Frame	*Ratio of Glass to Frame Area*	*Multiplying Factor for*	
		Single Glass	*Double Glass*
All Glass	100	1·00	1·00
Wood Frame	80	0·90	0·95
Wood Frame	60	0·80	0·85
Steel Frame	80	1·00	1·20
Aluminium Frame ...	80	1·10	1·30

CONDENSATION ON GLASS WINDOWS

The chart indicates the maximum heat transmission coefficient of the glass required for avoiding condensation at various outside temperatures and inside temperatures and relative humidities.

Example:

Assumptions—

Inside temp.=70° F.
Outside temp.=20° F.
Relative humidity inside=50%

From the chart the maximum permissible "U"-value is 0·6 B.t.u. per hour per sq. ft. °F.

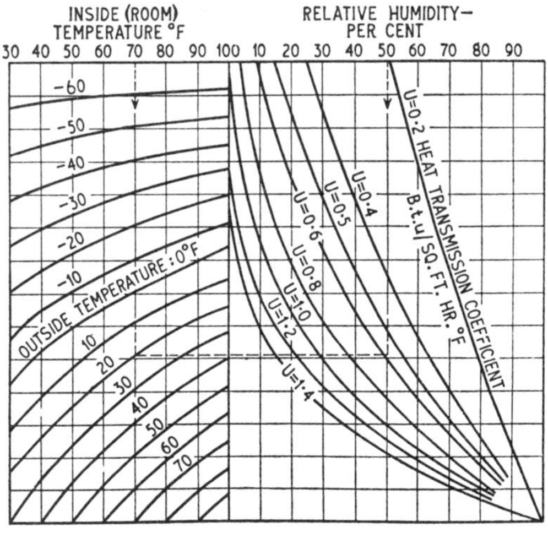

VERTICAL GLAZING — THERMAL TRANSMITTANCE ("U"-value) W./m.2 °C.

		Exposure to Wind																
Orientation	S	*Sheltered*			*Normal*			*Severe*										
	W, SW, E				*Sheltered*			*Normal*			*Severe*							
	NW							*Sheltered*			*Normal*			*Severe*				
	N, NE, E							*Sheltered*			*Normal*						*Severe*	
		A			B			C			D			E			F	
Single Glazing		4·0			4·5			5·0			5·7			6·5			7·4	
Double Multiple Glazing No. of Air Spaces		1	2	3	1	2	3	1	2	3	1	2	3	1	2	3	1	2	3
Air Space	19 mm.	2·32	1·64	1·19	2·49	1·76	1·30	2·66	1·81	1·36	2·83	1·87	1·42	3·0	1·98	1·47	3·17	2·04	1·47
	12 mm.	2·38	1·70	1·36	2·60	1·81	1·42	2·72	1·87	1·42	2·94	1·98	1 47	3·12	2·10	1·53	3·34	2·15	1·59
	10 mm.	2·49	1·81	1·42	2·66	1·93	1·47	2·89	1·98	1·53	3·06	2·10	1·59	3·28	2·21	1·64	3·51	2·32	1·70
	6 mm.	2·66	1·98	1·59	2·89	2·10	1·64	3·06	2·21	1·76	3·28	2·32	1·81	3·57	2·44	1·87	3·80	2·60	1·93
	3 mm.	2·94	2·32	1·93	3·23	2·49	2·10	3 45	2·66	2·15	3·85	2·83	2·26	4·13	3·06	2·44	4·47	3·23	2·55

HORIZONTAL AND SLOPING GLAZING THERMAL TRANSMITTANCE ("U"-value) W./m.2 °C.

Exposure		*Sheltered*			*Normal*			*Severe*		
Single Glazing		5·3			6·4			7·7		
Multiple Glazing No. of Air Spaces		1	2	3	1	2	3	1	2	3
Air Space ...	19 mm.	2·94	2·04	1·53	3·23	2·15	1·59	3·51	2·26	1·70
	12 mm.	3·00	2·04	1·59	3·23	2·15	1·64	3·57	2·32	1·70
	10 mm.	3·00	2·10	1·59	3·28	2·21	1·64	3·62	2·32	1·76
	6 mm.	3·17	2·26	1·70	3·45	2·38	1·81	3·80	2·55	1·87
	3 mm.	3·68	2·78	2·21	4·07	2·94	2·32	4·53	3·23	2·49

CONDENSATION ON GLASS WINDOWS

The chart indicates the maximum heat transmission coefficient of the glass required to avoid condensation at various outside temperatures and inside temperatures and humidities.

Example:

Inside temperature = 15°C.

Outside temperature = –5°C.

Relative humidity inside = 30%.

From the chart the maximum permissible "U"-value is:

7·0 W./m.2 deg. C.

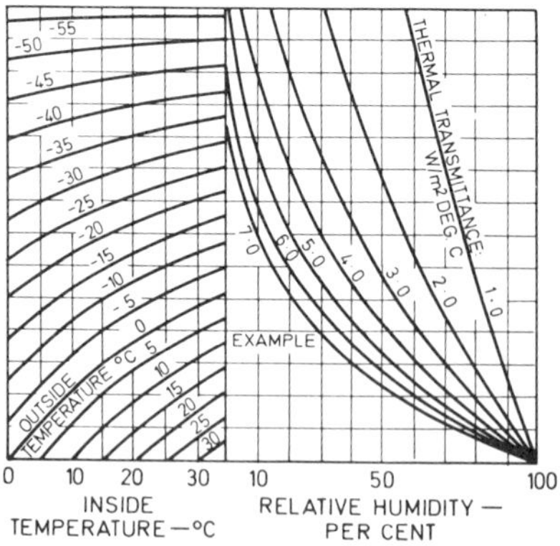

	Town	Country	SUMMER.				WINTER.		
			Average Max. Tem °F.	Mean Temp. °F.	Mean Humidity %	Corresp. Wet Bulb Temp. °F.	Mean Temp. to Feb. °F.		
EUROPE	Athens	Greece	100	78	49	64·4	50		
	Berlin	Germany	92	65	68	59	31		
	Budapest	Hungary	92	68	63	60	34		
	Bucarest	Rumania	96	71	56	61	29		
	Hamburg	Germany	91	—	—	—	30		
	Lisbon	Portugal	96	69	58	60	51		
	London	England	87	61	74	59	39		
	Oslo	Norway	87	61	69	57	24		
	Marseilles	France	93	70	64	60	44		
	Paris	France	93	64	73	62·5	37		
	Prague	Czechoslovakia	91	66	65	59	31		
	Rome	Italy	95	73	55	64	46		
	Seville	Spain	116	81	59	67	41		
	Sofia	Bulgaria	94	67	65	60	29		
	Vienna	Austria	91	66	70	55·5	30		
ASIA	Baghdad	Iraq	117	93	36		—		
	Djakarta	Java	93	79	78	74	—		
	Bombay	India	95	84	87	82	76		
	HongKong	China	97	82	84	78	62		
	Calcutta	India	102	86	89	83	68		
	Manila	Phillippine Is.	98	82	76	76	—		
	Singapore	Malaya	93	80	82	76	—		
	Tokyo	Japan	99	78	81	73·4	40		
AFRICA	Algiers	Algiers	106	78	72	71	57		
	Cairo	Egypt	109	84	70	76	55		
	Cape Town	South Africa	93	75	65	67	58		
	Katanga	Congo Rep	97	74	80	69	—		
	Tripoli	Libya	104	80	67	71	—		
AMERICA	Buenos Aires	Argentine	—	73	73	67	—		
	Caracas	Venezuela	87	69	78	64·5	—		
	Mexico	Mexico	86	65	71	60	—		
	Montevideo	Uruguay		71	66	64	—		
	New Orleans	U.S.A.	95	82	80	77	—		
	New York	U.S.A.	100	73	74	67	32		
	Rio de Janeiro	Brazil	—	78	79	73	—		
AUSTRALIA	Sydney	Australia	100	71	68	64	54		
	Melbourne	Australia	106	67	65	60	50		

NUMBER OF DEGREE-DAYS IN LONDON

	No.	%
September ...	177	4
October ...	473	10
November ...	561	12·5
December ...	618	14·5
January ...	821	18
February ...	652	14·5
March ...	536	12
April ...	420	9
May ...	245	5·5
Total ...	4,503	100

MEAN WINTER TEMPERATURE (AVERAGE) °F.

London	42·5
Scotland	40 to 41·0
England, N.E., E.	41·0
England, S.E.	43·5
England, N.W. and Wales	43·0
Ireland	43 to 44·5
England, S.W. and S. Wales	45·0

AVERAGE MINIMUM AND MEAN TEMPERATURES IN THE WINTER MONTHS IN THE UNITED KINGDOM

	Av. Min. Temp. °F.	Av. Mean Temp. °F
October	44	50
November ...	39	44
December ...	36	40
January	35	39
February ...	35	40
March	36	42
6 Months ...	37·3	42·6

The total efficiency "E" of a Central Heating Plant can be divided into the following partial efficiencies:

1. Boiler Efficiency. $E_1 = 60–78\%$.
2. Efficiency of Pipework (heat-loss of pipes). $E_2 = 75–90\%$.
3. Efficiency of Heaters (according to type of heaters). $E_3 = 90–100\%$.
4. Efficiency of Control (losses due to overheating). $E_4 = 80–95\%$.

$$E = E_1 \,.\, E_2 \,.\, E_3 \,.\, E_4 = 30 \text{ to } 65\%.$$

Fuel Consumption:

$$F = \frac{Hn\,(t_i - t_a)}{C.\,E\,(t_i - t_o)}$$

F = Fuel consumption during considered period kg.
H = Heat loss for a temperature difference $(t_i - t_o)$ W.
n = Heating time in seconds.
E = Efficiency of utilization of fuel %.
t_i = Inside temperature °C.
t_a = Average outside temperature during considered period °C.
t_o = Outside design temperature °C.
C = Calorific value of fuel J per kg.

Degree Day Method:

Number of Degree Days = Number of days × (60°F. — average outside temperature).

Fahrenheit Degree Days are listed to a base of 60°F. An agreed base for Celsius Degree Days has not yet been decided, and no Celsius Degree Days have yet been tabulated for the British Isles. The following data is therefore retained in Imperial units:

$$D = N\,(t_i - t_a).$$

D = Number of Degree Days.
N = Number of Days.
t_a = Average outside temperature °F.
t_i = Mean inside temperature during 24 hours = 60°F. for practical use.

For one day the number of Degree Days equals the temperature difference of inside and average outside.

Heat Loss per Degree Day:

$$h = \frac{24H}{t_i - t_a}$$

h = Heat Loss B.t.u. per degree day.
H = Heat Loss for temperature difference $t_i - t_a$, B.t.u.
t_i = Inside temperature °F.
t_a = Average outside temperature °F.

Fuel Consumption:

$$F = U \times D.$$

F = Fuel consumed lb.

U = Unit fuel consumption per degree day (to be found for the building concerned by test) lb. per degree day.

D = Number of Degree Days.

NUMBER OF DEGREE DAYS FOR VARIOUS TOWNS IN ENGLAND, SCOTLAND AND WALES

From 1st October to 31st May. Basis 60°F.

(See note about Celsius Degree Days on page VI. 20.)

Town	*Height in ft.*	*Degree Days*
Aberdeen...	37	4,250
Appleby (Westmorland)	440	4,460
Banff	130	4,237
Bellingham (Northumberland)... ...	849	4,845
Belper (Derbyshire)	222	4,072
Berwick-on-Tweed	76	4,168
Birmingham	535	3,920
Blackpool...	67	3,805
Brighton	32	3,415
Cambridge	41	3,810
Cardiff	202	3,650
Cranwell (Lincolnshire)	240	4,030
Crieff (Perthshire)	478	4,463
Deerness (Orkney Islands)	160	4,366
Douglas (Isle of Man)	284	3,700
Dover	22	3,550
Dublin	54	3,380
Dundee	147	4,300
Durham	336	4,640
Edinburgh	441	4,210
Fort Augustus (Invernessshire) ...	68	4,343
Fortrose (Ross and Cromarty)... ...	69	4,155
Glasgow	85	3,970
Hull	8	3,885
Kirkcaldy (Fife)	63	4,005

NUMBER OF DEGREE DAYS FOR VARIOUS TOWNS IN ENGLAND, SCOTLAND AND WALES (—*contd.*)

From 1st October to 31st May. Basis 60°F.

Town	*Height in ft.*	*Degree Days*
Lerwick (Shetland Islands)	156	4,367
Liverpool...	198	3,730
London, average...	—	3,615
Manchester	125	3,720
Mayfield (Staffordshire)...	374	4,305
Nairn	20	4,267
Newport (Isle of Wight)	48	3,525
Nottingham	192	3,905
Oxford	208	3,676
Plymouth...	117	3,135
Portsmouth	15	3,245
Renfrew	19	4,154
St. Andrew's (Fife)	13	4,200
Sheffield	428	3,865
Southampton	64	3,410
Stirling	151	4,275
Walton-on-Naze (Essex)	60	3,720
Wick (Caithness)	81	4,270
York	57	3,855

Section VII

HOT WATER HEATING

HOLDEN AND
BROOKE Pumps
We supply an extensive range of pumps which have been developed from our wide experience of the requirements of the heating and building services industries. Please send for technical literature.
HB
Holden and Brooke Limited
Sirius Works, Manchester M12 5JL
Telephone: 061-273 8262 Telex: 667318

THE HEATING MEDIUM IS WATER
carrying Heat through pipes from Boiler to Heaters.

Types of Hot Water Heating Schemes:

	Flow Temp. °F.	*Flow Temp.* °C.	*Temp. Drop* °F.	*Temp. Drop* °C.
(a) Low Pressure H.W.H.				
Forced circulation	up to	up to	20–30	10–15
Gravity	190	90	40	20
(b) Medium Pressure H.W.H.	190–250	90–120	30– 60	15–35
(c) High Pressure H.W.H.	250–400	120–200	50–150	27–85

Classification according to Pipe Schemes:

One-pipe or Two-pipe System.
Up-feed or Down-feed System. } Typical Schemes, see page VII. 4.

DESIGN OF HOT WATER HEATING SYSTEMS

1. Heat Losses of Heated Rooms.
2. Output of Boiler.
3. Heating Surface of Radiators or Output of Unit Heaters.
4. Type and Size of Circulating Pump-Circulating Pressure.
5. Pipe Scheme and Pipe Sizing.
6. Expansion Tank, Type and Size.

1. Heat Losses (see chapter VI).

2. Boiler:

$B = H(1 + X)$

B = Boiler rating watts (B.t.u./hr.).
H = Total heat loss of plant watts (B.t.u./hr.).
X = Margin for heating up = 0·10 to 0·15.

Boiler to be chosen from Maker's Catalogue.

3. Heater, Heating Surface of Radiators and Pipe Coils:

$$A = \frac{H_r}{K_r\left(\frac{t_1 + t_2}{2} - t_r\right)}$$

A = Heating surface of radiator, sq. ft. (m.^{2}).
H_r = Heat loss of room, B.t.u. per hr. (watts).
t_1 = Flow temperature, °F. (°C.).
t_2 = Return temperature, °F. (°C.).
t_r = Room temperature, °F. (°C.).

K_r = Heat transmission of heater, B.t.u. per sq. ft. per hr. °F. (watts/m.^{2} deg. C.). (See Tables IV. 15–18.)

Type and size of Heaters to be chosen from Maker's Catalogue.

4. Circulating Pressure — Pump Size:

Circulating Pressure for Gravity Systems—

$H = h(d_2 - d_1)$

H = Pressure, ft. or m.

h = height between middle of boiler and radiator, ft. or m.

$d_1\ d_2$ = density of water in flow and return, lb./ft.^{3} or kg./m.^{3}.

Volume of Water Handled—

$$Q = \frac{H}{600\,(h_1 - h_2) \times d}$$

Q = Volume of water in gal./min.

H = Total heat-loss of plant, B.t.u./hr.

$h_1\ h_2$ = Heat of water in flow and return respectively, B.t.u./lb.

d = relative density of water in pump.

$$Q' = \frac{H'}{(h_1' - h_2') \times d'}$$

Q' = Volume of water in m.^{3}/s.

H' = Total heat loss of plant in J/s.

$h_1'\ h_2'$ = Heat of water in flow and return respectively in J/kg.

d = Density of water in pump in kg./m.^{3}

For water temperatures up to 100°C. (212°F.):

$$Q = \frac{H}{600\,(t_1 - t_2)}$$

t_1, t_2 = Flow and return temperatures respectively, °F.

$$Q' = \frac{H' \times 10^{-6}}{4{\cdot}185\,(t_1' - t_2')}$$

t_1', t_2' = Flow and return temperatures respectively, °C.

Pump Head—

	Pump Head		*Friction Resistance* millinches per ft.	mm. per m.
For low pressure H.W.H.	1 to 6 m.	3 to 20 ft.	100 to 300	8 to 25
For high pressure H.W.H.	6 to 25 m.	20 to 72 ft.	120 to 400	10 to 30

5. Pipe Sizing — (see Tables VII. 11 and Chart No. 1):

6. Expansion Tank:

Open Expansion Tanks (for Low Pressure H.W.H.)—

Expansion of water from 7°C. to 100°C. (45°F. to 212°F.) = approximately 4 per cent.

Requisite volume of expansion tank = 0·08 × water volume of system.

For estimating:

Volume of expansion tank = 1 litre for every 1 m.^{2} of radiator surface (approximately. 2 gal. for every 100 ft.^{2}).

Closed Expansion Tanks (for Medium and High Pressure H.W.H.)—

$V_1 = V(d_1 - d_2)$

$V_E P_1 = (V_E - V_1) P_2$

$$V_E = \frac{V_1 P_2}{(P_2 - P_1)}$$

V_1 = Water volume expanded by warming, litres.
d_1 = Density of water at 7°C.
d_2 = Density of water at max. flow temp.
V_E = Volume of expansion tank, litres.
P_1 = Pressure at tank connection at 7°C.
P_2 = Pressure of tank connection at max. flow temperature, N/m.^{2}

RECOMMENDED FLOW TEMPERATURES FOR LOW TEMPERATURE HOT WATER HEATING PLANTS

Outside temperature, °C. ...	0	2	4	7	10
Flow temperature at Boiler, °C.	80	70	56	45	37
Outside temperature, °F. ...	30	35	40	45	50
Flow temperature at Boiler, °F.	180	160	135	115	100

SAFETY VALVES

Safety Valve setting = Head of water in metres + 1·5.
or Head of water in feet + 5.

To prevent leakage due to shocks in the system, it is recommended that the load should not be less than 240 kN/m.^{2} (about 35 lb./in.^{2}).

Valves shall have clearances to allow a lift of D/5.

In systems with accelerated circulation, the load should be 70 kN/m.^{2} (about 10 lb./in.^{2}) in excess of the pressure on the outlet side of the pump, but should otherwise conform to the foregoing rules.

WATER HEATING BOILERS

Output up to		*Minimum Clear Bore of Safety Valve and Open Vents*	
B.t.u./hr.	*W.*	*in.*	*mm.*
900,000	275,000	one ¾	one 20
1,200,000	350,000	,, 1	,, 25
1,500,000	440,000	,, 1¼	,, 32
1,800,000	530,000	one 1½	one 40
3,000,000	880,000	two 1½	two 40
5,000,000	1,500,000	3 to 5	80 to 150

STEAM HEATING BOILERS

(Working Pressure up to 500 N/m.^{2}, or 10 lb./sq. ft.)

Output up to		*Minimum Clear Bore of Safety Valve*	
B.t.u./hr.	*W.*	*in.*	*mm.*
80,000	24,000	one ¾	one 20
150,000	44,000	,, 1	,, 25
250,000	75,000	,, 1¼	,, 32
350,000	100,000	,, 1½	,, 40
650,000	200,000	one 2	one 50
950,000	275,000	,, 2½	,, 65
1,500,000	440,000	two 2	two 50

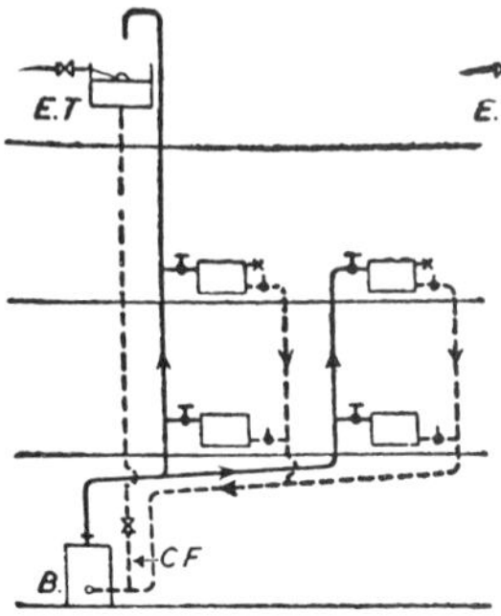

TWO PIPE UPFEED SYSTEM.

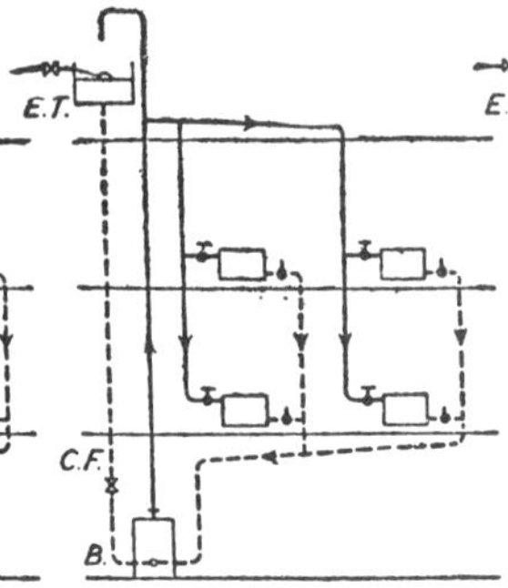

TWO PIPE DROP SYSTEM.

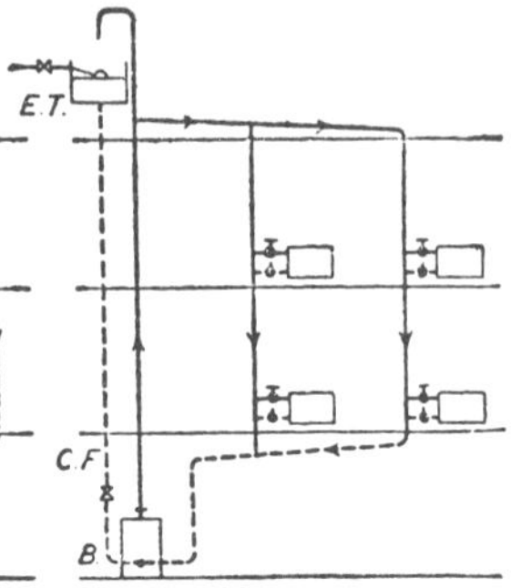

ONE PIPE DROP SYSTEM.

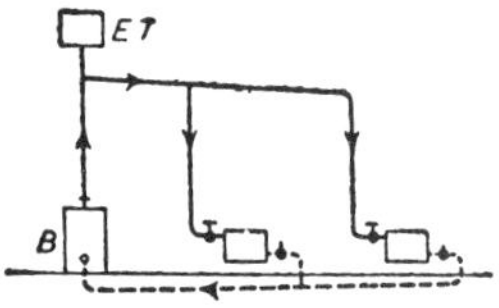

TWO PIPE DROP SYSTEM. BOILER & RADIATORS ON SAME LEVEL

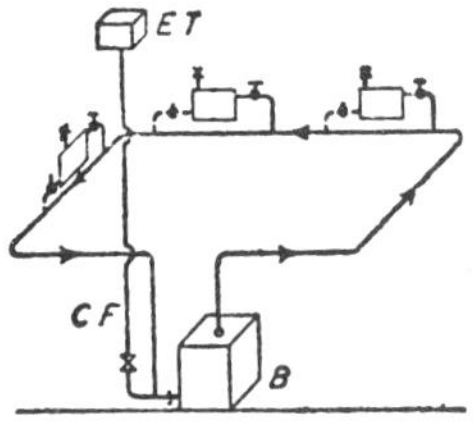

ONE PIPE RING MAIN SYSTEM.

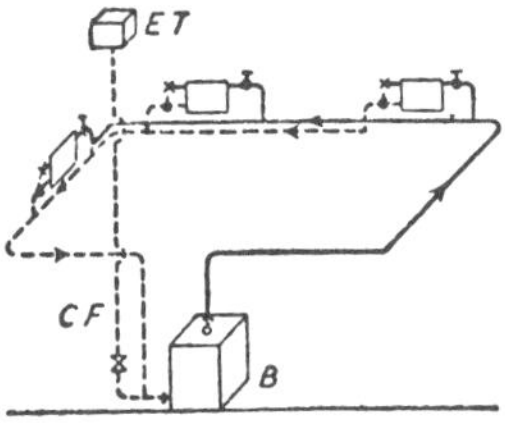

TWO PIPE RING MAIN SYSTEM. REVERSED RETURN

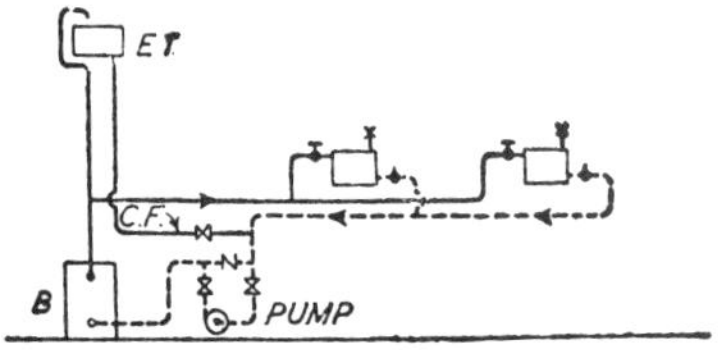

FORCED CIRCULATION, DIRECT RETURN SYSTEM.

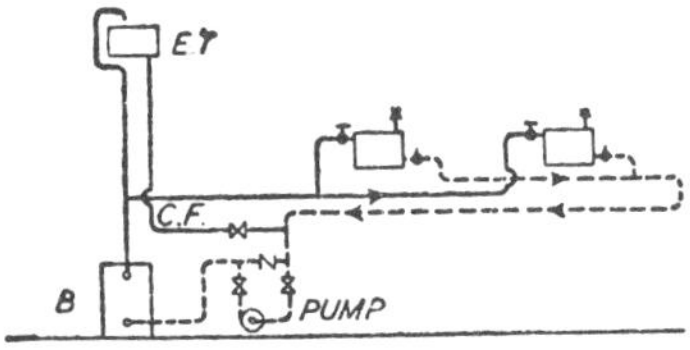

FORCED CIRCULATION, REVERSED RETURN SYSTEM.

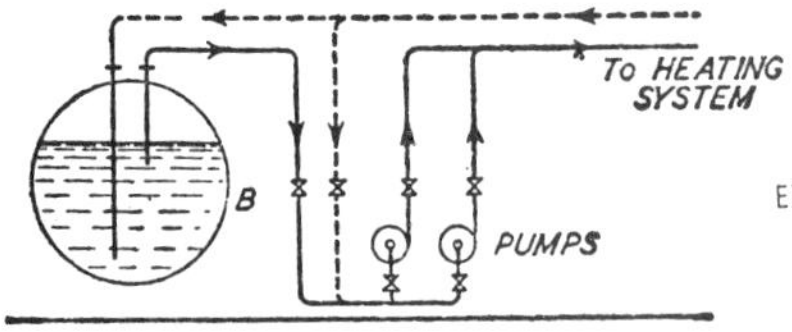

HIGH TEMPERATURE HOT WATER HEATING SYSTEM.

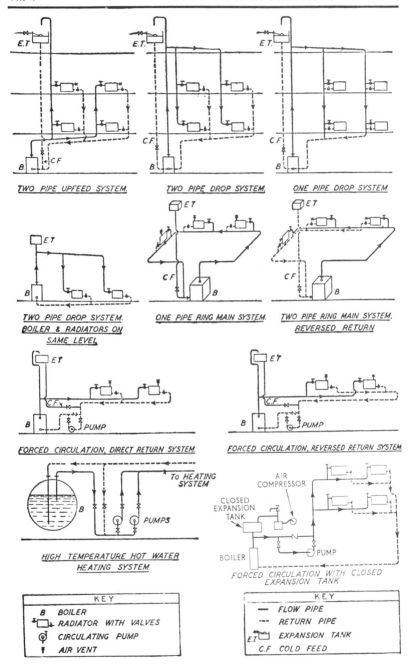

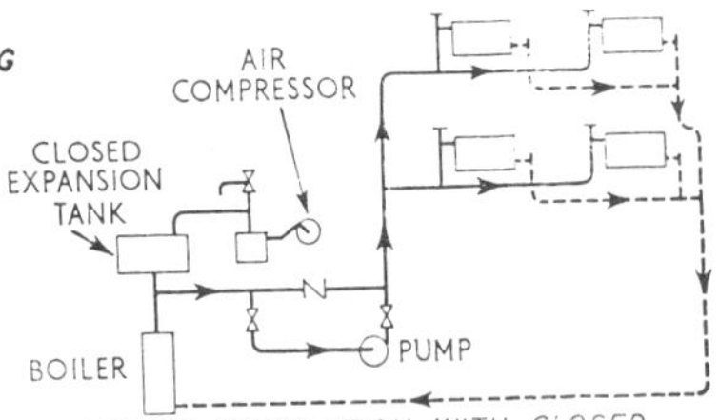

FORCED CIRCULATION WITH CLOSED EXPANSION TANK

KEY	
B	BOILER
(symbol)	RADIATOR WITH VALVES
(symbol)	CIRCULATING PUMP
(symbol)	AIR VENT

KEY	
——	FLOW PIPE
- - -	RETURN PIPE
E.T	EXPANSION TANK
C.F	COLD FEED

CIRCULATING PRESSURES FOR GRAVITY HEATING

mm. water gauge per metre circulating height.

Return Temp. °C.	*Flow Temperature °C.*							
	95	90	85	80	75	70	65	60
50	26·2	22·7	19·4	16·2	13·2	10·3	7·5	4·0
55	23·8	20·4	17·1	13·9	10·8	9·9	5·1	2·5
60	21·3	17·9	14·6	11·4	8·4	5·4	2·7	—
65	18·7	15·3	11·9	8·9	5·7	2·8	—	—
70	15·9	12·5	9·2	6·0	2·9	—	—	—
75	13·0	9·6	6·2	3·1	—	—	—	—
80	10·0	6·5	3·2	—	—	—	—	—
85	6·7	3·3	—	—	—	—	—	—

Inches per foot of circulating height.

Return Temp. °F.	*Flow Temperature °F.*						
	200	190	180	170	160	150	140
120	0·324	0·277	0·230	0·187	0·145	0·104	0·068
130	0·293	0·244	0·198	0·153	0·111	0·070	0·035
140	0·258	0·210	0·163	0·118	0·077	0·036	—
150	0·221	0·172	0·126	0·081	0·040	—	—
160	0·181	0·133	0·086	0·040	—	—	—
170	0·140	0·090	0·044	—	—	—	—
180	0·096	0·046	—	—	—	—	—
190	0·048	—	—	—	—	—	—

Density and Specific Gravity of Water at Various Temperatures, see IV. 29 and IV. 30.

Boiler and Radiator on Same Level

Circulation pressure in millimetres for 90°C. flow temperature 70°C. return downcomers bare, without covering. (From Rietschel-Brabée.)

Horizontal Extent of Plant in metres	*Horizontal distance of downcomer from main riser in metres*						
	5	*5—10*	*1—15*	*15—20*	*20—30*	*30—40*	*40—50*
up to 10	7	18	—	—	—	—	—
10 to 25	7	11	15	20	25	—	—
25 to 50	5	8	11	14	18	24	30

In milli-inches for flow 195°F., return 160°F.

Horizontal Extent of Plant in feet	*Horizontal distance of downcomer from main riser in feet*						
	16	*16—32*	*32—48*	*48—64*	*64—96*	*96—125*	*125—160*
32	275	710	—	—	—	—	—
32—82	275	430	600	800	1000	—	—
82—164	200	310	430	550	710	950	1180

Invisible Panel Heating Installations are heating systems using as heating surfaces pipe coils embedded in the concrete structure. The coils are generally constructed from black, mild steel tubes to B.S. 1387: 1967 $\frac{1}{2}$ or $\frac{3}{4}$ in. nom. bore. Coils embedded in concrete or plaster to be tested to a hydraulic pressure of 3,400 kN/m.^{2} or 500 lb f./in.^{2} for half an hour.

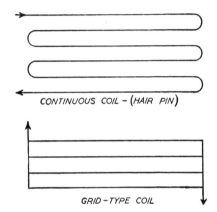

Heating Coils in form of Hair Pins are generally used, maximum length. of $\frac{1}{2}$ in. nom. bore coils approximately 180 ft. lin. or 55 m.

CONTINUOUS COIL – (HAIR PIN)

Grid Coils have less resistance, are more rigid, but there is a danger of short circuiting, therefore they are seldom used.

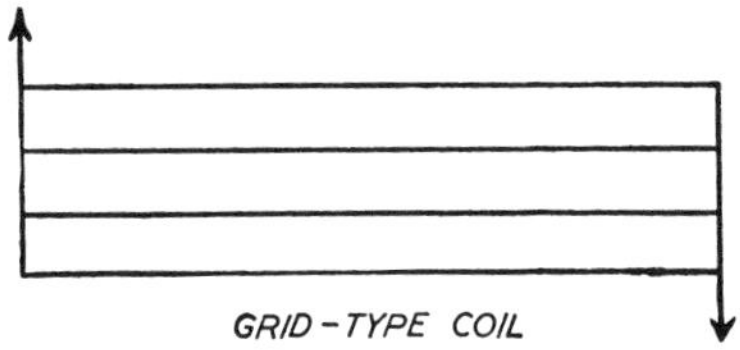

GRID-TYPE COIL

Advantages of Panel Heating Systems:

Saving of floor space.
No visible heaters and tubing, no dust collection.
No staining of walls, cleanliness and low maintenance.
More even temperature distribution, no excessive temperature at ceiling height.
Lower air temperature, therefore lower heat loss and saving of fuel.

Disadvantages:

More expensive first cost than radiator and convector heating.
Special building structure sometimes required.
Time lag for heating up.

Invisible Heating Panels can be arranged as:

A. Ceiling Panels.
B. Floor Panels.
C. Wall Panels.

The advantages of one type over another are small; advantages and disadvantages can be summarized as shown on page VII. 8.

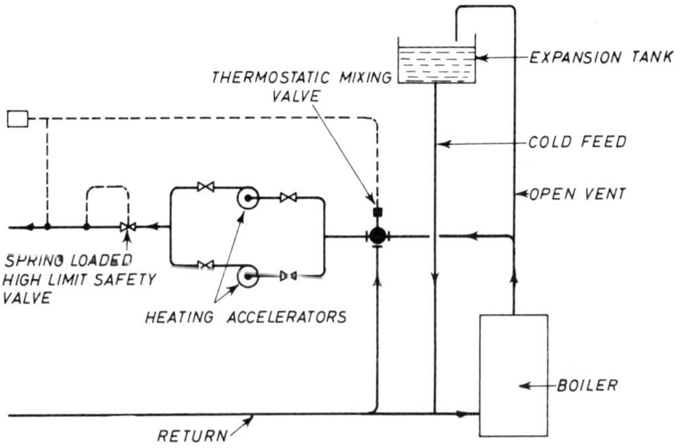

SCHEME OF MIXING ARRANGEMENT FOR PANEL HEATING INSTALLATION

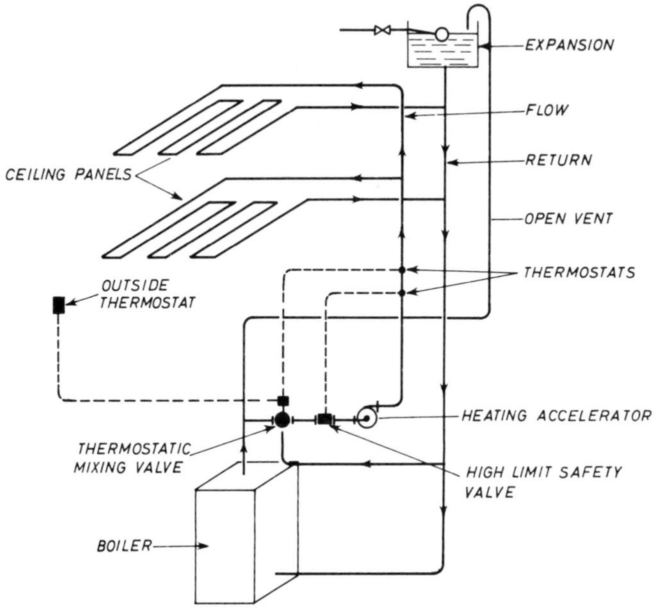

PIPE SCHEME OF PANEL HEATING INSTALLATION

CEILING PANELS:

Advantages:

Never obstructed by furniture. Comfortable conditions.
Higher permissible surface temperature than for floor panels, therefore smaller heating surface and lower cost.

Disadvantages:

Not suitable for low rooms. Feet shaded under desk in schools.
Not advisable for rooms with very cold floors. Special ceiling structure required.

Applications:

Hospitals, Schools, Flats, Offices.

Data:

Coils: ½ in. nom. bore tubes at 150 mm. centres.

Flow temp.	54 °C.(130°F.)	Output to below 205 W./m.² (65 B.t.u./hr.ft.2)	
Return temp.	46°C. (115°F.)	Loss to above 32 W./m.² (10 B.t.u./hr. ft.2)	
Temp. drop	8°C. (15°F.)	Emission by radiation	70%
Panel surface temp.	43°C. (110°F.)	Emission by convection	30%.

FLOOR PANELS:

Advantages:

Heat transfer partly by convection, comfort for feet.
No special building structure required, coils embedded in floor slab or screed.

Disadvantages:

Permissible surface temperature is low, therefore larger heating surfaces are required than for ceiling panels, and the first costs are higher.
Risk of panels being covered by furniture.

Applications:

Schools, Nurseries (especially where children play on floor), Single-floor buildings, Churches, Factories.

Data:

Coils: ½ in. (15 mm.) nom. bore or ¾ in. (20 mm.) nom. bore tubes at approximately 230 mm. centres.

Flow temp.	32°C. (90°F.)	Output 125 W./m.² (40 B.t.u./hr. ft.2)	
Return temp.	24°C. (75°F.)	Loss to below 19 W./m.² (6 B.t.u./hr. ft.2)	
Temp. drop	8°C. (15°F.)	Emission by radiation	55%
Panel surface temp.	24°C. max. (75°F.).	Emission by convection	45%.

WALL PANELS:

Advantages:

No special wall structure is required, panels are embedded in plaster. Wall panels are useful under windows as an addition to floor panels to cover the heat loss of windows.

Disadvantages:

Difficulty in arranging pipe connections.
Danger of covering wall panels by furniture.

Applications:

Halls, Staircases.

Data:

Coils: ½ in. (15 mm.) nom. bore tubes at 150 mm. centres.

Flow temp.	32° to 38° C. (90° to 100°F.)	Output 142 W./m.² (45 B.t.u./hr. ft.2)	
Return temp.	24° to 32°C. (75° to 90°F.)	Loss to outside 19 W./m.² (6 B.t.u./hr. ft.2)	
Temp drop	8°C. (15°F.)	Emission by radiation	65%
Panel surface temp.	26° to 29°C. (78° to 85°F.).	Emission by convection	35%.

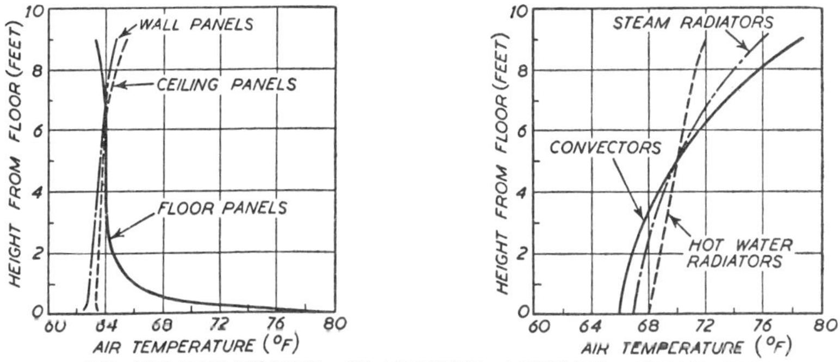

AIR TEMPERATURES AT VARIOUS HEIGHTS ABOVE FLOOR LEVEL FOR PANEL AND RADIATOR HEATING

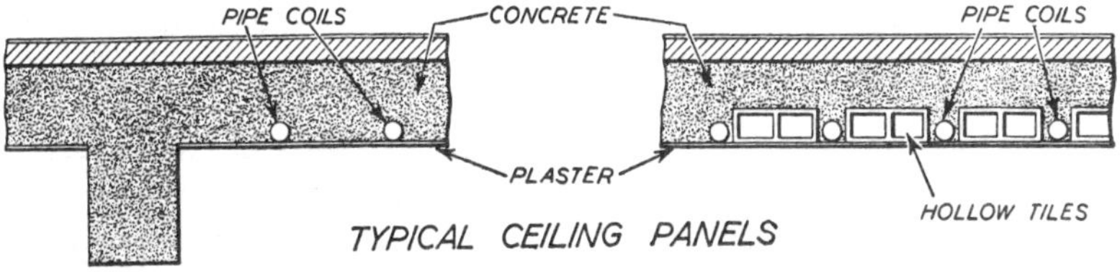

TYPICAL CEILING PANELS

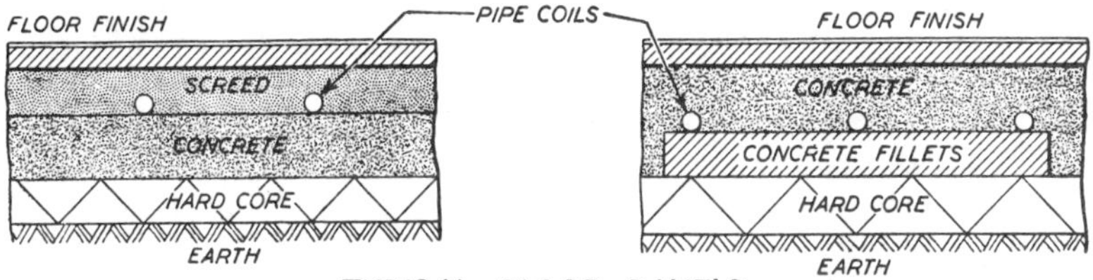

TYPICAL FLOOR PANELS

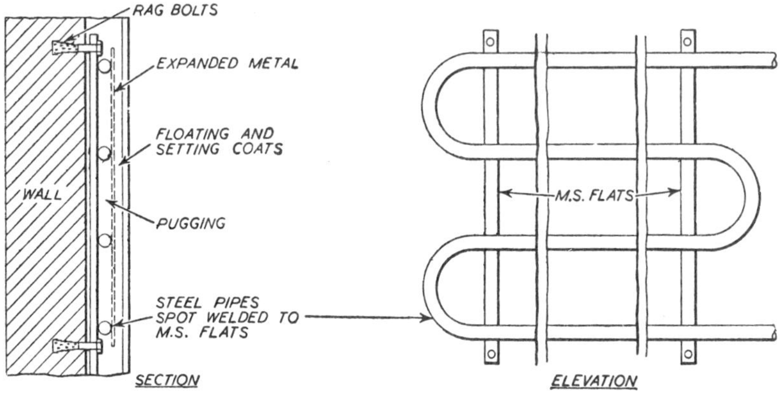

TYPICAL WALL PANEL

PLASTER FOR INVISIBLE HEATING PANELS

Heating coils embedded in concrete, wired to cork slabs or to suspended ceilings require careful treatment as to floating and finishing.

The suspended ceilings are covered with expanded metal lathing and pricked up with lime, sand and cement. When the pricking-up coat is stiff enough it should be well scratched with a wire comb ready to receive the floating coat.

Where time does not permit the rendering coat with cement to become thoroughly dry the metal lathing can be covered with gauged coarse stuff (lime and sand).

Sometimes heating panels are covered with lathwork and can be treated as previously mentioned if time is short.

A good mixture for lath or expanded metal is 4 parts of sand, 1 part of lime, and 1 part of cement; all floating coats should be lime 1, sand 3, with the usual hair added to this mix. About 30 per cent of plaster of Paris is added, floating the area to be covered to the ordinary thickness of the other part of the plastering surface, usually about 16 mm. thick.

The finishing coat should be applied within 24 hours and composed of setting stuff, 3 parts of washed sand and 2 parts of lime putty. Two parts of this mixture is gauged on the board to 1 part of plaster.

This is applied to the floating coat as evenly as possible, following on with a covering of hessian scrim about 760 mm. wide, pressing it well into the plaster, making a surface if possible without the addition of any extra application of plaster.

The canvas should overlap the panel about 150 mm. with the ends of the canvas unthreaded or opened out.

General Formulae for Pipe Sizing:

$$H_T = \Sigma H_1 + \Sigma H_2$$
$$= \Sigma iL + \Sigma F \frac{v^2 \rho}{2g}$$
$$H_1 = iL; \quad i = \frac{H_1}{L}$$
$$H_2 = F \frac{v^2 \rho}{2g}$$

H_T = Total loss of head.
H_1 = Loss of head due to friction.
H_2 = Loss of head due to fittings.
i = Resistance per unit of length of pipe.
L = Length of pipe (index circuit) ft. or m.
v = Velocity of water, ft. per sec. or m/s.
ρ = Density of water.
F = Coefficient of resistance.

or—

$$H_T = I(L + L_E) = iL_T$$
$$i = \frac{H_T}{L_T}$$

L_E = Equivalent length for the resistance of fittings, see Table XI. 5.
L_T = $(L + L_E)$ = total length.

(Graph for Pipe Sizing, see Chart No. 1).

Ratio of Resistances of Fittings (H_2) to the total resistance of the circuit in per cent:

Heating installations in buildings	40 to 50
District heating mains	10 to 30
Heating main lines in boiler houses	70 to 90

Resistance of Fittings for Hot Water Heating Schemes.
Values of "F" for various kind of Fittings.

Radiators	3·0	Tee, Straightway ...	1·0
Boilers	2·5	Branch ...	1·5
Abrupt velocity change	1·0	Counter current	3·0
Cross-over... ...	0·5	Double branch	1·5

VALUES FOR "F" FOR VARIOUS FITTINGS (see XI. 5)
(Dimensionless)

Type of Fitting	*½ in.*	*¾ in.*	*1 in.*	*1¼ in.*	*1½ in.*	*2 in.*
Radiator Valve, Angle	7	4	4	4	—	—
Radiator Valve, Straight	4	2	2	2	—	—
Valve, Steam Globe ...	16	10	9	9	8	7
Gate Valve, Screwed	1·5	0·5	0·5	0·5	0·5	0·5
Gate Valve, Flanged ...	0	0	0	0	0	0
Elbow	2	2	1·5	1·5	1·0	1·0
Bend	1·5	1·5	1·0	1·0	0·5	0·5

LOSS OF HEAD H_2 DUE TO FITTINGS FOR VARIOUS VELOCITIES AND VALUES OF F

1. Loss of Head in mm. W.G.:

Velocity	*Values of F.*						
m./s.	1	2	3	4	5	10	15
0·24	2·9	5·7	8·6	11·5	14·4	28·5	43
0·50	12·5	25	37·5	50	62	125	187
0·75	28·0	56	84	112	140	280	420
1·0	50	100	150	200	250	500	750
1·2	72	144	215	285	360	720	1,080
1·5	112	225	335	450	560	1,120	1,680
1·7	144	290	430	580	720	1,440	2,160
2·0	200	400	600	800	1,000	1,990	2,990
2·4	285	570	860	1,150	1,440	2,870	4,310
2·8	390	780	1,170	1,560	1,950	3,910	5,860
3·0	450	900	1,350	1,790	2,240	4,490	6,730

2. Loss of Head in inches W.G.

Velocity	*Values of F.*						
ft./sec.	1	2	3	4	5	10	15
1	0·18	0·36	0·54	0·72	0·90	1·8	2·7
1·5	0·22	0·45	0·67	0·90	1·12	2·25	3·37
2	0·72	1·5	2·2	3·0	3·6	7·2	10·8
3	1·6	3·3	4·9	6·6	8·2	16·3	24·4
4	3·0	5·8	8·7	11·6	14·5	28·9	43·4
5	4·5	9·0	13·6	18·0	22·6	45·2	67·8
6	6·5	13·0	19·6	26·0	32·6	65·2	97·8
7	8·9	17·7	26·6	35·4	44·3	88·6	133·0
8	11·6	23·2	34·7	46·4	57·9	116·0	174·0
9	14·7	29·3	44·0	58·6	73·3	147·0	220·0
10	18·0	36·2	54·3	72·4	90·5	181·0	272·0

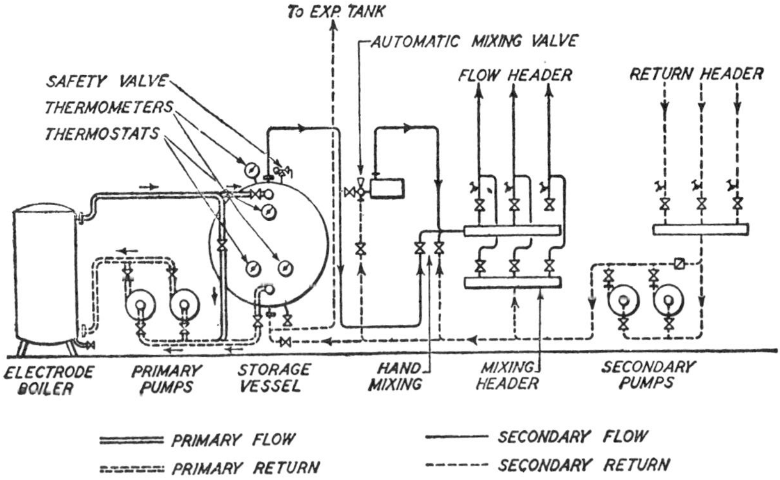

SCHEME OF HOT WATER HEATING PLANT WITH ELECTRODE BOILER

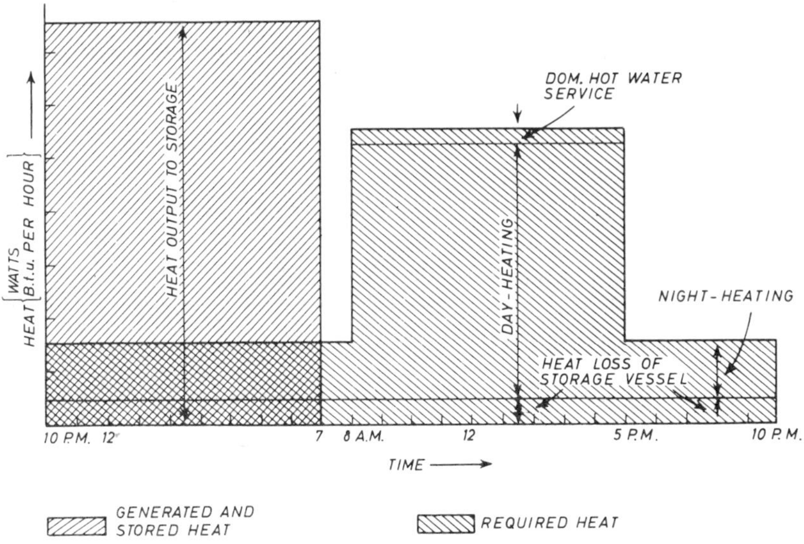

DIAGRAM OF HEAT STORAGE

The Safe Storage Temperature is approximately 10°C. (or about 20°F.) below the boiling temperature of water under the pressure of the respective water head.

Total Heat to be Stored (H) Joules:

H = Heat demand for heating (watts) × Time (secs.).
+ Heat demand for hot water supply (watts) × Time (secs.).
+ Heat loss of storage vessel (watts) × Time (secs.).

Capacity of Storage Vessel:

$$V = \frac{H}{d_s (t_s - t_2)}$$

$$V' = \frac{H'}{4{\cdot}2\, d_s^{\,2} (t_s' - t_2')}$$

V = Capacity of storage vessel, cu. ft.
V = ,, ,, ,, ,, litres.
t_s = Storage temperature, °F.
t_s' = ,, ,, °C.
t_2 = Return temperature, °F.
t_2' = ,, ,, °C.
d_s = Density of water at storage temp. lb./ft.^{2}
d_s' = ,, ,, ,, ,, ,, ,, l./kg.
H = Total heat to be stored, B.t.u.
H' = ,, ,, ,, ,, ,, kJ.
(= total heat in Joules × 10^{-3}).

Heat Input to Storage:

$$H_s = \frac{H}{n}$$

H_s = Heat input to storage.
n = Time for loading of storage vessel.

If H is in B.t.u. and H_5 in B.t.u./hr., then n must be expressed in hours.
If H is in Joules and H_5 in watts, then n must be expressed in seconds.

Load of Electrode Boiler:

$$kW = \frac{V d_r (t_s - t_2)}{n \times 3{,}415}$$

$$kW = \frac{4{\cdot}2\, V' d_r' (t_s' - t_2')}{n'}$$

kW = load in kW.
d_r = density of water at return temp. lb./ft^2.
d_r' = ,, ,, ,, ,, ,, ,, l./kg.
n = time of loading, hours.
n' = ,, ,, ,, seconds.

HEAT LOSS THROUGH LAGGING

Material	*Loss in Watts through 76 mm. (3 in.) thickness per 55°C. (100°F.) difference between faces*	
	W./m.^{2}	W./ft.^{2}
Asbestos	75	7
Cork	32	3
Sawdust	54	5
Soft Wood 25 mm. (1 in.) thick	215	20

Loss from bare metal for 55°C. difference is approximately 485 W/m.^{2} (45 W/ft.^{2}).

HEAT TRANSMISSION COEFFICIENTS FOR UNCOVERED TUBES
(Water to Air) in B.t.u. per lineal ft. per hour per °F.

Pipe Size ins.	*Mean Temperature Difference — Water-Air °F.*									*Pipe Size ins.*
	175	200	225	250	275	300	350	400	500	
½	0·65	0·69	0·72	0·75	0·79	0·82	0·89	0·97	1·19	½
¾	0·81	0·85	0·89	0·92	0·98	1·01	1·09	1·19	1·36	¾
1	0·99	1·04	1·09	1·13	1·18	1·24	1·35	1·46	1·68	1
1¼	1 20	1·27	1·33	1·37	1·45	1·53	1·64	1·79	2·06	1¼
1½	1·35	1·43	1·48	1·55	1·63	1 70	1·85	2·02	2·31	1½
2	1·65	1·75	1·83	1·91	2·00	2·10	2·28	2·50	2·86	2
2½	1·95	2·06	2·16	2·25	2·37	2·50	2·70	2·96	3·40	2½
3	2·34	2·46	2·90	2·68	2·84	3·00	3·24	3·54	4·08	3
4	2·92	3·08	3·24	3·38	3·56	3·73	4·06	4·46	5·20	4
5	3·50	3·68	3·87	4·04	4·27	4·46	4·86	5·36	6·17	5
6	4·06	4·28	4·50	4·69	4·98	5·14	5·67	6·25	7·20	6
7	4·51	4·96	5·22	5·44	5·76	6·04	6·58	7·10	8·40	7
8	5·15	5·45	5·72	5·96	6·33	6·64	7·25	7·96	9·25	8
9	5·70	6·00	6·30	6·60	7·00	7·30	8·00	8·16	10·35	9
10	6·26	6·61	6·98	7·25	7·70	8·05	8·82	9·74	11·40	10
12	7·30	7·72	8·15	8·50	9·00	9·40	10·00	11·40	13·30	12

HEAT TRANSMISSION COEFFICIENTS FOR UNCOVERED TUBES
(Water to Air) in Watts per metre per °C.

Pipe Size (mm.)	*Mean Temperature Difference — Water-Air °C.*									*Pipe Size in.*
	98	110	125	140	150	165	195	225	280	
15	1·13	1·20	1·25	1·30	1·37	1·42	1·45	1·68	2·06	½
20	1·40	1·47	1·54	1·59	1·70	1·75	1·89	2·06	2·35	¾
25	1·71	1·80	1·89	1·96	2·04	2·15	2·34	2·53	2·91	1
32	2·08	2·20	2·31	2·37	2·51	2·65	2·84	3·10	3·57	1¼
40	2·34	2·48	2·56	2·68	2·82	2·94	3·20	3·50	4·00	1½
50	2·86	3·03	3·17	3·31	3·46	3·63	3·95	4·33	4·95	2
65	3·38	3·57	3·74	3·89	4·10	4·33	4·67	5·12	5·88	2½
80	4·05	4·26	4·50	4·64	4·92	5·19	5·61	6·13	7·07	3
100	5·05	5·33	5·61	5·85	6·16	6·46	7·03	7·72	9·00	4
125	6·06	6·38	6·70	6·99	7·39	7·72	8·41	9·298	10·68	5
150	7·03	7·41	7·79	8·12	8·62	8·90	9·81	10·8	12·5	6
175	7·81	8·59	9·04	9·42	9·97	10·5	11·4	12·3	14·7	7
200	8·92	9·44	9·90	10·3	11·0	11·5	12·5	13·8	16·0	8
225	9·86	10·4	10·9	11·4	12·1	12·6	13·8	14·1	17·9	9
250	10·8	11·4	12·1	12·5	13·3	13·9	15·3	16·8	14·7	10
300	12·6	13·4	14·1	14·7	15·6	16·3	17·3	19·7	23·0	12

Return Temp. °C.	Flow Temperature °C. 210	200	195	190	185	180	175	170	165	160	155	150	145	140	135	130	125	120	115	110	105	100
75	569	538	516	493	471	449	427	405	383	351	340	318	297	275	254	232	220	190	170	147	126	105
80	547	516	494	471	449	427	405	383	361	339	318	296	275	253	232	210	198	168	145	125	104	83
85	527	496	474	451	429	407	385	363	341	319	300	278	255	233	212	190	178	148	127	105	84	63
90	506	475	453	430	408	396	374	342	320	298	277	255	234	212	191	169	158	127	106	85	63	42
95	495	454	432	409	387	365	342	321	299	277	256	234	213	191	170	148	136	106	85	63	42	21
100	464	433	411	388	366	344	322	300	278	256	235	213	192	170	149	127	115	85	64	42	21	
105	443	412	390	367	345	323	301	279	257	235	214	192	171	149	128	106	94	64	43	21		
110	422	391	369	346	324	302	280	258	236	214	193	171	150	128	107	85	73	43	22			
115	400	369	347	324	302	280	258	236	214	192	171	150	128	106	85	63	51	21				
120	379	348	327	303	281	259	237	215	193	171	150	128	107	85	64	42	31					
125	349	317	296	273	251	229	207	185	163	141	120	98	77	55	33	12						
130	337	306	284	259	239	217	195	173	151	129	108	86	65	43	22							
135	315	284	262	239	217	195	173	151	129	108	86	65	43	22								
140	294	263	241	218	196	174	151	130	108	86	65	43	22									
145	272	241	219	196	174	152	130	108	86	64	43	21										
150	251	220	198	175	153	131	109	87	65	43	22											
155	229	198	176	153	131	109	87	65	43	21												
160	208	177	155	132	110	88	66	44	22													
165	186	155	133	110	88	66	44	22														
170	124	133	111	88	66	44	22															
175	142	111	89	66	44	22																
180	122	89	67	43	22																	
185	98	67	45	22																		
190	76	45	23																			
195	53	22																				
200	31																					

Example: Flow Temperature = 180°C
Return Temperature = 130°C.
Heat given up by 1 kg. of water = 217 kJ.

Heat in kJ. given up by 1 kg. of water for various Temperature Drops.

Return Temp. °F.	Flow Temperature °F.																			
	400	**390**	**380**	**370**	**360**	**350**	**340**	**330**	**320**	**310**	**300**	**290**	**280**	**270**	**260**	**250**	**240**	**230**	**220**	**210**
170	237·3	226·3	215·5	204·9	194·3	183·7	173·1	162·6	152·2	141·8	131·5	121·2	110·9	100·7	90·5	80·4	70·3	60·2	50·1	40
180	227·3	216·3	205·5	194·9	184·3	173·7	163·1	152·6	142·2	131·8	121·5	111·2	100·9	90·7	80·5	70·4	60·3	50·2	40·1	30
190	217·3	206·3	195·5	184·9	174·3	163·7	153·1	142·6	132·2	121·8	111·5	101·2	90·9	80·7	70·5	50·4	50·3	40·2	30·1	20
200	207·3	196·3	185·5	174·9	164·3	153·7	143·1	132·6	122·2	111·8	101·5	91·2	80·9	70·7	60·5	50·4	40·3	30·2	20·1	10
210	197·3	186·3	175·5	164·9	154·3	143·7	133·1	122·6	112·2	101·8	91·5	81·2	70·9	60·7	50·5	40·4	30·3	20·2	10·1	
220	187·3	176·2	165·4	154·8	144·2	133·6	123·0	112·5	102·1	91·7	81·4	71·1	60·8	50·6	40·4	30·3	20·2	10·1		
230	177·1	166·1	155·3	144·7	134·1	123·6	112·9	102·4	92·0	81·6	71·3	60·0	50·7	40·5	30·3	20·2	10·1			
240	167·0	156·0	145·2	134·6	124·0	113·4	102·8	92·3	81·9	71·5	61·2	50·9	40·6	30·4	20·2	10·1				
250	156·3	145·9	135·1	124·5	113·9	103·3	92·7	82·2	71·8	61·4	51·1	40·8	30·5	20·3	10·1					
260	146·7	135·7	124·9	114·3	103·7	93·1	82·5	72·0	61·6	51·2	40·9	30·6	20·3	10·2						
270	136·6	125·6	114·8	104·2	93·6	83·0	72·4	61·9	51·5	41·1	30·8	20·5	10·2							
280	126·4	115·4	104·6	94·0	83·4	72·8	61·2	51·9	41·3	36·9	20·6	10·3								
290	116·1	105·1	94·3	83·7	73·1	62·5	51·9	41·4	31·0	20·6	10·3									
300	105·8	94·3	84·0	73·4	62·8	52·2	41·6	31·1	20·7	10·8										
310	95·5	84·5	73·7	63·1	52·5	41·9	31·3	20·8	10·4											
320	85·0	74·5	63·2	51·7	41·1	30·5	20·9	10·4												
330	74·7	63·7	52·9	42·3	31·7	21·1	10·5													
340	64·1	53·1	42·3	31·8	21·2	10·6														
350	53·6	42·6	31·8	21·2	10·6															
360	42·0	31·0	21·4	10·6																
370	32·4	21·0	10·6																	
380	21·8	10·8																		
390	11·0																			

Example: Flow Temperature = 350°F.
Return Temperature = 240°F.
Heat given up by 1 lb. of water is **113·4 B.t.u. per lb.**

Heat in B.t.u. given up by one lb. of Water for various Temperature Drops

HEAT PUMPS

The Heat Pump is a common Refrigeration Unit arranged in such a way that it can be used for both cooling and heating, or for heating only. The initial cost of the installation is high, and savings and advantages are achieved mainly when heating and cooling are required in winter and summer respectively.

Operation of the Heat Pump:

Referring to the scheme drawing below, the Heat Pump consists of the following parts:

Compressor, with driving motor, for raising the pressure and temperature of the refrigerant vapour.

Condenser, for extracting heat from the refrigerant.

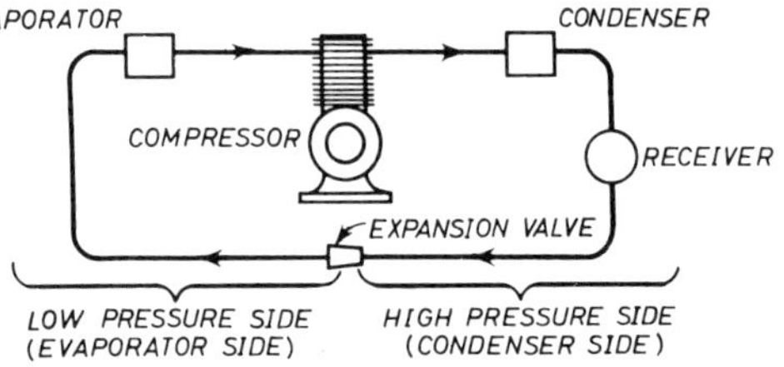

SCHEME OF HEAT PUMP SYSTEM

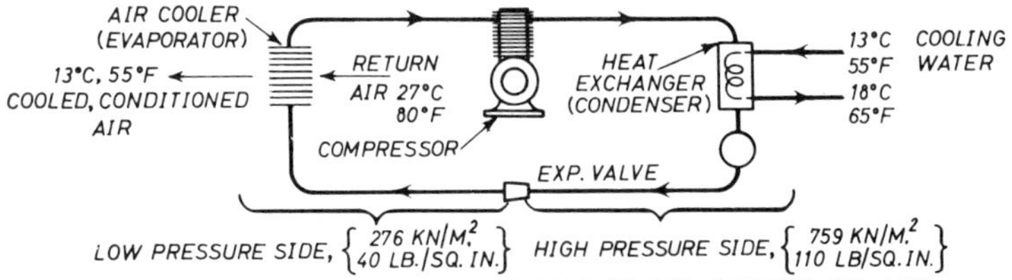

COOLING CYCLE OF THE HEAT PUMP (WATER TO AIR)

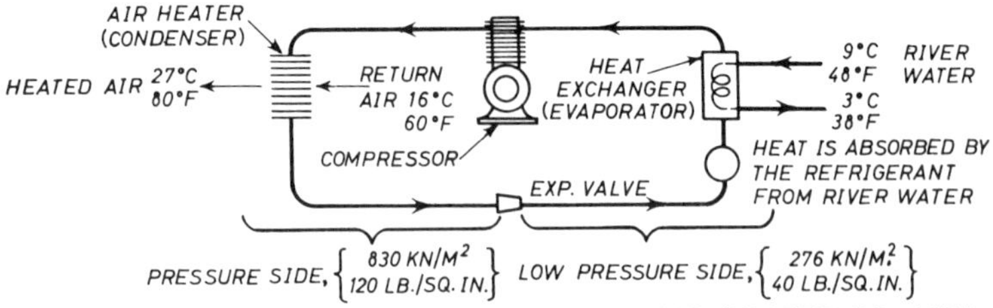

HEATING CYCLE OF THE HEAT PUMP (WATER TO AIR)

Receiver (Storage Tank) to hold the liquid refrigerant in the high pressure side before it passes the expansion valve.

Expansion Valve, for causing expansion of the refrigerant and for lowering the pressure from the high pressure to the low pressure side of the system.

Evaporator, in which heat is absorbed by the refrigerant from some source. Water, earth or air can be used as the source of heat.

A Commercial Refrigeration Unit and a Heat Pump consist of the same units and the same plant can be used either for cooling or heating.
The changing of the system from cooling to heating can be carried out by either of the following methods:

(*a*) Leave the flow of the refrigerant unchanged and change the circuit of the heat source and the medium to be heated.

(*b*) Leave the heat source and the medium to be heated unchanged and reverse the flow of the refrigerant by a suitable pipe and valve scheme.

Schemes for a Heat Pump indicating suitable temperatures when used for cooling and heating are shown, the data being chosen for the purpose of illustration only.

Section VIII

STEAM HEATING

THE HEATING MEDIUM IS STEAM

(Carrying Heat through Pipes from Boiler to Heaters)

Types of Steam Heating Schemes:

(a) High Pressure Steam Heating System.
(b) Low Pressure Steam Heating System.
(c) Vacuum System.

Classification according to method of Returning Condensate:
Gravity System.
Mechanical System.

Classification according to Pipe Scheme:
One-pipe or Two-pipe System.
Up-feed or Down-feed System.

Design of Steam Heating Systems:

1. Heat Losses of Heated Rooms.
2. Type and Capacity of Boilers.
3. Heating Surface of Radiators or Output of Unit Heaters.
4. Pipe Scheme and Sizing of Steam and Condense Pipes.

1. Heat Losses (see Section VI.):

2. Boiler:

$$B = H_t (1 + X)$$

$$S = \frac{H_t}{H}$$

B = Boiler rating, watts (or B.t.u./hr.).
H_t = Total Heat Loss of plant watts, (or B.t.u./hr.)
X = Margin for heating up = 0·10 to 0·15.
S = Steam consumption kg./sec. (or lb./hr.).
H = Heat utilized from unit mass of steam, J/kg. (or B.t.u./lb.).

Boiler Type and Capacity to be chosen from Maker's Catalogue.

3. Heater, Heating Surface of Radiators and Pipe Coils:

$$A = \frac{H_R}{K (t_S - t_R)}$$

A	= Heating surface	m.^{2}	(or ft.^{2}).
H_R	= Heat loss of room	watts	(or B.t.u./hr.).
t_S	= Temperature of Steam	°C.	(or °F.).
t_R	= Room Temperature	°C.	(or °F.).
K	= Heat transmission of heater	W/m.^{2} deg. C.	(or B.t.u./°F. ft.^{2} hr.)

Type and Size of Heater to be chosen from Maker's Catalogue.

STEAM HEATING SYSTEMS

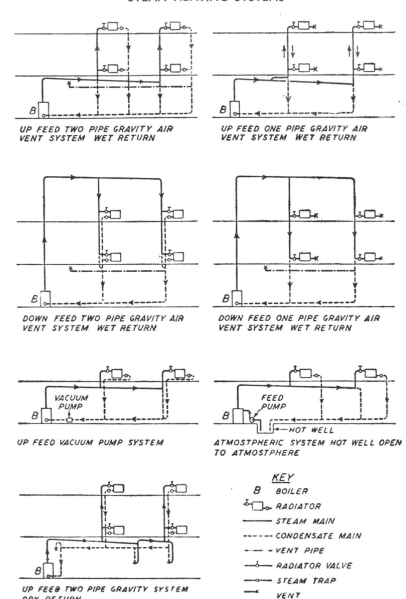

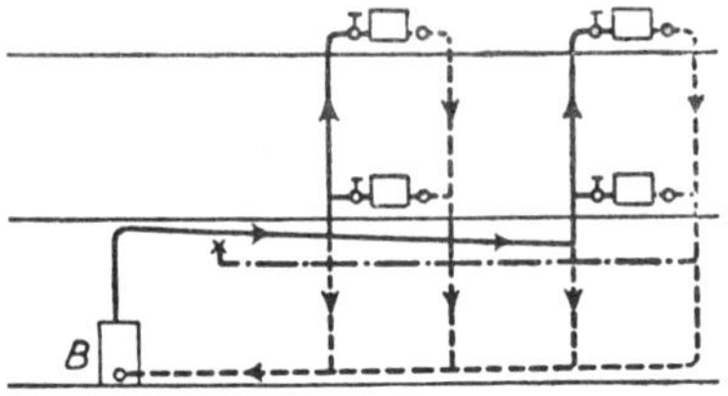

UP FEED TWO PIPE GRAVITY AIR VENT SYSTEM WET RETURN

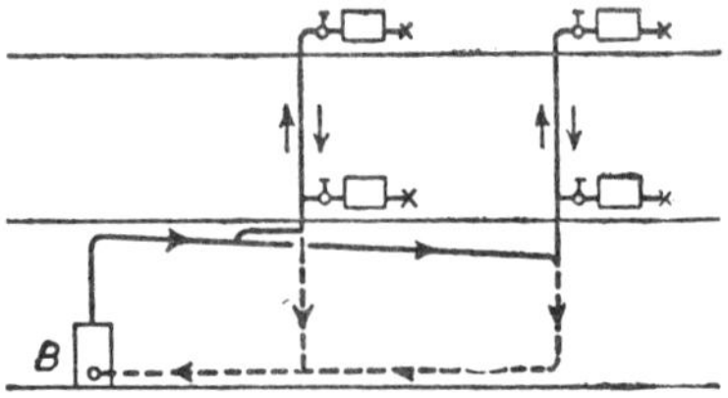

UP FEED ONE PIPE GRAVITY AIR VENT SYSTEM WET RETURN

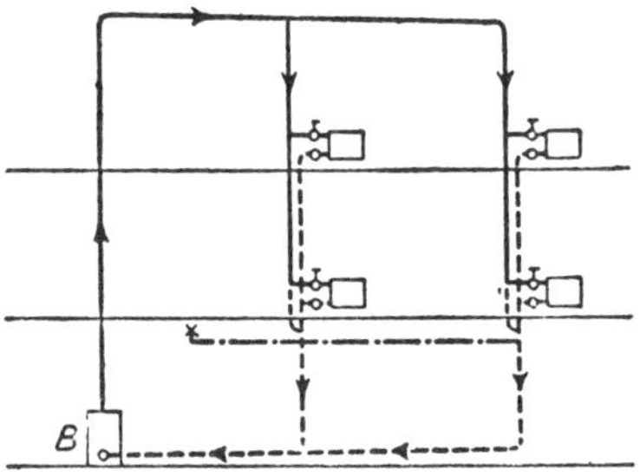

DOWN FEED TWO PIPE GRAVITY AIR VENT SYSTEM WET RETURN

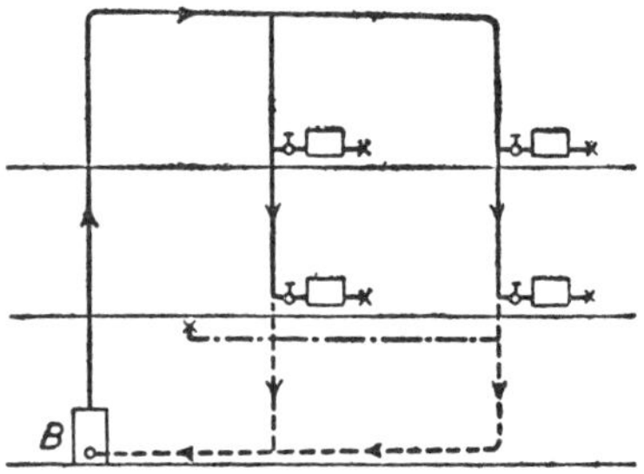

DOWN FEED ONE PIPE GRAVITY AIR VENT SYSTEM WET RETURN

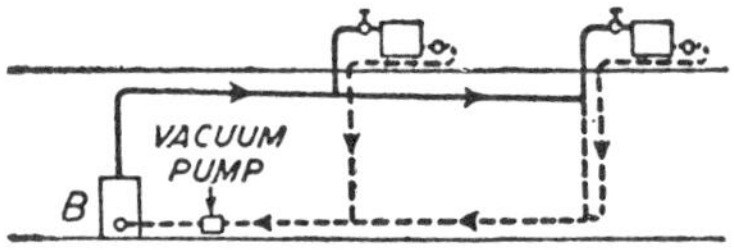

UP FEED VACUUM PUMP SYSTEM

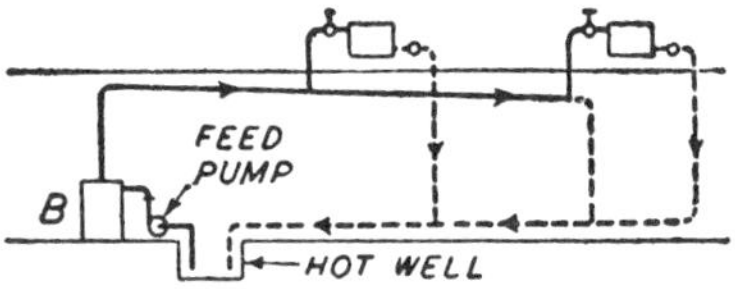

ATMOSTPHERIC SYSTEM HOT WELL OPEN TO ATMOSPHERE

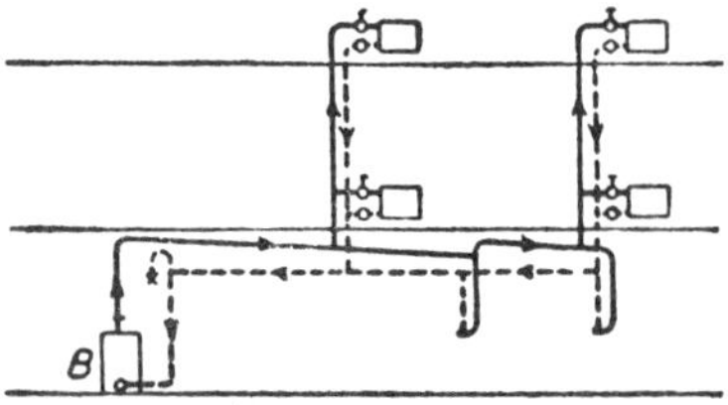

UP FEEB TWO PIPE GRAVITY SYSTEM DRY RETURN

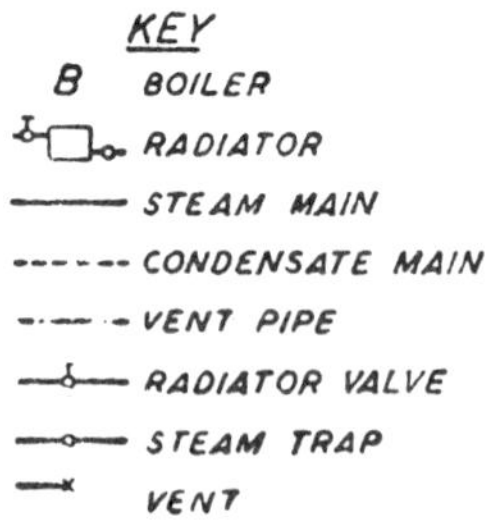

CAPACITIES OF CONDENSATE PIPES IN WATTS

Nominal Pipe Size		Wet Main	Dry Main with Gradient			Vent Pipes
in.	mm.		1 in 200	1 in 600.	Vertical	
½	15	30,000	10,000	6,000	10,000	12,000
¾	20	70,000	30,000	18,000	30,000	47,000
1	25	120,000	50,000	34,000	50,000	94,000
1¼	32	300,000	120,000	80,000	120,000	211,000
1½	40	420,000	176,000	117,000	176,000	293,000
2	50	760,000	350,000	225,000	350,000	530,000
2½	65	1,900,000	800,000	510,000	800,000	1,200,000
3	80	2,700,000	1,200,000	740,000	1,200,000	1,870,000

CAPACITIES OF CONDENSATE PIPES IN B.t.u./Hr.

Nominal Pipe Size		Wet Main	Dry Main with Gradient			Vent Pipes
in.	mm.		$\frac{3}{16}$ in. per yd.	$\frac{1}{16}$ in. per yd.	Vertical	
½	15	100,000	40,000	24,000	40,000	40,000
¾	20	240,000	108,000	68,000	108,000	160,000
1	25	400,000	192,000	120,000	192,000	320,000
1¼	32	1,000,000	440,000	280,000	440,000	720,000
1½	40	1,440,000	600,000	400,000	600,000	1,000,000
2	50	2,600,000	1,120,000	700,000	1,120,000	1,800,000
2½	65	6,400,000	2,800,000	1,760,000	2,800,000	4,000,000
3	80	9,600,000	4,000,000	2,520,000	4,000,000	6,400,000

SAFETY VALVES FOR STEAM HEATING
(Working pressure = 70 kN/m.^{2} or 10 lb./sq. in.).

Output Watts	Minimum Clear Bore	
	in.	mm.
24,000	¾	20
44,000	1	25
73,000	1¼	32
100,000	1½	40
230,000	2	50
275,000	2½	65
440,000	Two 2	Two 50

Output B.t.u./hr.	Minimum Clear Bore	
	in.	mm.
80,000	¾	20
150,000	1	25
250,000	1¼	32
350,000	1½	40
800,000	2	50
950,000	2½	65
1,500,000	Two 2	Two 50

RECOMMENDED WORKING PRESSURE FOR LOW PRESSURE STEAM HEATING PLANTS

Length of Index Circuit in m.	up to 180	275	450
Working Pressure in kN/m.^{2}	7–10	14	20
Length of Index Circuit in ft.	up to 600	900	1,500
Working Pressure in lb./in.^{2}	1–1·5	2	3

GENERAL FORMULAE FOR PIPE SIZING

$$H_T = H_1 + H_2$$
$$= pL + \Sigma F \frac{v^2\rho}{2}$$
$$H_T = P_1 - P_2$$
$$H_1 = pL, \ p = \frac{H_1}{L}$$
$$H_2 = F \frac{v^2\rho}{2}$$

P_1 = Initial steam pressure at boiler.
P_2 = Final steam pressure at heater valve.
H_T = Total pressure drop.
H_1 = Pressure drop due to friction.
H_2 = Pressure drop due to fittings.
p = Pressure drop per unit length of pipe.
L = Length of pipe
v = Velocity of steam
ρ = Density of steam = w/g.
F = Coefficient of resistance.
w = Weight of steam.

or—

$$H_T = p\,(L + L_E) = pL_T$$

L_E = Equivalent length for the resistance of fittings (see Table VIII. 7).

$$p = \frac{H_T}{L_T}$$

$L_T = (+ L_E)$ = total length.

(Graph for Pipe Sizing, see Chart No. 2).

Ratio of the Resistance of Fittings (H_2) to the Total Resistance of the Circuit is about 33 per cent.

Resistance of Fittings at Low Pressure Steam Heating.

Values of "F" for various kind of Fittings:

Radiator ... F = 1·5
Abrupt Velocity Change 1·0
Cross-over ... ... 0·5
Long Sweep Elbow ... 1·5
Tee, Straightway ... 1·0
Branch ... 1·5
Counter Current 3·0
Double Branch ... 1·5

Kind of Fitting.	½ in.	¾ in.	1 in.	1¼ in.	1½ in.	2 in.
Frictional Angle Valve	9	9	9	9		
Fractional Globe Valve ...	15	17	19	30		
Angle Cock	7	4	4	4		
Straight Cock	4	2	2	2		
Gate Valve, Screwed	1·5	0·5	0·5	0·5	0·5	0·5
Gate Valve, Flanged	0	0	0	0	0	0
Damper	3·5	2	2	1·5	1·5	1
Elbow Standard, G.F.	2	2	1·5	1·5	1	1
Long Sweep Elbow, G.F. ...	1·5	1·5	1	1	0·5	0·5
Bend, Short Radius	2	2	2	2	2	2
Bend, Long Radius	1	1	1	1	1	1
Standard Pipe Coupling, G.F.	0·5	0	0	0	0	0

CHART Ia. PIPE SIZES FOR HOT WATER HEATING

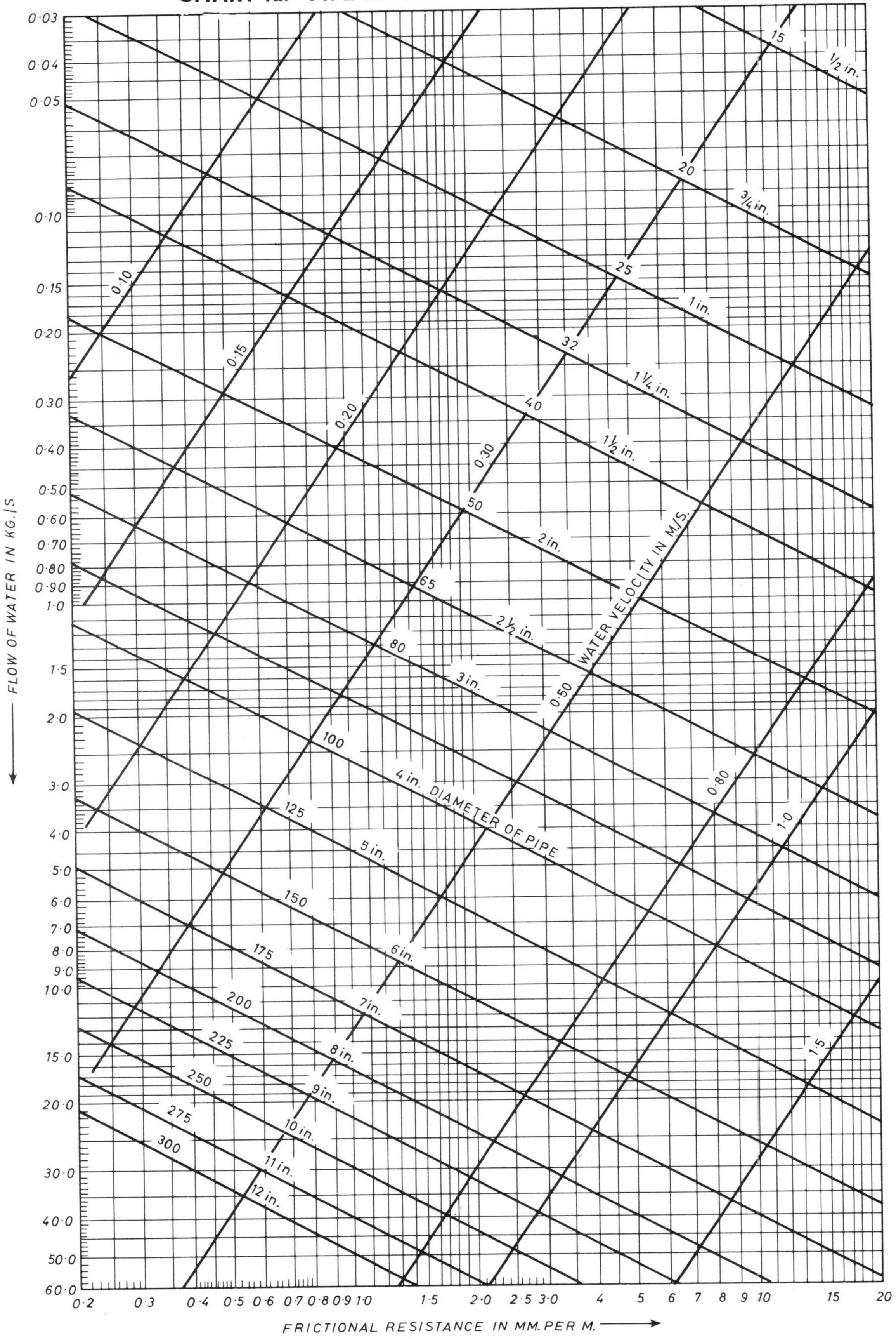

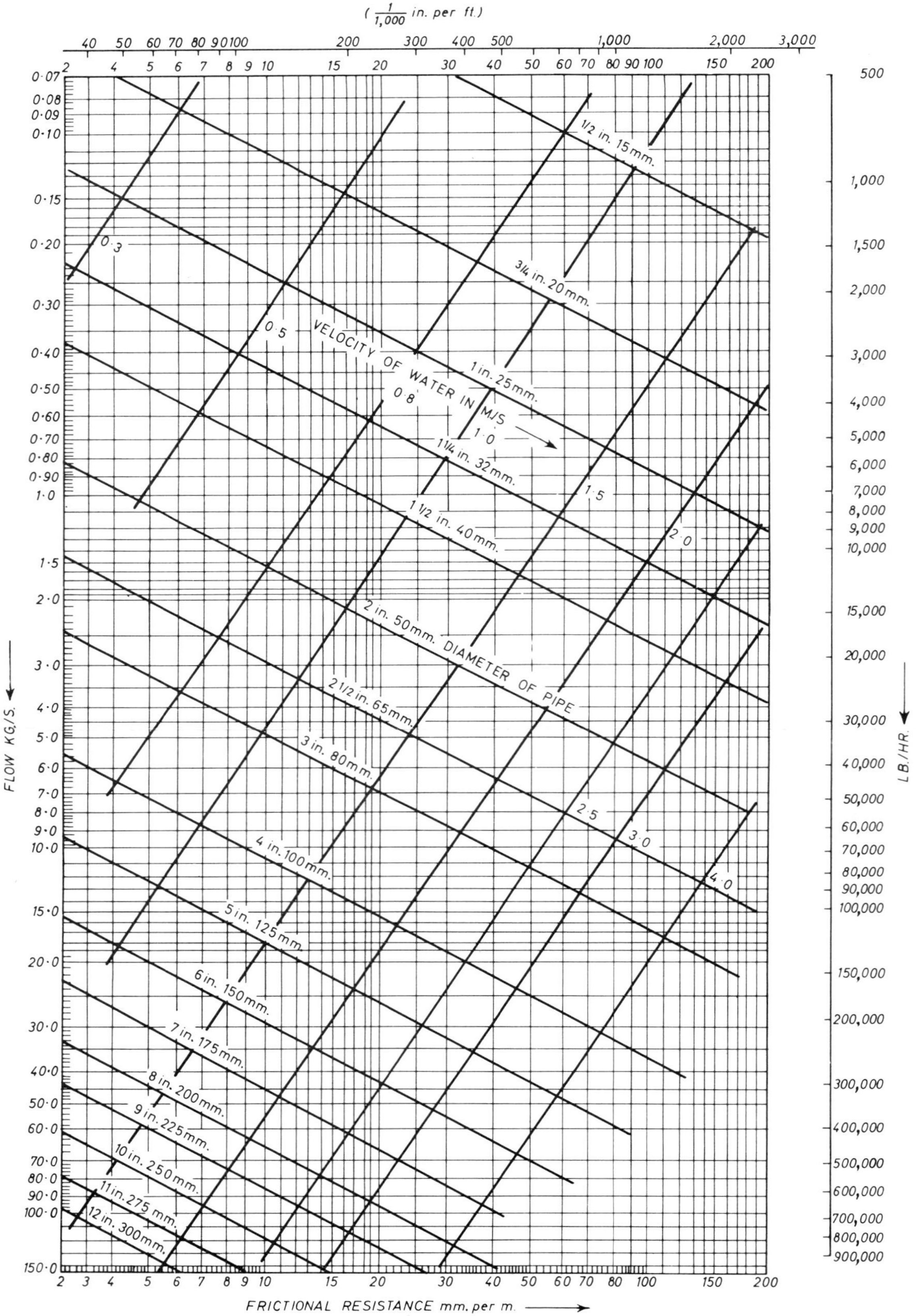
(1/1,000 in. per ft.)
FLOW KG/S.
LB./HR.
FRICTIONAL RESISTANCE mm. per m.
VELOCITY OF WATER IN M/S
DIAMETER OF PIPE
1/2 in. 15 mm.
3/4 in. 20 mm.
1 in. 25 mm.
1 1/4 in. 32 mm.
1 1/2 in. 40 mm.
2 in. 50 mm.
2 1/2 in. 65 mm.
3 in. 80 mm.
4 in. 100 mm.
5 in. 125 mm.
6 in. 150 mm.
7 in. 175 mm.
8 in. 200 mm.
9 in. 225 mm.
10 in. 250 mm.
11 in. 275 mm.
12 in. 300 mm.
0·3
0·5
0·8
1·0
1·5
2·0
2·5
3·0
4·0

CHART 3. PIPE SIZES FOR HIGH PRESSURE STEAM HEATING

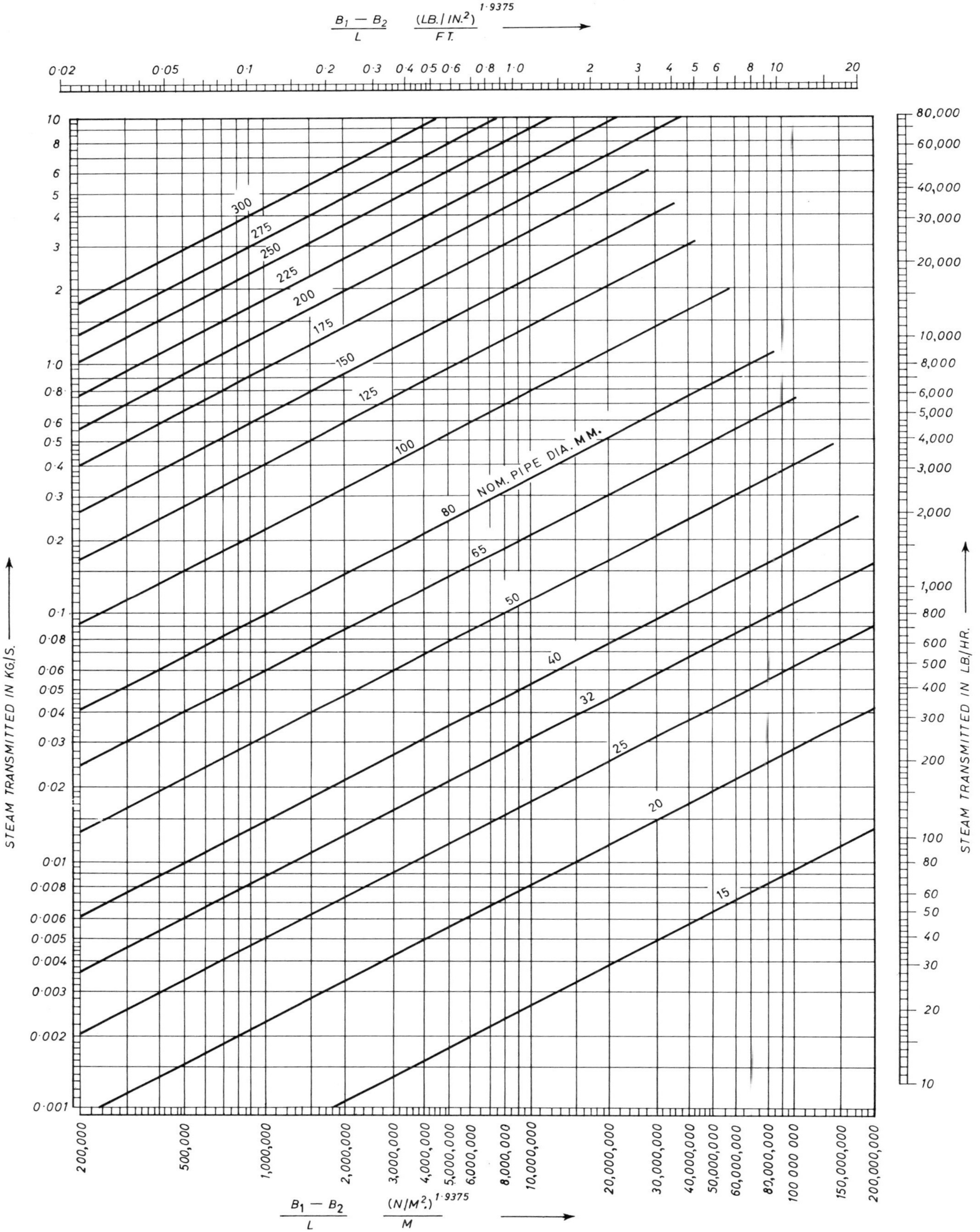

Explanation see page VIII 5

CHART 2. PIPE SIZES FOR LOW PRESSURE STEAM HEATING

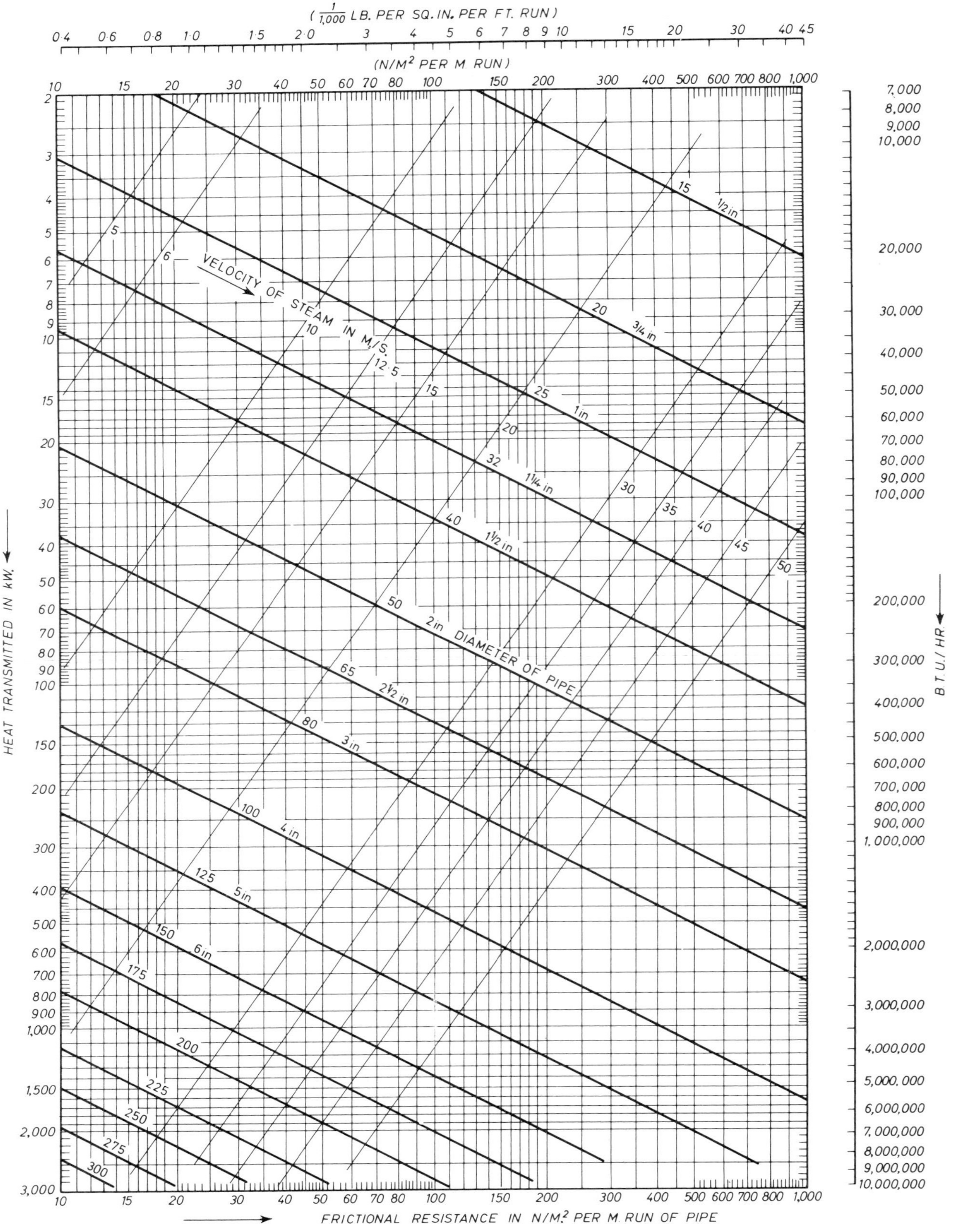

PRESSURE DROP (H_2) DUE TO FITTINGS FOR VARIOUS VALUES OF VELOCITY AND "F"

1. Pressure drop in N/m.^{2}:

Steam Velocity m./s.	*Value of "F"*						
	1	2	3	4	5	10	15
4	5·2	10·4	15·6	20·8	26·0	52	78
6	11·7	23·4	35·1	46·8	58·5	117	176
9	26·3	52·6	78·9	105·2	131·5	263	395
12	46·8	93·6	140·4	177·2	234·0	468	702
15	73·1	146·2	219·3	292·4	365·5	731	1,097
18	105·3	210·6	315·9	421·2	526·5	1,053	1,580
21	143·4	286·8	430·2	573·6	717·0	1,434	2,151
24	187·2	374·4	561·6	748·8	936·0	1,872	2,808
27	236·9	473·8	710·7	947·6	1,185	2,369	3,554
30	292·5	585·0	877·5	1,170	1,463	2,925	4,388

2. Pressure drop in lb./in.^{2}:

Steam Velocity ft./s.	*Value of "F"*						
	1	2	3	4	5	10	15
14	0·000874	0·00175	0·00262	0·00350	0·00437	0·00874	0·0131
20	0·00178	0·00357	0·00535	0·00714	0·00892	0·0178	0·0268
30	0·00401	0·00803	0·0120	0·0160	0·0201	0·0401	0·0602
40	0·00714	0·0143	0·0214	0·0286	0·0357	0·07154	0·107
50	0·0112	0·0223	0·0335	0·0446	0·0558	0·112	0·167
60	0·0161	0·0321	0·0482	0·0642	0·0803	0·161	0·241
70	0·0219	0·0437	0·0696	0·0874	0·109	0·219	0·328
80	0·0285	0·0571	0·0856	0·1142	0·143	0·285	0·428
90	0·0361	0·0723	0·108	0·1446	0·181	0·361	0·542
100	0·0446	0·0892	0·134	0·1784	0·223	0·446	0·669

TABLES FOR PIPE SIZING (see Chart 3)

Table for $B = p^{1\cdot9375}$

p	B	*p*	B	*p*	B	*p*	B	*p*	B
14	167	29	680	44	1,540	68	3,550	98	7220
15	190	30	720	45	1,610	70	3,750	100	7,500
16	215	31	775	46	1,680	72	3,960	105	8,250
17	240	32	825	47	1,750	74	4,180	110	9,020
18	270	33	875	48	1,820	76	4,400	115	9,830
19	300	34	925	49	1,890	78	4,630	120	10,680
20	330	35	980	50	1,960	80	4,870	125	11,500
21	365	36	1,035	52	2,120	82	5,110	130	12,470
22	400	37	1,090	54	2,280	84	5,350	135	13,420
23	435	38	1,150	56	2,440	86	5,590	140	14,390
24	470	39	1,210	58	2,610	88	5,850	145	15,400
25	510	40	1,270	60	2,790	90	6,110	150	16,450
26	550	41	1,330	62	2,970	92	6,380		
27	590	42	1,400	64	3,150	94	6,660		
28	635	43	1,470	66	3,350	96	6,940		

DETERMINATION OF STEAM MAINS

The Available Pressure Drop is given by the initial pressure P_1 (Boiler pressure or pressure on header) and the required final pressure P_2.

The total pressure drop H is made up by:

1. Pressure drop due to friction H_1
2. Pressure drop due to fittings H_2

$$H = H_1 + H_2$$

The formula $p = \frac{P_1 - P_2}{L}$ expressing the pressure drop due to friction for low pressure steam pipes is to be superseded for high pressure steam pipes by formula:

$$p = \frac{B_1 - B_2}{L}$$

L = length of pipe line

$B_1 = P_1^{1 \cdot 9375}$ } Auxiliary values

$B_2 = P_2^{1 \cdot 9375}$

The pressure drop due to fittings is:

$$H = \Sigma F \frac{v^2 \rho}{2g}$$

F = coefficient of resistance

v = velocity of steam

ρ = density of steam

The pressure drop due to fittings (resistance), can also be expressed as an equivalent length of straight pipe, see Table VIII. 7.

The pipe sizing is to be carried out by using the auxiliary value B_1 and B_2 and Chart No. 3.

Pressure Drop generally kept about 2 per cent of the initial pressure per 100 ft. of pipe or 6 per cent. per 100 metres.

Advisable Velocity of Steam:

For exhaust steam	20 to 30 m./s. 70 to 100 ft./s.
For saturated steam	30 to 40 m./s. 100 to 130 ft./s.
For superheated steam	40 to 60 m./s. 130 to 200 ft./s.

RESISTANCE OF VALVES AND FITTINGS TO FLOW OF STEAM

Expressed as an Equivalent Length of Straight Pipe

Nom. Bore of Pipe		Bends of Standard Radius				Barrel of Tee				Branch of Tee		Valves						Lyre Expansion Bends	
		90°		45°		Plain		Reduced 25%				Through		Angle		Globe			
in.	mm.	ft.	m.	ft.	m.	ft.	m.	ft.	m.	ft.	m.	ft.	m.	ft.	m.	ft.	m.	ft.	m.
1	25	0·5	0·15	0·4	0·12	0·5	0·15	0·7	0·21	2·2	0·67	0·4	0·12	1·5	0·46	3·3	1·0	2·2	0·67
1¼	32	0·7	0·21	0·5	0·15	0·7	0·21	0·9	0·27	2·9	0·89	0·5	0·15	2·0	0·61	4·3	1·3	2·9	0·88
1½	40	0·9	0·27	0·7	0·21	0·9	0·27	1·1	0·33	3·6	1·1	0·7	0·21	2·4	0·73	5·4	1·6	3·6	1·1
2	50	1·3	0·40	1·0	0·30	1·3	0·40	1·6	0·49	5·1	1·6	1·3	0·40	3·4	1·0	7·6	2·3	5·1	1·6
2½	65	1·6	0·49	1·2	0·37	1·6	0·49	2·1	0·64	6·6	2·0	1·6	0·49	4·5	1·4	10·0	3·0	6·6	2·0
3	80	2·1	0·64	1·6	0·49	2·1	0·64	2·6	0·80	8·3	2·5	2·1	0·64	5·6	1·7	12·0	3·7	8·3	2·5
4	100	2·9	0·88	2·2	0·67	2·9	0·88	3·7	1·1	12·0	3·7	2·2	0·67	7·9	2·4	18·0	5·5	12·0	3·7
5	125	3·8	1·2	2·9	0·88	3·8	1·2	4·8	1·5	15·0	4·6	2·9	0·89	10·0	3·0	23·0	7·0	15·0	4·6
6	150	4·7	1·4	3·6	1·1	4·7	1·4	6·0	1·8	19·0	5·8	3·6	1·1	13·0	4·0	29·0	8·8	19·0	5·8
7	175	5·7	1·7	4·3	1·3	5·7	1·7	7·2	2·2	23·0	7·0	4·3	1·3	15·0	4·6	34·0	10	23·0	7·0
8	200	6·7	2·0	5·0	1·5	7·6	2·0	8·5	2·6	27·0	8·2	5·0	1·5	18·0	5·5	40·0	12	27·0	8·2
9	225	7·7	2·3	5·8	1·8	7·7	2·3	9·8	3·0	31·0	10	5·8	1·7	21·0	6·4	46·0	14	31·0	9·5
10	250	8·7	2·7	6·6	2·0	8·7	2·7	11·0	3·4	35·0	11	6·6	1·8	24·0	7·3	53·0	16	35·0	11

SUCTION LIFT OF BOILER FEED PUMPS FOR VARIOUS WATER TEMPERATURES

Temp. of Feed Water °F.	*Max. Suction Lift Feet*	*Minimum Pressure Head Feet*
130	10	
150	2	
170	7	
175	0	0
190		5
200		10
210		15
212		17

Temp. of Feed Water °C.	*Max. Suction Lift Metres*	*Minimum Pressure Head Metres*
55	3	
65	2	
77	0·6	
80	0	0
87·5		1·5
95		3·5
99		4·5
100		5·0

QUANTITIES OF FLASH STEAM

Condensate		*Percentage of Condensate Flashed off at Reduction of Pressure to kN/m.² absolute*					
Absolute Pressure kN/m.²	*Temperature °C.*	*400*	*260*	*170*	*101·33*	*65*	*35*
1500	198·3	11·3	14·0	16·4	18·9	20·4	23·2
1150	186·0	8·7	11·5	13·9	16·5	18·4	20·9
800	170·4	5·5	8·2	10·8	13·4	15·4	17·9
650	162·0	3·7	6·5	9·1	11·8	13·7	16·3
500	151·8	1·6	4·6	7·1	9·8	11·8	14·4
400	143·6	—	3·0	5·5	8·3	10·3	12·9
260	128·7	—	—	2·6	5·4	7·5	10·2
170	115·2	—	—	—	2·8	5·0	7·7
101·33	100	—	—	—	—	2·2	4·9

Condensate		*Percentage of Condensate Flashed off at Reduction of Pressure to lb./in.² gauge or in. Hg. vacuum*					
Gauge Pressure lb./in.²	*Temperature °F.*	*40*	*20*	*10*	*0*	*10 in.*	*20 in.*
200	388	11·5	14·3	16·2	18·8	20·5	23·2
150	366	9·0	11·8	13·0	16·4	18·2	20·9
100	338	5·8	8·6	10·6	13·3	15·1	17·9
80	324	4·2	7·1	9·1	11·9	13·7	16·5
60	308	2·3	5·2	7·3	10·0	11·8	14·7
40	287	—	3·0	5·0	7·8	9·7	12·6
20	259	—	—	2·1	5·0	6·8	9·8
10	240	—	—	—	2·9	4·8	7·8
0	212	—	—	—	—	1·9	5·0

SCHEME OF FLASH STEAM RECOVERY

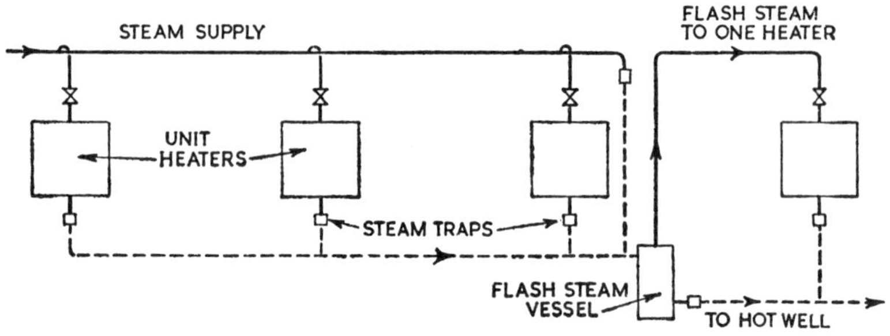

In vacuum steam heating systems, a partial vacuum is maintained in the return line by means of a vacuum pump. The vacuum maintained is approx. 3 to 10 in. mercury = approx. 75 to 250 mm. mercury.

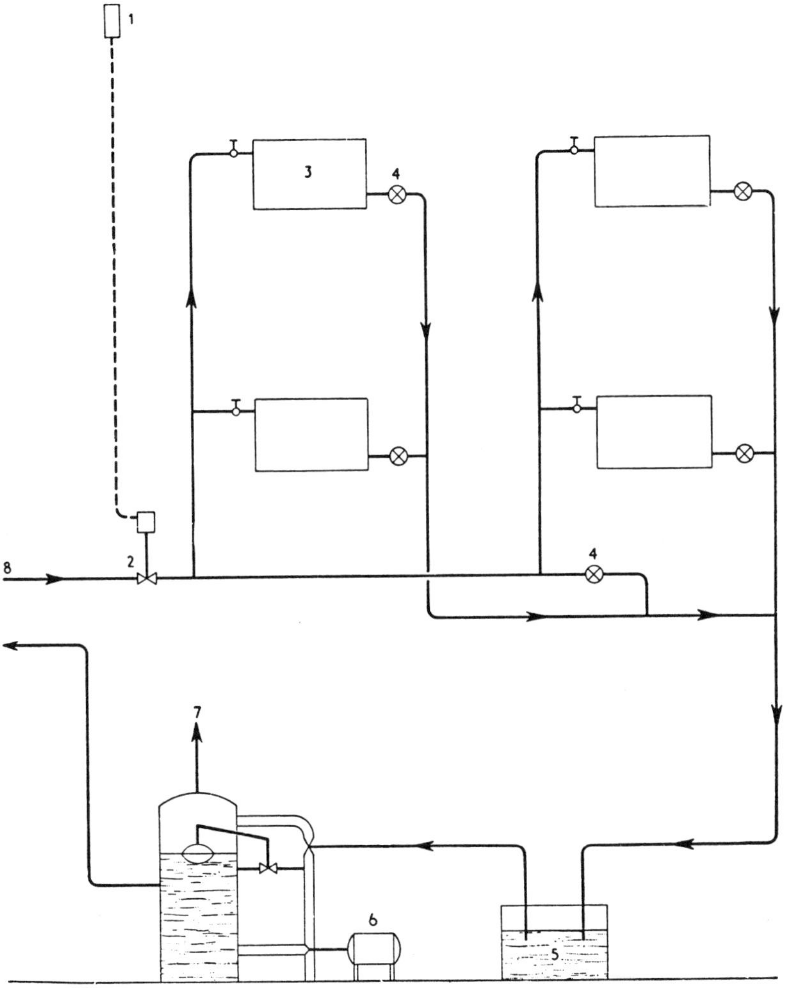

1 OUTSIDE THERMOSTAT
2 CONTROL VALVE
3 RADIATORS
4 STEAM TRAPS
5 CONDENSE RECEIVER
6 VACUUM PUMP
7 VENT
8 STEAM SUPPLY

VACUUM DIFFERENTIAL HEATING SYSTEM

Section IX

DOMESTIC HOT WATER SUPPLY AND GAS SUPPLY

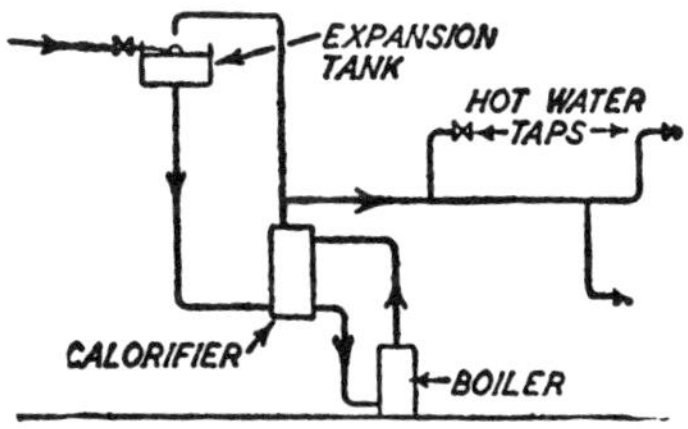

Direct Hot Water Service System

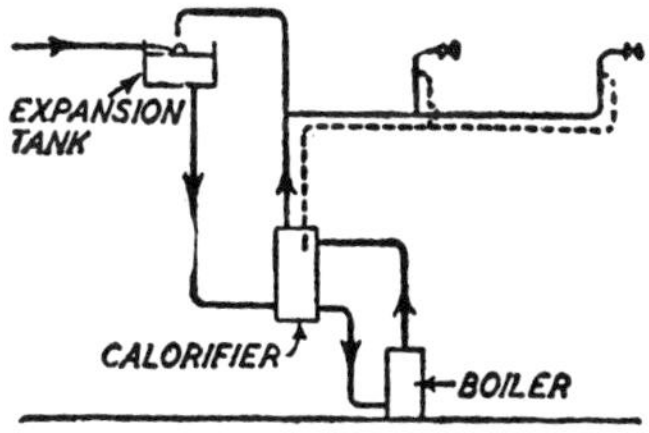

Direct Hot Water Service System with Secondary Circulation.

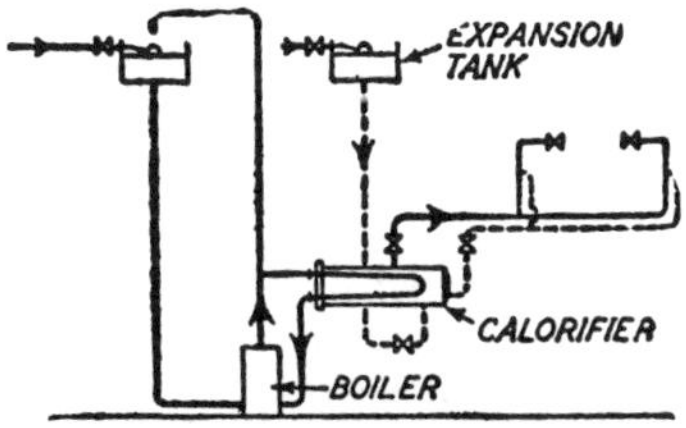

Indirect Hot Water Service System with Secondary Circulation.

Classification of Hot Water Supply schemes:

- Direct system.
- Indirect system.

- Closed system.
- Open system.

Design of Hot Water Systems:

1. Determination of demand of hot water, quantity and temperature.
2. Design, type, capacity and output of hot water calorifier.
3. Design, type and size of boiler.
4. Design, arrangement of boiler, calorifier, automatic control and pipe scheme.
5. Determination of primary and secondary mains.

Requisite Output of Calorifier:

$$H = \frac{Q\,(t_2 - t_1)}{n}.$$

H = Output of H.W. calorifier (B.t.u./hr.)
Q = Required quantity of hot water (gal.).
n = Time for warming up (hr.)
t_2 = Temperature of hot water (°F.)
t_1 = Temperature of cold water (°F.)
7·2 m.³/s.

$$H = \frac{4{\cdot}2\,Q\,(t_2 - t_1)}{n}.$$

H in kW.
Q in litres.
n in seconds.
t_2 °C.
t_1 °C.

Heating Surface of Indirect Calorifier:

$$A_c = \frac{H}{k\left(t_m - \frac{t_e + t_1}{2}\right)}$$

A_c = Heating surface of coil (sq. ft.).
t_m = Mean temperature of heating medium °F.
Logarithmic mean temperature difference, see IV. 10, 14.
k = Coefficient of heat transmission for heating coil (B.t.u. per sq. ft. per hr.). per °F.
See IV. 15 Heat Transmission Coefficients for Metals.

Heating Surface of Boiler:

$$A = \frac{1{\cdot}1H}{k_b}$$

A = Heating surface of boiler (sq. ft.).
k_b = Rating of boiler in B.t.u. per sq. ft. per hr.
= 3,800 to 4,400 B.t.u./sq. ft. hr. for cast-iron indirect boilers.
= 10,000 B.t.u./sq. ft. hr. for cast-iron direct boilers.

TEMPERATURE DROP IN PIPES IN °F. PER FOOT OF BARE PIPE

Weight of water, lb. per hour	*½ in.*	*¾ in.*	*1 in.*	*1¼ in.*	*1½ in.*	*2 in.*	*2½ in.*	*3 in.*	*4 in.*
100	0·45	0·6	0·65	0·8	0·9	1·1	1·25	1·5	1·9
120	0·38	0·5	0·54	0·62	0·75	0·92	1·04	1·21	1·58
140	0·32	0·43	0·46	0·57	0·64	0·79	0·89	1·07	1·36
160	0·28	0·37	0·41	0·5	0·56	0·69	0·78	0·94	1·19
180	0·25	0·33	0·36	0·44	0·5	0·61	0·69	0·83	1·06
200	0·22	0·3	0·33	0·4	0·45	0·55	0·63	0·75	0·95
250	0·18	0·24	0·26	0·32	0·36	0·44	0·5	0·6	0·76
300	0·15	0·2	0·22	0·27	0·3	0·37	0·42	0·5	0·63
350	0·13	0·17	0·19	0·23	0·26	0·31	0·36	0·43	0·54
400	0·11	0·15	0·165	0·2	0·225	0·275	0·315	0·375	0·475
450	0·1	0·133	0·14	0·177	0·2	0·244	0·277	0·33	0·422
500	0·09	0·12	0·13	0·16	0·18	0·22	0·25	0·3	0·38
600	0·075	0·1	0·11	0·135	0·15	0·185	0·21	0·25	0·365
700	0·065	0·085	0·095	0·125	0·13	0·155	0·18	0·215	0·27
800	0·055	0·075	0·083	0·1	0·113	0·138	0·158	0·189	0·238
900	0·05	0·066	0·07	0·089	0·1	0·122	0·139	0·165	0·211
1000	0·045	0·060	0·065	0·08	0·09	0·11	0·125	0·15	0·19

TEMPERATURE DROP IN PIPES IN °C. PER METRE OF BARE PIPE

Weight of Water kg./s.	½ *in.* 15 *mm.*	¾ *in.* 20 *mm.*	1 *in.* 25 *mm.*	1¼ *in.* 32 *mm.*	1½ *in.* 40 *mm.*	1½ *in.* 50 *mm.*	2½ *in.* 65 *mm.*	3 *in.* 80 *mm.*	4 *in.* 100 *mm.*
0·010	1·03	1·37	1·49	1·83	2·06	2·52	2·88	3·44	4·35
0·012	0·86	1·14	1·24	1·54	1·72	2·10	2·40	2·87	3·63
0·014	0·74	0·98	1·06	1·31	1·45	1·80	2·06	2·43	3·11
0·016	0·65	0·86	0·93	1·14	1·29	1·57	1·80	2·14	2·72
0·018	0·57	0·76	0·83	1·02	1·14	1·40	1·60	1·91	2·42
0·020	0·52	0·69	0·74	0·92	1·03	1·26	1·44	1·77	2·68
0·025	0·41	0·55	0·60	0·73	0·82	1·01	1·16	1·37	1·78
0·030	0·34	0·45	0·50	0·61	0·69	0·84	0·96	1·15	1·45
0·035	0·29	0·39	0·43	0·52	0·59	0·72	0·82	0·98	1·24
0·040	0·26	0·34	0·39	0·46	0·52	0·63	0·72	0·86	1·09
0·045	0·23	0·30	0·33	0·41	0·46	0·56	0·64	0·76	0·97
0·050	0·21	0·27	0·30	0·37	0·41	0·50	0·57	0·69	0·87
0·060	0·17	0·23	0·25	0·31	0·34	0·42	0·48	0·57	0·76
0·070	0·15	0·20	0·21	0·26	0·29	0·36	0·41	0·49	0·62
0·080	0·13	0·17	0·19	0·23	0·26	0·32	0·36	0·47	0·54
0·090	0·11	0·15	0·17	0·20	0·23	0·27	0·32	0·43	0·48
0·100	0·10	0·14	0·15	0·18	0·21	0·25	0·29	0·34	0·44

HOT WATER CONSUMPTION PER FITTING (at 65°C.)

Fitting	*Consumption*	
	litres/hour	*gal./hour*
Basin (Private)	14	3
Basin (Public)	45	10
Shower	180	40

Fitting	*Consumption*	
	litres/hour	*gal./hour*
Sink	45—90	10—20
Bath	90—180	20—40

HOT WATER CONSUMPTION AND STORAGE PER OCCUPANT (at 65°C.)

Type of Building	*Consumption per Occupant*		*Peak Consumption per Occupant*		*Storage per Occupant*	
	Gal./Day	*l./Day*	*Gal./Hour*	*l./Hour*	*Gal.*	*litres*
Schools (Boarding)	20	90	4	18	5	22
Blocks of Flats	20—35	90—160	10	45	7	30
Hotels	20—35	90—160	10	45	7	30
Factories, no process	5—10	22—45	2	9	1	5
Blocks of Offices	5	22	2	9	1	5
Hospitals, Infection	50	220	10	45	10	45
Hospitals, Sick	35	160	7	30	6	27
Hospitals, Mental	25	110	5	22	6	27
Hospitals, Maternity	40	180	8	35	7	30

CONTENTS OF DIFFERENT FITTINGS

Fitting	*Contents*	
	Gal.	*Litres*
Lavatory Basin, Normal	1	5
Lavatory Basin, Full	2	9
Sink, Normal	5	22
Sink, Full	10	45
Bath	30—40	135—180

Fitting	*Consumption*	
	gal./min.	*l./s.*
Shower Bath, "Winns"	1—1½	0·07—0·12
Shower Bath 6—7 Rose	7—8	0·5—0·6

SIZES OF WATER METER PITS:

Minimum 2 ft. 0 in. × 2 ft. 0 in. × 3 ft. 3 in. deep
or 0·6 m. × 0·6 m. × 1·0 m. deep.
For 3 in. main — 3 ft. 0 in. × 2 ft. 0 in. × 3 ft. 3 in. deep
or 0·9 m. × 0·6 m. × 1·0 m. deep.

PIPE DIAMETERS FOR DOMESTIC COLD AND HOT WATER SERVICE

Nominal Pipe Diameter		*Maximum Number of Taps*		
		Flow Pipes		*Return Pipes*
in.	*mm.*	*Head up to 70 ft. or 20 m.*	*Head over 70 ft. or 20 m.*	
$\frac{1}{2}$	15	1	1 to 2	1 to 8
$\frac{3}{4}$	20	2 to 4	3 to 9	9 to 29
1	25	5 to 8	10 to 19	30 to 66
$1\frac{1}{4}$	32	9 to 24	20 to 49	67 to 169
$1\frac{1}{2}$	40	25 to 49	50 to 79	170 to 350
2	50	50 to 99	80 to 153	—
$2\frac{1}{2}$	65	100 to 200	154 to 300	—

When using the above Table one bath is to be taken into account as two taps. Sinks, Lavatory Basins, Showers, Slop Sinks = one tap.

HOT WATER SUPPLY BY ELECTRICITY.

(*Immersion Heater with Thermostat.*)
One electrical unit (kWh) = 1,000 watt hours = 3,412 B.t.u.

Electric Load

$$E = \frac{10\, Q_1\, \triangle t_1}{3{\cdot}412\, n} \qquad \text{or } E = \frac{4{\cdot}2\, Q_2\, \triangle t_2}{N}.$$

where E = Electric load in kW.
Q_1 = Quantity of water heated in gallons.
Q_2 = ,, ,, ,, ,, ,, kilogrammes.
$\triangle t_1$ = required temperature rise in °F.
$\triangle t_2$ = ,, ,, ,, ,, °C.
n = number of hours for warming up (usually 3 to 4 hours).
N = ,, ,, seconds ,, ,, ,,

Recommended Loads

Capacity of Hot Water Supply Tank		*Load of Immersion Heater*
gal.	*litres*	*kW.*
20	90	2
30	135	$2\frac{1}{2}$
40	180	3

GAS CONSUMPTION OF EQUIPMENT (NATURAL GAS)

	ft.³/h.	*m.³/s.*	*l./s.*
10 gal. Boiling Pan	45	350×10^{-6}	0·35
20 gal. Boiling Pan	60	475×10^{-6}	0·48
30 gal Boiling Pan	75	600×10^{-6}	0·60
40 gal. Boiling Pan	90	700×10^{-6}	0·70
4 ft. Hot Cupboard	48	375×10^{-6}	0·38
6 ft. Hot Cupboard	54	425×10^{-6}	0·43
Steaming Oven	40 to 50	300 to 400×10^{-6}	0·30 to 0·40
Double Steaming Oven ...	100	800×10^{-6}	0·80
2-tier Roasting Oven	50	400×10^{-6}	0·40
Double Oven Range	400	3200×10^{-6}	1·6
Roasting Oven...	30	240×10^{-6}	0·24
Gas Cooker	75	600×10^{-6}	0·30
Hot Cupboard...	17	140×10^{-6}	0·14
Drying Cupboard	5	40×10^{-6}	0·04
Gas Iron Heater	5	40×10^{-6}	0·04
Washing Machine	20	150×10^{-6}	0·15
Wash Boiler	30 to 50	230 to 400×10^{-6}	0·23 to 0·40
Bunsen Burner	3	20×10^{-6}	0·02
Bunsen Burner, full on ...	10	80×10^{-6}	0·08
Glue Kettle	10	80×10^{-6}	0·08
Forge	15	115×10^{-6}	0·12
Brazing Hearth	30	230×10^{-6}	0·23

Flow of Gas in Steel Tubes:

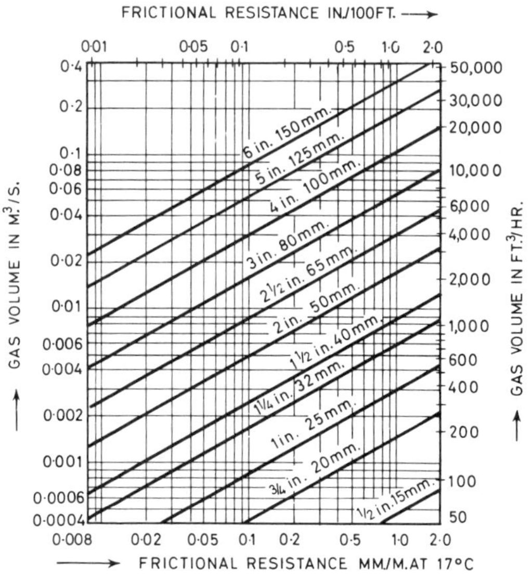

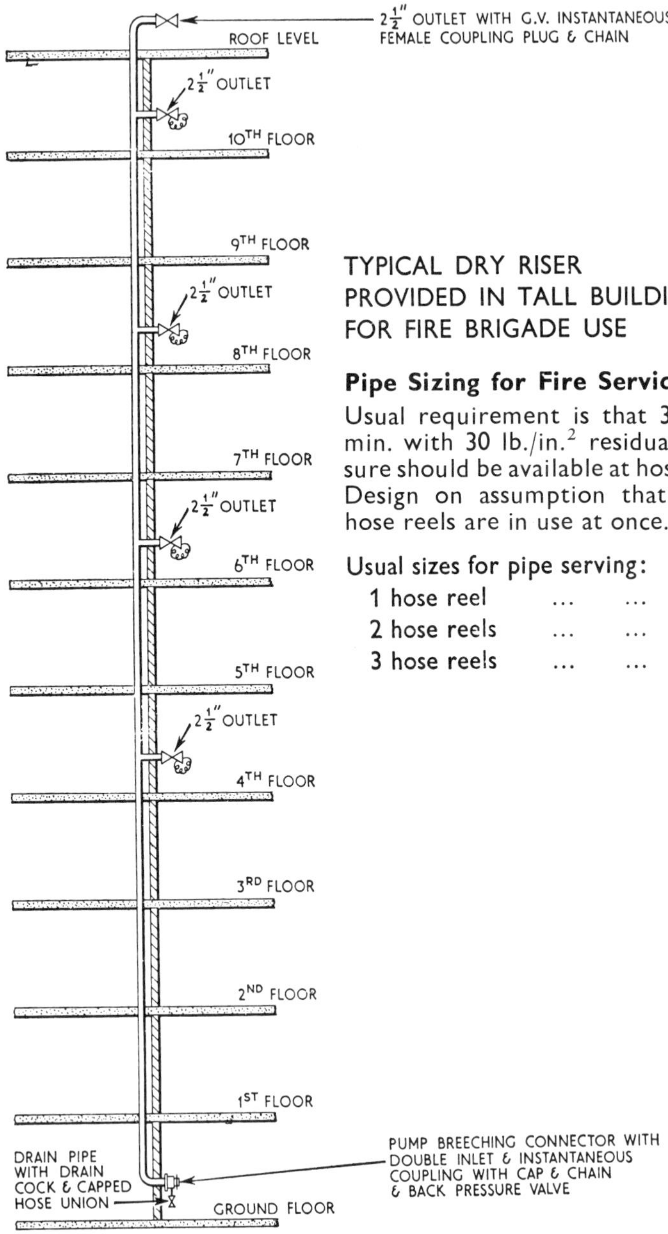

TYPICAL DRY RISER PROVIDED IN TALL BUILDINGS FOR FIRE BRIGADE USE

Pipe Sizing for Fire Service:

Usual requirement is that 30 gal./min. with 30 lb./in.^{2} residual pressure should be available at hose reel. Design on assumption that three hose reels are in use at once.

Usual sizes for pipe serving:

1 hose reel	1¼ in.
2 hose reels	1½ in.
3 hose reels	2 in.

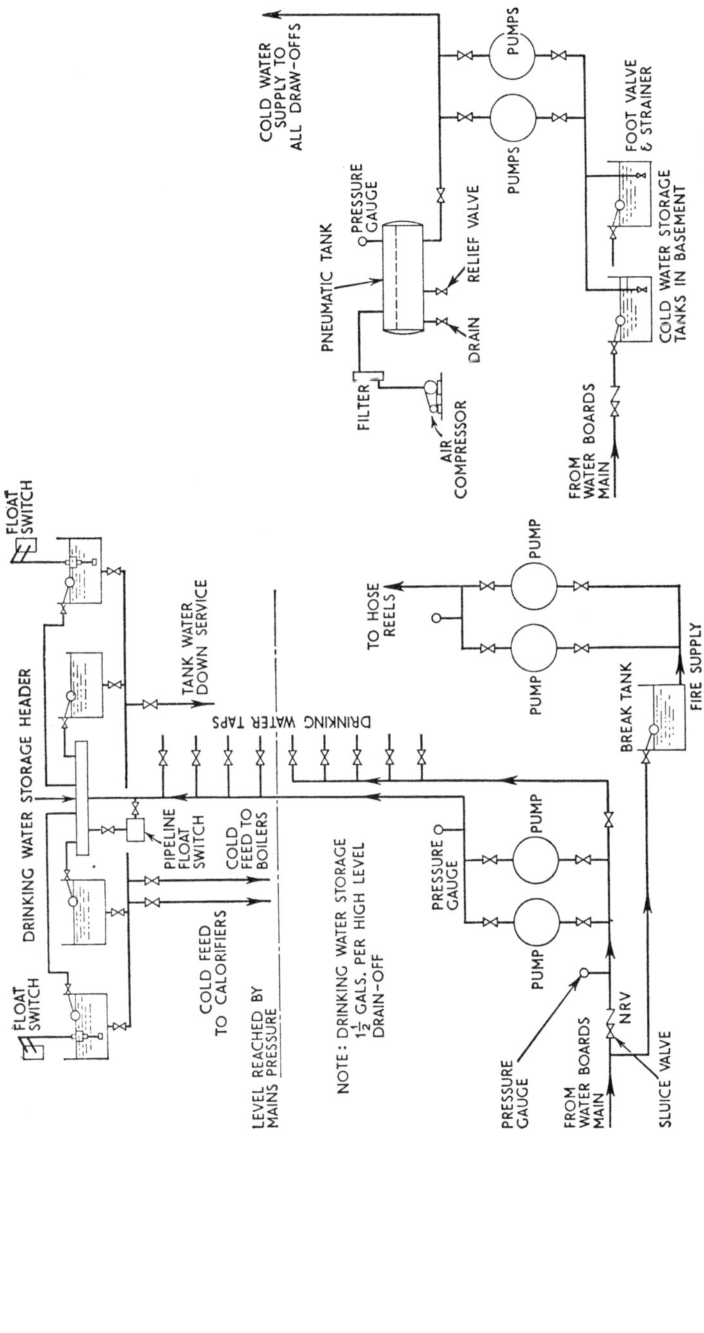

ROOF STORAGE

BASEMENT STORAGE

COLD WATER STORAGE SYSTEMS FOR TALL BUILDINGS

Section X

VENTILATION AND AIR CONDITIONING

SYSTEMS OF VENTILATION

General Types: (a) Central Conditioning System.
(b) Unit Conditioner.

(a) *Split System of Heating and Ventilating*—Heat-losses from building are supplied by direct radiators and ventilation or air conditioning delivers air at room temperature.

(b) *Combination System*—Entire operation of heating and ventilating is handled by central system.

Arrangement of Heating Unit and Fan:

(a) Draw-through, heating unit on suction side.

(b) Blow-through, heating unit on discharge.

Schemes of Air Distribution:

Fig. 1. Upward flow system.
,, 2. Downward flow system.
,, 3. High supply and return openings.
,, 4. Low supply and return openings.
,, 5. Ejector system.

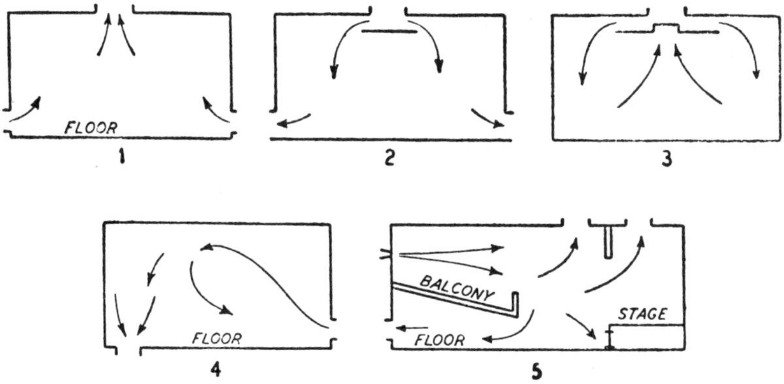

Diagrammatic Views (in Elevation) showing how various systems of Air Distribution are applied in buildings.

Dust Load for Air Filters:

	grains/1,000 cu. ft.	mg./m.³
Rural and suburban districts ...	0·2–0·4	0·45–1·00
Metropolitan districts	0·4–0·8	1·00–1·8
Industrial districts	0·8–1·5	1·8 –3·5

Classification of Air Filters:

(a) *Air Washers*

Overall length, approximately 20 m. (6 to 7 ft.).
Air velocity in washer, 2·5 m/s. (500 ft./min.).
Water quantity required 0·5–0·8 l. per $m.^3$ air (3 to 5 gal. per 1,000 $ft.^3$).
Water pressure for spray nozzles 140–170 $kN/m.^2$ (20–25 $lb./in.^2$).
Water pressure for flooding nozzles, 35–70 $kN/m.^2$ (5–10 $lb./in.^2$).

(b) *Viscous Air Filters* — Plates coated with a viscous fluid.

1. Unit type:
   Cartridges about 500 × 500 mm. (20 × 20 in.).
   Air Volume 0·3 to 0·8 $m.^3/s.$ (600 –1,600 $ft.^3/min.$).
2. Automatic Type:
   Entrance velocity, 2·5 m/s. (500 ft./min.).
   Resistance 4·5 to 10·0 mm. W.G. ($\frac{3}{16}$ to $\frac{3}{8}$ in. W.G.).

(c) *Dry Filters*

Felt, cloth, cellulose, glass, silk, etc., without adhesive liquid.
Air velocity 0·05 to 0·25 m/s. (10 to 50 ft./min.).
Resistance of clean filter approximately 2·5 mm. W.G. (0·1 in. W.G.).

Testing and Rating of Air Cleaning Devices:
(Standard Code, A.S.H.R.A.E.)

1. Air capacity.
2. Resistance.
3. Dust precipitation.
4. Reconditioning power, energy necessary for automatic mechanism.
5. Dust-holding capacity, for non-automatic devices.

Design of Plenum Heating and Air Conditioning Systems:

1. Calculate heating load or cooling load:
   (*a*) Sensible heat.
   (*b*) Latent heat.
2. Calculate, or assume, temperature of air leaving grilles.
3. Calculate weight of air.
4. Estimate temperature loss in duct system.
5. Calculate output of heaters and washers, and select type and size.
6. Calculate total heat required and select type and size of boiler.
7. Design duct system, calculate duct size.

Air Temperature at Supply Grilles:

		°C.	°F.
For heating	...	38–50	100–120
For cooling, inlets near to occupied zone	Below room temp.	6–8	10–15
For cooling, high velocity jets, diffusing nozzles	Below room temp.	17	30

Quantity of Air required:

When temperature is limiting factor:

$$W = \frac{H_s}{C\,(t - t_i)} = Vd.$$

$$V = \frac{H}{d.\,C.\,(t - t_i)}$$

			S.I. units	*Imperial units*
W	=	Mass of air	kg./s.	lb./hr.
H_s	=	Sensible heat loss or gain	kW.	B.t.u./hr.
C	=	Specific heat capacity of air at constant pressure	= 1·01 kJ/kg. deg. C.	= 0·24 B.t.u./lb. deg. F.
t	=	Outlet temperature at grilles	°C.	°F.
t_i	=	Room temperature	°C.	°F.
V	=	Volume of air	$m^3/s.$	$ft.^3/hr.$
d	=	Density of air	= 1·21 $kg./m.^3$ at 20°C.	= 0·075 $lb./ft.^3$ at 70°F.

When moisture content is limiting factor:

$$W = \frac{M}{(w_2 - w_1)} = Vd.$$

$$V = \frac{M}{d.\,(w_2 - w_1)}$$

M	=	Moisture to be absorbed	gramme/s.	grains/hr.
w_1	=	Specific humidity of air supply	gramme/kg.	grains/lb.
w_2	=	Desired specific humidity in room	gramme/kg.	grains/lb.

Heat required for Ventilation:

$$H = W.C.\,(t_2 - t_1) = V.d.c.\,(t_2 - t_1).$$

t_1	=	Initial temperature	°C.	°F.
t_2	=	Final temperature	°C.	°F.

Requisite Temperature of Incoming Air:

$$t = t_i + \frac{H_S}{d.c.V.}$$

Temperature Drop in Ducts:

		S.I. units	*Imperial units*
Let: W	= Weight of air flowing.	kg./s.	lb./hr.
A	= Area of duct walls	m.^{2}	ft.^{2}
s	= Specific heat capacity of air	1×10^3 J/kg. deg. C.	0·24 B.t.u./lb. deg. F.
k	= Heat loss coefficient of duct walls	= 5·68 W/m.^{2} deg. C. for sheet metal ducts.	= 1 B.t.u./ft.^{2} hr. deg. F. for sheet metal ducts.
		= 2·3 W/m.^{2} deg. C. for insulated ducts.	= 0·24 B.t.u./ft.^{2} hr. deg. F. for insulated ducts.
t_2	= Initial temperature in duct	°C.	°F.
t_1	= Final temperature in duct	°C.	°F.
t_R	= Air temperature outside duct	°C.	°F.

$$W.s. (t_2 - t_1) = A.k. \left(\frac{t_1 + t_2}{2} - t_R\right).$$

$$(t_2 - t_1) = A.k. \left(\frac{t_1 + t_2}{2} - t_R\right) \div W.s.$$

$$\text{or } t_2 = \frac{(2\,W\,s + A.k.)\,t_1 - 2\,A.k.\,t_R}{2\,W.s. - A.k.}$$

For high temperature falls the logarithmic mean temperature is to be used.

$$W.s. (t_2 - t_1) = \frac{A.k. \,[(t_2 - t_R) - (t_1 - t_R)]}{\log_e \dfrac{t_2 - t_R}{t_1 - t_R}}$$

Allowance to be made for Altered Surface Conditions of Ducts:

New brass seamless drawn tubing, clean, smooth ...	1·00
New galvanized iron pipe ...	1·15
High pressure steam lines, flanged ...	1·30
High pressure water lines, flanged ...	1·30
Sheet iron ducts ...	1·50
Concrete, very smooth ...	4·00
New asphalted cast-iron pipe ...	6·00
Concrete, ordinary ...	7·00
New black wrought iron pipe ...	1·80

Desirable Temperatures and Humidities for Industrial Processing (A.S.H.V.E. Guide)

Industry.	*Process.*	*Temperature °F.*	*Relative Humidity %*
Textile ...	Cotton—Carding	75–80	50
	Spinning	60–80	60–70
	Weaving	68–75	70–80
	Rayon—Spinning	70	85
	Twisting	70	65
	Silk—Spinning	75–80	65–70
	Weaving	75–80	60–70
	Wool—Carding	75–80	65–70
	Spinning	75–80	55–60
	Weaving	75–80	50–55
Tobacco ...	Cigar and Cigarette making...	70–75	55–65
	Softening	90	85
	Stemming and Strigging ...	75–85	70
Paint ...	Drying of Oil Paints	60–90	25–50
	Brush and Spray Painting ...	60–80	25–50
Paper ...	Binding, Cutting, Drying, Folding, Gluing	60–80	25–50
	Storage of Paper	60–80	35–45
Hospitals...	Operating Theatre	75–80	40–65
Libraries...	Book Storage...	65–70	38–50
Printing ...	Binding	70	45
	Folding	77	65
	Press Room/General ...	75	60–78
Photographic	Development of Film ...	70–75	60
	Drying	75–80	50
	Printing	70	70
	Cutting	72	65
Fur ...	Storage of Furs	28–40	25–40
	Drying of Furs	110	—

Air Change for Rooms, Occupancy known:

Type of Building	*Air Changes per cu. ft. per person, per hour*
Hospitals—Ordinary	2,500
Surgical Cases	3,000
Contagious Diseases	6,000
Schools, Theatres, Prisons, Assembly Halls	1,800
Factories, Shops	2,000
Factories, Unhealthy Trades	3,500
London County Council Regulations	1,000 min.
Code, A.S.H.V.E.	600 min.

Air Change for Rooms, Occupancy unknown:

Type of Building	*Air Changes per hour*
Cinemas, Theatres	5–10
Assembly Rooms	5–10
Kitchens, large	10–20
Kitchens, small	20–40
Lavatories	5–10
Restaurants	5–10
Offices	3–8
Baths	5–8
Garages	5–6
Boiler Houses, Engine Rooms	4

Garage Ventilation:

Allow 6 air changes per hour.
Two-thirds total extracted at high level; one-third total extracted at low level.
Two fans should be provided one serving as a stand by.

Bathroom and W.C. Ventilation:

Allow 40 ft^3/min per room, or 6 air changes per hour, or 0·018 $m.^3$/s. per room.

P.V.C. ducting is often employed for quietness and durability. To provide a standby service two fans with an automatic change-over switch are installed.

Typical Schemes:

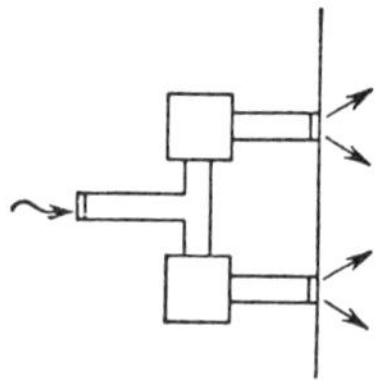

CENTRIFUGAL FANS WITH COMMON INLET & SEPARATE DISCHARGE

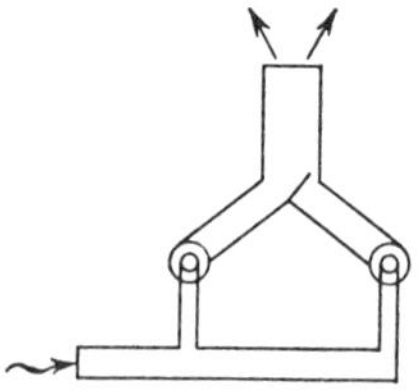

CENTRIFUGAL FANS WITH COMMON INLET & DISCHARGE

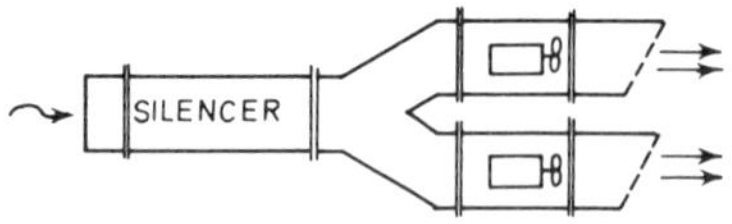

AXIAL FLOW FANS WITH SILENCER

THEORETICAL VELOCITY OF AIR (Due to Natural Draught)

$$V = 4{\cdot}48 \sqrt{\frac{h\,(t_c - t_o)}{273 + t_o}}$$

V = Theoretical Velocity, m./s.
h = Height of flue, m.
t_c = Temperature of warm air column, °C.
t_o = Temperature of outside air, °C.

Height of Flue m.	*Excess of Temperature in Flue above Outside Air °C. (for t_o = 2°C.)*								
	2	4	6	8	10	15	30	50	80
0·3	0·21	0·30	0·36	0·42	0·47	0·57	0·81	1·0	1·3
1·0	0·39	0·54	0·66	0·77	0·86	1·0	1·48	1·91	2·42
3·0	0·66	0·94	1·15	1·32	1·48	1·81	2·57	3·31	4·19
5·0	0·87	1·22	1·50	1·73	1·93	2·37	3·35	4·31	5·46
6·0	0·94	1·32	1·62	1·87	2·10	2·57	3·63	4·68	5·93
7·5	1·07	1·51	1·85	2·13	2·38	2·92	4·13	5·32	6·75
10	1·21	1·71	2·10	2·42	2·73	3·32	4·69	6·04	7·66
15	1·49	2·10	2·57	2·96	3·31	4·06	5·74	7·40	9·37
20	1·71	2·42	2·97	3·42	3·83	4·69	6·63	8·55	10·8
25	1·91	2·71	3·32	3·83	4·28	5·24	7·42	9·55	12·1
30	2·10	2·96	3·63	4·19	4·69	5·74	8·12	10·5	13·2
35	2·27	3·20	3·92	4·53	5·07	6·20	8·77	11·3	14·3
40	2·42	3·42	4·20	4·84	5·41	6·63	9·33	12·1	15·3
45	2·57	3·63	4·45	5·13	5·74	7·03	9·95	12·8	16·2
50	2·71	3·82	4·69	5·41	6·05	7·41	10·5	13·5	17·1

$$V = 8{\cdot}02 \sqrt{\frac{h\,(t_c - t_o)}{460 + t_o}}$$

V = in ft./s.
h = in ft.
t_c = in °F.
t_o = in °F.

Height of Flue ft.	*Excess of Temperature in Flue above Outside Air °F. (for ta = 35°F.)*								
	5	10	15	20	25	30	50	100	150
1	0·8	1·1	1·4	1·6	1·8	2·0	2·5	3·6	4·4
5	1·8	2·5	3·1	3·6	4·0	4·5	5·6	8·1	9·9
10	2·6	3·6	4·4	5·1	5·7	6·6	8·1	11·4	14·0
15	3·1	4·4	5·4	6·3	7·0	7·7	9·9	14·0	17·1
20	3·6	5·1	6·3	7·2	8·1	8·8	11·4	16·1	19·8
30	4·4	6·3	7·8	8·8	9·9	10·9	14·0	19·8	24·2
40	5·1	7·3	8·9	10·2	11·4	12·5	16·1	22·8	27·9
50	5·7	8·1	9·9	11·4	12·8	14·0	18·0	25·5	31·1
60	6·3	8·8	10·8	12·6	14·0	15·3	19·8	27·8	33·3
70	6·8	9·5	11·7	13·6	15·2	16·5	21·4	30·0	36·1
80	7·3	10·2	12·5	14·4	16·2	18·7	22·9	32·2	38·9
90	7·7	10·8	13·3	15·3	17·2	18·8	24·3	34·2	41·6
100	8·1	11·4	14·0	16·2	17·8	19·8	25·6	36·0	45·2
125	9·1	12·8	15·6	18·1	20·1	22·1	28·7	40·3	49·3
150	9·9	14·0	17·2	19·8	22·2	24·3	31·4	44·3	54·3

AIR VELOCITIES AND EQUIVALENT PRESSURES FOR STANDARD AIR

$$V = 4{\cdot}41 \sqrt{\frac{h}{P}} = 4{\cdot}01 \sqrt{h}.$$

V = Velocity of air m./s.
h = Velocity head mm. W.G.
p = Density of air = 1·2 kg./m.^{3}

V (m./s.)	h (mm.)	V (m./s.)	h (mm.)	V (m./s.)	h (mm.)
0·25	0·0038	5·0	1·56	17·5	19·1
0·5	0·0156	5·67	2·00	20·0	24·9
0·75	0·0350	6	2·24	22·5	31·5
1·0	0·0622	7	3·05	25·0	38·9
1·25	0·0972	8	3·98	27·5	47·1
1·5	0·140	9	5·04	28·3	50·0
2·0	0·249	10	6·22	30·0	56·0
2·5	0·389	11	7·53	32·5	62·5
3·0	0·560	12	8 96	34·7	75·0
3·3	0·764	13	10·5	35·0	76·0
4·0	0·995	14	12·2		
4·01	1·000	15	14·0		

$$V = 1096{\cdot}2 \sqrt{\frac{h}{P}} = 4{,}000 \sqrt{h}.$$

V = in ft./min.
h = in inches w.g.
p = 0·075 lb./ft.^{3}

V (ft./min.)	h (in.)	V (ft./min.)	h (in.)	V (ft./min.)	h (in.)
60	0·00023	1,200	0·0915	3,430	0·750
120	0·00092	1,250	0·1000	3,600	0·824
180	0·00206	1,500	0·1430	4,000	1·000
240	0·00366	1,770	0·2000	4,200	1·121
360	0·00824	1,800	0·2059	4,800	1·464
480	0·01464	2,000	0·2500	4,850	1·500
600	0·0229	2,100	0·2803	5,400	1·853
720	0·0329	2,400	0·366	5,600	2·000
840	0·0448	2,700	0·463	6,000	2·288
960	0·0586	2,800	0·500	6,260	2·500
1,080	0·0741	3,000	0·572	6,870	3·00

GENERAL FORMULA FOR AIR FLOW

$$Q = AV; \; A = \frac{Q}{V}; \; V = \frac{Q}{A}.$$

Q = Air volume m^3/s.
A = Area of duct $m.^2$
V = Air velocity m/s.

Total Loss of Head (H_t) = Loss of head due to friction in duct (H_1).
+ Loss of head due to fittings (H_2).
+ Loss of head in apparatus (Filters, Heaters) (H_3).

$$H_t = H_1 + H_2 + H_3.$$

1. Loss of Head due to Friction:

for Circular Ducts —

$$H_1 = f\frac{2V^2}{gD} \times L = h_1.L$$

H_1 = Loss of head due to friction mm. W.G.
h_1 = ditto per unit length mm./m.
L = Length of duct m.
V = Velocity of air m/s.
D = Diameter of circular duct m.
g = Gravity of earth = 9·81 $m/s.^2$
f = Friction factor, which is a function of Reynold's number.

for Rectangular Ducts —
The diameter of the above formula is to be replaced by the equivalent diameter (see Table X. 12).

$$D_E = 1{\cdot}265 \sqrt[6]{\frac{(ab)^3}{(a+b)}}$$

D_E = Equivalent diameter.
a, b = Sides of rectangular duct.

(For sizing Air Ducts, use Chart No. 4).

Addition to Friction Factor —
For smooth ducts of brickwork or plaster ... 100 per cent.
For rough ducts ... ... 200 per cent.

2. Loss of Head due to Fittings:

$$H_2 = \Sigma F \frac{V^2\rho}{2g}$$

H_2 = Total loss of head due to fittings.
V = Air velocity.
F = Coefficient of resistance.
ρ = Density of air.
(Values of F, see Table X. 10.)

for Standard Air —

$$H_2 = F\left(\frac{V}{4{\cdot}0}\right)^2.$$

H_2 in mm. W.G.
V in m/s.

or

$$H_2 = F\left(\frac{V}{4{,}000}\right)^2.$$

H_2 in inches w.g.
V in ft./min.

RESISTANCE H_2 FOR VARIOUS VELOCITIES AND VALUES OF F.

Air Velocity m./s.	Resistance H_2 in mm. W.G. for F.								
	1	2	3	4	5	6	7	8	9
2·0	0·249	0·498	0·747	0·996	1·25	1·49	1·74	1·99	2·24
2·5	0·389	0·778	1·17	1·56	1·95	2·33	2·72	3·11	3·50
3·0	0·560	1·12	1·68	2·24	2·80	3·36	3·92	4·48	5·13
3·5	0·764	1·53	2·29	3·06	3·72	4·58	5·35	6·11	6·97
4·0	0·995	2·00	2·99	3·98	4·98	5·97	6·97	7·96	8·96
4·5	1·20	2·41	3·61	4·82	6·02	7·22	8·43	9·63	10·8
5·0	1·56	3·12	4·68	7·44	7·80	9·36	10·9	12·5	14·0
6·0	2·24	4·48	6·72	8·96	11·2	13·4	15·7	17·9	20·2
7·0	3·05	6·10	9·15	12·2	15·3	18·3	21·4	24·4	27·5
8·0	3·98	7·96	11·9	15·9	19·9	23·9	27·9	31·8	35·8
9·0	5·04	10·1	15·1	20·2	25·2	30·2	35·3	40·3	45·4
10·0	6·22	12·4	18·7	24·9	31·1	37·3	43·5	49·8	65·0
12·5	9·72	19·4	29·2	38·9	48·6	58·3	68·0	77·8	87·5
15·0	14·0	28·0	42·0	56·0	70·0	84·0	98·0	112	126
17·5	19·1	39·1	57·2	78·2	95·3	114	113	152	171
20	24·9	49·8	74·7	99·6	125	149	174	199	224
25	38·9	79·8	116·7	156	195	233	272	311	350

Air Velocity ft./min	Resistance H_2 in inches W.G. for F.								
	1	2	3	4	5	6	7	8	9
400	0·0099	0·0198	0·0297	0·0396	0·0496	0·0595	0·0694	0·0793	0·0892
500	0·0155	0·0310	0·0465	0·0620	0·0775	0·0929	0·108	0·124	0·139
600	0·0223	0·0446	0·0669	0·0892	0·112	0·134	0·156	0·178	0·201
700	0·0304	0·0607	0·0911	0·121	0·152	0·182	0·213	0·243	0·273
800	0·0396	0·0793	0·119	0·159	0·198	0·238	0·277	0·317	0·357
900	0·0502	0·100	0·151	0·201	0·251	0·301	0·351	0·401	0·452
1,000	0·0619	0·124	0·186	0·248	0·310	0·372	0·434	0·496	0·557
1,200	0·0892	0·178	0·268	0·357	0·446	0·535	0·624	0·714	0·803
1,400	0·121	0·243	0·364	0·486	0·607	0·728	0·850	0·974	1·09
1,600	0·159	0·317	0·476	0·634	0·793	0·952	1·11	1·27	1·43
1,800	0·201	0·401	0·602	0·803	1·00	1·20	1·41	1·61	1·81
2,000	0·248	0·496	0·743	0·991	1·24	1·49	1·73	1·98	2·23
2,400	0·357	0·714	1·07	1·43	1·78	2·14	2·50	2·85	3·21
2,800	0·486	0·971	1·46	1·94	2·43	2·91	3·40	3·88	4·37
3,200	0·634	1·27	1·90	2·54	3·17	3·81	4·44	5·07	5·71
3,600	0·803	1·61	2·41	3·21	4·01	4·82	5·62	6·42	7·22
4,000	0·991	1·98	2·97	3·96	4·96	5·95	6·94	7·93	8·92

COEFFICIENTS OF RESISTANCE. For Fittings of Ventilating Systems

ITEM	RESISTANCE	F
1	ELL 90°	1·5
2	ELL 90°, ROUNDED	0·5
3	LONG SWEEP ELL 90°, R=2D	0·1
4	ELL 45°	0·5
5	ELL 45°, ROUNDED	0·2
6	LONG SWEEP ELL 45°, R=2D	0·05
7	GRADUALLY INCREASED AREA, FOR $\alpha \leq 8°$; A_1, A_2 = SECTIONAL AREA	$0{\cdot}15\left[1-\left(\frac{A_1}{A_2}\right)\right]^2$
	FOR $\alpha > 8°$	$\left[1-\left(\frac{A_1}{A_2}\right)\right]^2$
8	ABRUPT ENLARGEMENT OF AREA A_1 → A_2	$\left[1-\left(\frac{A_1}{A_2}\right)\right]^2$
9	FLOW FROM DUCT INTO ROOM	1·0
10	GRADUAL REDUCING	0
11	ABRUPT REDUCING	0 TO 0·35
12	FLOW FROM ROOM INTO DUCT	0 TO 0·35
13	BRANCH $A_1+A_2=A_0$ (SEE No. 6, SEE No. 10, SEE No. 7)	–

14 GRILLS					
FREE AREA DIVIDED BY TOTAL SURFACE	0·6	0·5	0·4	0·3	0·2
WIRE GAUZE	2	3	5	8	17
SHEET METAL	4	6	10	20	50

RATIO OF THE RESISTANCE OF FITTINGS TO THE TOTAL RESISTANCE OF THE CIRCUIT, IN PER CENT.

Inside Dimensions of Duct						
	in.	2–4	4–12	8–24	16–45	> 40
	mm.	5–10	10–30	20–60	40–120	> 100
Sheet Metal Ducts		40	60	80	90	95
Brick Ducts		30	50	70	80	85

3. Loss of Head in Apparatus (usually given by Manufacturers):

Average Values	*mm. W.G.*	*in. W.G.*
Filters	5 to 10	$\frac{3}{16}$ to $\frac{3}{8}$
Air Washers	5 to 10	$\frac{3}{16}$ to $\frac{3}{8}$
Heating Batteries	3 to 10	$\frac{1}{8}$ to $\frac{3}{8}$

CIRCULAR EQUIVALENTS OF RECTANGULAR DUCTS FOR EQUAL FRICTION

$$d = 1{\cdot}265 \sqrt[5]{\frac{(a\,b)^3}{a + b}}$$

Side Rect. Duct.	5	6	7	8	9	10	11	12	13	14	15	16	18	20	22	24	26	28	30	32	34	36
5	5·5																					
6	6·0	6·6																				
7	6·5	7·0	7·7																			
8	6·9	7·6	8·2	8·8																		
9	7·3	8·0	8·7	9·3	9·9																	
10	7·7	8·4	9·2	9·8	10·4	11·0																
11	8·1	8·8	9·6	10·2	10·9	11·5	12·1															
12	8·4	9·2	10·0	10·7	11·4	12·0	12·6	13·2														
13	8·8	9·6	10·4	11·1	11·9	12·5	13·3	13·7	14·3													
14	9·0	9·9	10·8	11·5	12·3	12·9	13·6	14·3	14·9	15·4												
15	9·2	10·2	11·1	11·9	12·7	13·4	14·1	14·7	15·3	16·0	16·5											
16	9·6	10·5	11·5	12·3	13·1	13·8	14·5	15·2	15·8	16·5	17·1	17·6										
17	9·8	10·8	11·9	12·6	13·3	14·2	14·9	15·7	16·3	17·0	17·5	18·2										
18	10·1	11·1	12·1	12·9	13·9	14·6	15·4	16·1	16·8	17·4	18·1	18·7	19·8									
19	10·3	11·4	12·4	13·3	14·2	15·0	15·8	16·5	17·2	18·0	18·7	19·2	20·4									
20	10·5	11·6	12·7	13·6	14·5	15·4	16·2	17·0	17·6	18·4	19·1	19·7	20·9	22·0								
22	11·0	12·1	13·2	14·2	15·1	16·1	16·9	17·8	18·5	19·2	20·0	20·6	21·9	23·1	24·2							
24	11·4	12·6	13·7	14·8	15·8	16·8	17·8	18·5	19·3	20·0	20·8	21·5	22·8	24·0	25·2	26·4						
26	11·8	13·1	14·2	15·4	16·4	17·3	18·2	19·2	20·0	20·8	21·6	22·3	23·8	25·1	26·3	27·5	28·6					
28	12·2	13·5	14·7	15·9	16·9	18·0	18·9	19·8	20·7	21·5	22·4	23·1	24·6	26·0	27·3	28·5	29·7	30·8				
30	12·6	13·9	15·2	16·4	17·4	18·7	19·5	20·5	21·4	22·2	23·1	23·9	25·4	26·8	28·2	29·5	30·7	32·4	33·0			
32	12·9	14·3	15·6	16·9	18·0	19·1	20·1	21·1	22·0	22·9	23·8	24·6	26·2	27·7	29·1	30·5	31·7	33·0	34·1	35·2		
34	13·2	14·7	16·1	17·3	18·5	19·6	20·7	21·6	22·6	23·5	24·4	25·3	26·9	28·5	30·0	31·3	32·7	34·0	35·1	36·3	37·4	
36	13·6	15·1	16·5	17·7	19·1	20·1	21·2	22·2	23·2	24·2	25·2	26·0	27·7	29·3	30·8	32·2	33·6	34·8	36·1	37·3	38·5	39·6
38	13·9	15·4	16·8	18·2	19·4	20·6	21·7	22·9	23·8	24·8	25·8	26·7	28·4	30·0	31·5	33·1	34·5	35·9	37·1	38·4	39·5	40·7
40	14·3	15·7	17·2	18·6	19·9	21·1	22·2	23·3	24·4	25·4	26·4	27·3	29·1	30·8	32·4	33·9	35·3	36·7	38·0	39·3	40·5	41·7
42	14·5	16·1	17·6	19·0	20·3	21·6	22·7	23·8	24·9	25·9	26·9	27·9	29·8	31·4	33·0	34·5	36·0	37·6	39·0	40·3	41·5	42·7
44	14·8	16·4	18·0	19·4	20·7	22·0	23·2	24·2	25·4	26·5	27·5	28·5	30·3	32·1	33·7	35·2	36·9	38·5	39·9	41·2	42·5	43·7
46	15·1	16·7	18·3	19·8	21·1	22·4	23·6	24·8	25·9	27·0	28·1	29·1	31·0	32·8	34·6	36·2	37·8	39·3	40·8	42·2	43·5	44·8
48	15·4	17·0	18·6	20·1	21·5	22·8	24·1	25·2	26·4	27·5	28·6	29·6	31·6	33·4	35·2	37·0	38·5	40·0	41·5	43·0	44·4	45·6
50	15·7	17·3	19·0	20·4	21·9	23·2	24·5	25·7	26·9	28·0	29·1	30·3	32·2	34·1	35·9	37·6	39·2	40·8	42·3	43·8	45·2	46·5
54	16·1	17·9	19·6	21·1	22·6	24·0	25·3	26·6	27·8	29·0	30·1	31·2	33·4	35·3	37·2	38·9	40·7	42·4	44·0	45·5	47·0	48·2
60	16·8	18·7	20·4	22·1	23·6	25·1	26·5	27·8	29·1	30·5	31·6	32·7	34·9	37·1	39·1	40·9	42·7	44·5	46·1	47·8	49·3	50·9
64	17·3	19·2	21·0	22·7	24·3	25·9	27·3	28·6	29·9	31·3	32·6	33·7	35·9	38·2	40·2	42·2	44·0	45·8	47·5	49·2	50·9	52·4
70																	46·0	47·8	49·5	51·3	52·9	54·5
	5	6	7	8	9	10	11	12	13	14	15	16	18	20	22	24	26	28	30	32	34	36

RECOMMENDED VELOCITIES FOR VENTILATING SYSTEMS

Service	Public Buildings		Industrial Plants	
	m./s.	*ft./min.*	*m./s.*	*ft./min.*
Air intake from outside ...	2·5–4·5	500–900	5–6	1,000–1,200
Air washers	2·5	500	2·5–3·0	500– 600
Heater connection to Fan ...	3·5–4·5	700–900	5–7	1,000–1,400
Main Ducts	5·0–8·0	1,000–1,500	6–12	1,200–2,400
Branch Ducts and Risers ...	2·5–3·0	500–600	4·5–9	900–1,800
Supply Registers and Grilles ...	1·2–2·3	250–450	—	—
Supply Openings	—	—	1·5–2·5	350–500
Supply Grilles near the Floor ...	0·8–1·2	150–250	—	—

RECOMMENDED VELOCITIES FOR EXHAUST VENTILATING SYSTEMS

Service	Public Buildings		Industrial Plants	
	m./s.	*ft./min.*	*m./s.*	*ft./min.*
Risers	2·5–3·0	500–600	4·5–9·0	900–1,800
Main Ducts	4·5–8·0	900–1,500	6–12	1,200–2,400

Air Velocities in gravity exhaust Systems are 1·0 to 3·0 m/s. (200 to 600 ft./min.).

SIZE OF DUCTS

	Longest Side or Diameter		*Thickness S.W.G.*
	mm.	*inches*	
Rectangular ...	UP to 150	Up to 6	24
	150–300	6–12	22
	300–450	12–18	20
	450–900	18–36	18
	900–1,400	36–54	16
	Above 1,400	Above 54	14
Circular ...	Up to 300	Up to 12	22
	300–500	12–20	20
	500–1,000	20–42	18
	Above 1,000	Above 42	16

Ducts outside Buildings exposed to atmosphere are one gauge heavier.

Ducts over 750 mm. longest side (30 in.) to have flanged joints.

Minimum radius of Duct Bends, $R_{min} = 1{\cdot}5$ D.

Weight of Air to be circulated:

$$W = \frac{X}{w_2 - w_1}$$

W = Mass of air to be circulated kg./s.
X = Mass of water to be evaporated kg./s.
w_1 = Absolute humidity of entering air kg./kg
w_2 = Absolute humidity of leaving air kg./kg.

The relative humidity of the air leaving the dryer is usually kept below 75 per cent.

Heat Amount:

Total Heat Amount = 1. Heat for evaporating moisture.
2. Heat for heating of stock.
3. Heat-loss due to air change.
4. Heat transmission loss of drying chamber.

WATER CONTENT OF VARIOUS MATERIALS

Material	*Original per cent.*	*Final per cent.*	*Material*	*Original per cent.*	*Final per cent.*
Bituminous Coal ...	40–60	8–12	Hides...	45	0
Earth	45–50	0	Glue	80–90	0
Earth, Sandy... ...	20–25	0	Glue, Air Dried ...	15	0
Grain...	17–23	10–12	Macaroni	35	0
Rubber Goods ...	30–50	0	Soap	27–35	25–26
Green Hardwood ...	50		Starch	38–45	12–14
Green Softwood ...	30–50	10–15	Starch, Air Dried ...	16–20	12–14
Air Dried Hardwood	17–20		Peat	85–90	30–35
Air Dried Softwood	10–15		Yarn, Washing ...	40–50	0
Cork	40–45	10–15			

DRYING TEMPERATURES AND TIME FOR VARIOUS MATERIALS

Material	*Temperature*		*Time*
	°C.	*°F.*	*Hr.*
Bedding	66–88	150–190	
Cereals	43–66	110–150	
Coconut	63–68	145–155	4–6
Coffee	71–82	160–180	24
Cores, Oil Sand	150	300	0·5
Films, Photo	32	90	
Fruits, Vegetable	60	140	2–6
Furs	43	110	
Glue	21–32	70–90	2–4
Glue Size on Furniture	54	130	4
Gut	66	150	
Gypsum Wall Board			
Start Wet	175	350	
Finish	88	190	
Gypsum Blocks	175–88	350–180	8–16
Hair Goods	66–88	150–190	
Hats, Felt	60–82	140–180	
Hops	49–82	120–187	

Material	*Temperature*		*Time*
	°C.	*°F.*	*Hr.*
Hides, Thin	32	90	2–4
Ink, Printing	21–150	70–300	
Knitted Fabrics	60–82	140–180	
Leather, Thick Sole	32	90	4–6
Lumber:			
Green, Hardwood	38–82	100–180	3–180
Green Softwood	71–105	160–220	24–350
Macaroni	32–43	90–110	
Matches	60–82	140–180	
Milk	120–150	250–300	
Paper Glued	54–150	130–300	
Paper Treated	60–93	140–200	
Rubber	27–32	80–80	6–12
Soap	52	125	12
Sugar	66–93	150–200	0·3–0·5
Tannin	120–150	250–300	
Terra Cotta	66–93	150–200	12–96

CHART 5. PSYCHROMETRIC CHART FOR AIR

Barometric Pressure = 1·013 bar

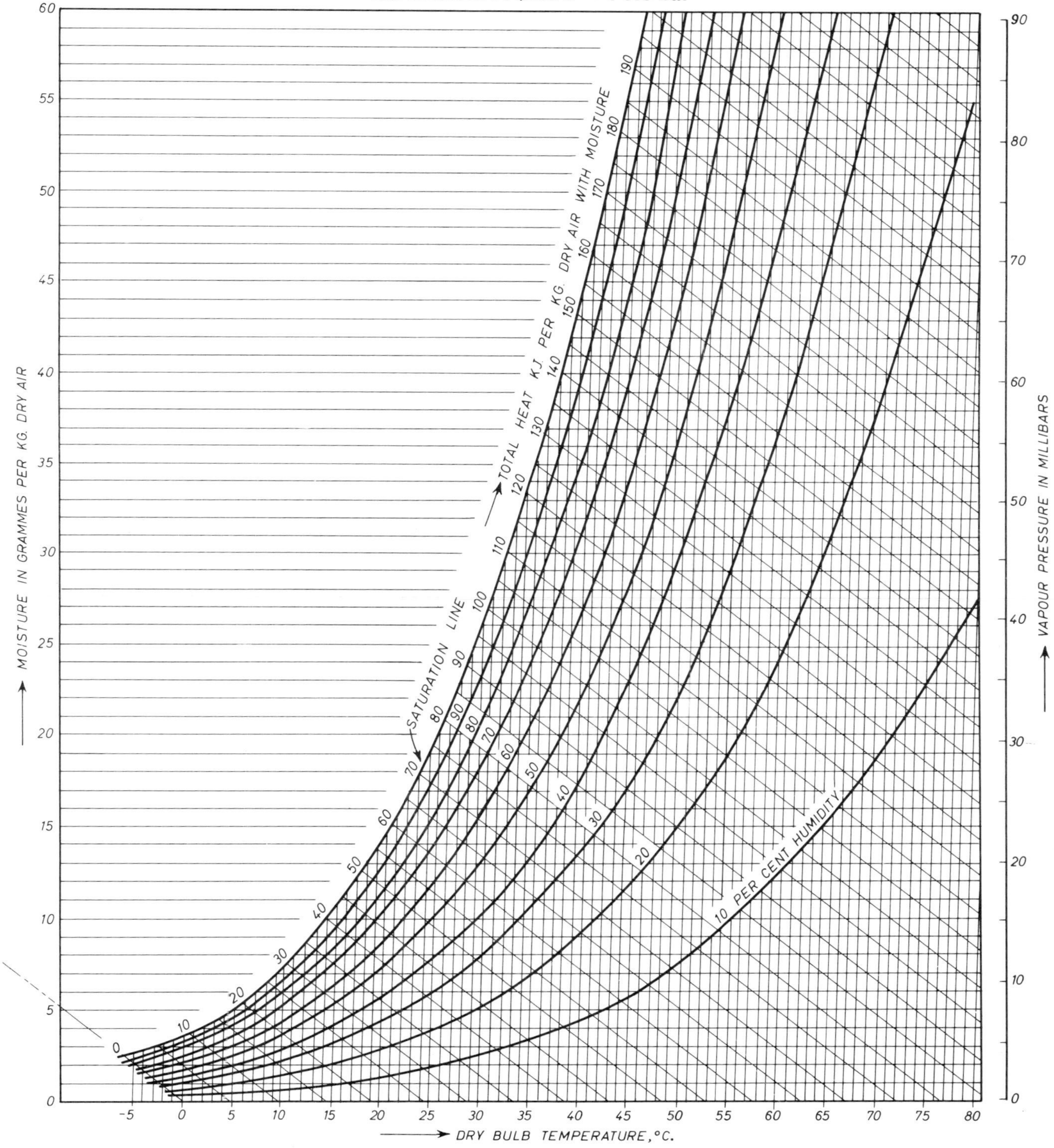

The defogging of rooms is carried out by blowing in dry, hot air and exhausting humidified air.

Mass of Water Evaporated from Open Vats:

$$W = 1{\cdot}25 \times 10^{-2} \times Ac\frac{(P_S - P_A)}{P}$$

W = Mass of water evaporated kg./s.
A = Surface of vats, m.^{2}
P_S = Partial pressure of water vapour of saturated air at the temperature of the water, mm. Hg.
P_A = Partial pressure of water vapour of surrounding air, mm. Hg.
P = Atmospheric pressure mm. Hg.
c = 0·55 for still air.
0·71 for slight air movement.
0·86 for fast air movement.

$$W = 9{\cdot}4\, Ac\frac{(P_S - P_A)}{P}$$

W = Weight of water evaporated lb./hr.
A = Surface of vats, ft.^{2}
P_S = Partial pressure of water vapour of saturated air at the temperature of the water, in Hg.
P_A = Partial pressure of water vapour of surrounding air, in Hg.
P = Atmospheric pressure, in. Hg.
c = 0·55 for still air.
0·71 for slight air movement.
0·86 for fast air movement.

Mass of Air to be circulated:

$$G = \frac{W}{(w_2 - w_1)}$$

G = Mass of air, kg./s.
W = Mass of water vapour to be removed, kg./s.
w_1 = Original absolute humidity of air, kg./kg.
w_2 = Final absolute humidity of air, kg./kg.

Amount of Heat:

$$H = Gc\,(t_i - t_o)$$

H = Amount of heat, without fabric loss of room or other losses, W.
G = Mass of air (see above), kg./s.
t_i = Inside air temperature, °C.
t_o = Outside air temperature, °C.
c = Specific heat capacity of air = $1{\cdot}012 \times 10^3$ J/kg. deg. C.

Air Change for Good Conditions:

Laundries	10–15 air changes per hour
Large Kitchens	10–20 ,, ,, ,, ,,
Dyeing Shops	10–20 ,, ,, ,, ,,
Swimming Baths	5–10 ,, ,, ,, ,,

Principle of Air Conditioning: Air conditioning is the control of atmospheric conditions within an enclosure with reference to temperature, humidity, air motion, and cleanliness. Often combined with removing of bacteria, odours, toxic gases, and with ionization of air.

Requirements:

1. *For Human Comfort.*—Human comfort depends on cooling of the body surface and takes place partially by evaporation, so that both temperature and humidity, as well as air velocity of surrounding air affect comfort and health.

   Thermo-Equivalent Conditions are combinations of temperature, humidity and air movement which produce the same feeling of warmth. (See pages V. 15 and V. 16.)

2. *Industrial Purposes.*—Hygroscopic materials such as textile fibres, paper, timber, leather, etc., take up or give off moisture to the atmosphere to an extent governed by the temperature and humidity of the surrounding air. Eventually the water content of the material approaches an ultimate value termed the Equilibrium Water Content. Therefore many process works have to be carried out in air-conditioned rooms. (See page X. 5.)

Air Conditioning Equipment consists of all apparatus and installations used for ventilation and plenum heating plants (such as Fans, Heater Batteries, Mixing Chambers with Dampers, Distribution Ducting, etc.) with the addition of an Air Washer for humidification or dehumidification, and a Refrigeration Plant for cooling in summer, if required, and an Automatic Control System.

CHART 4. DUCT SIZING FOR VENTILATION SCHEMES

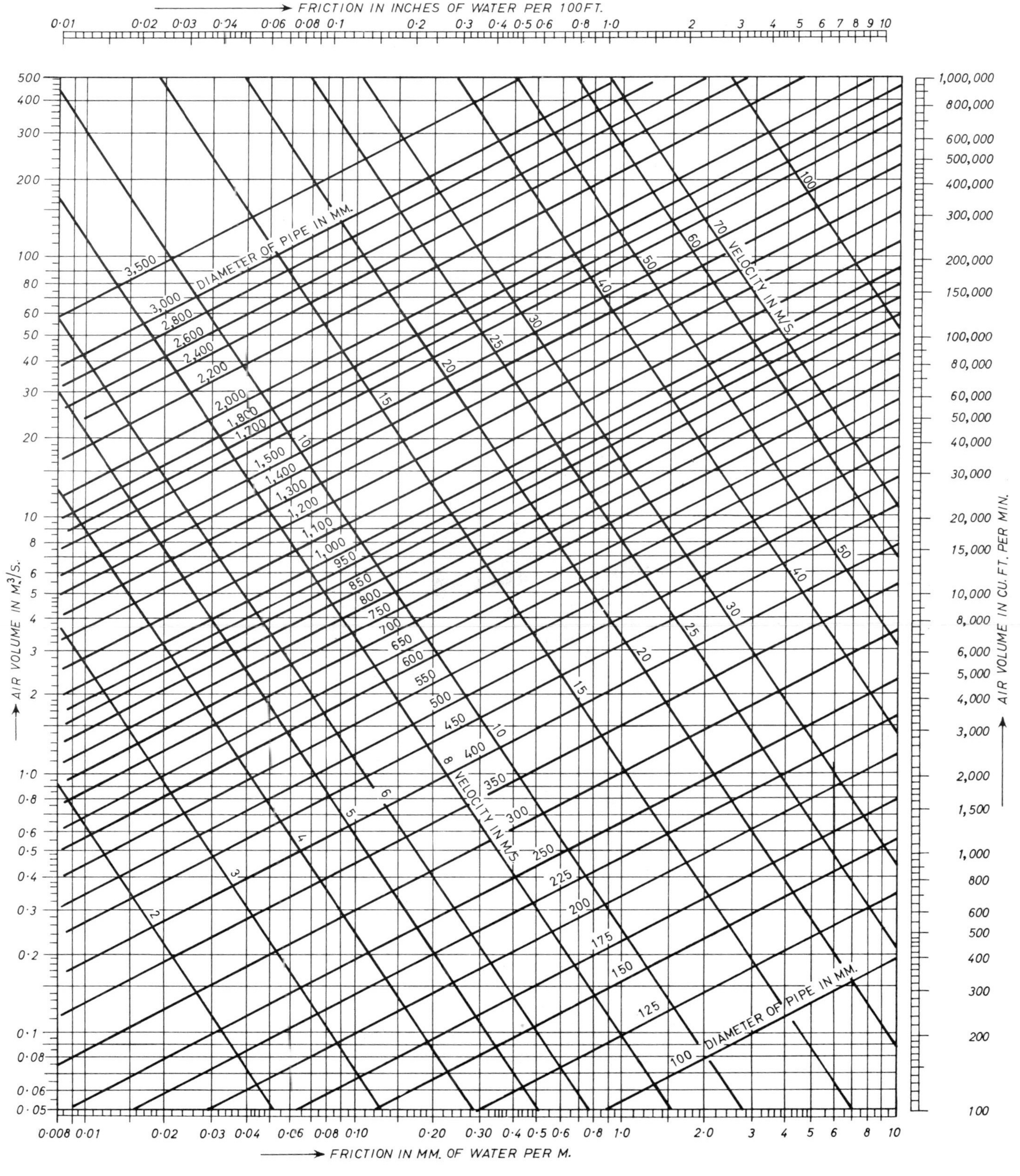

CHART 4a. DUCT SIZING FOR HIGH AIR VELOCITIES

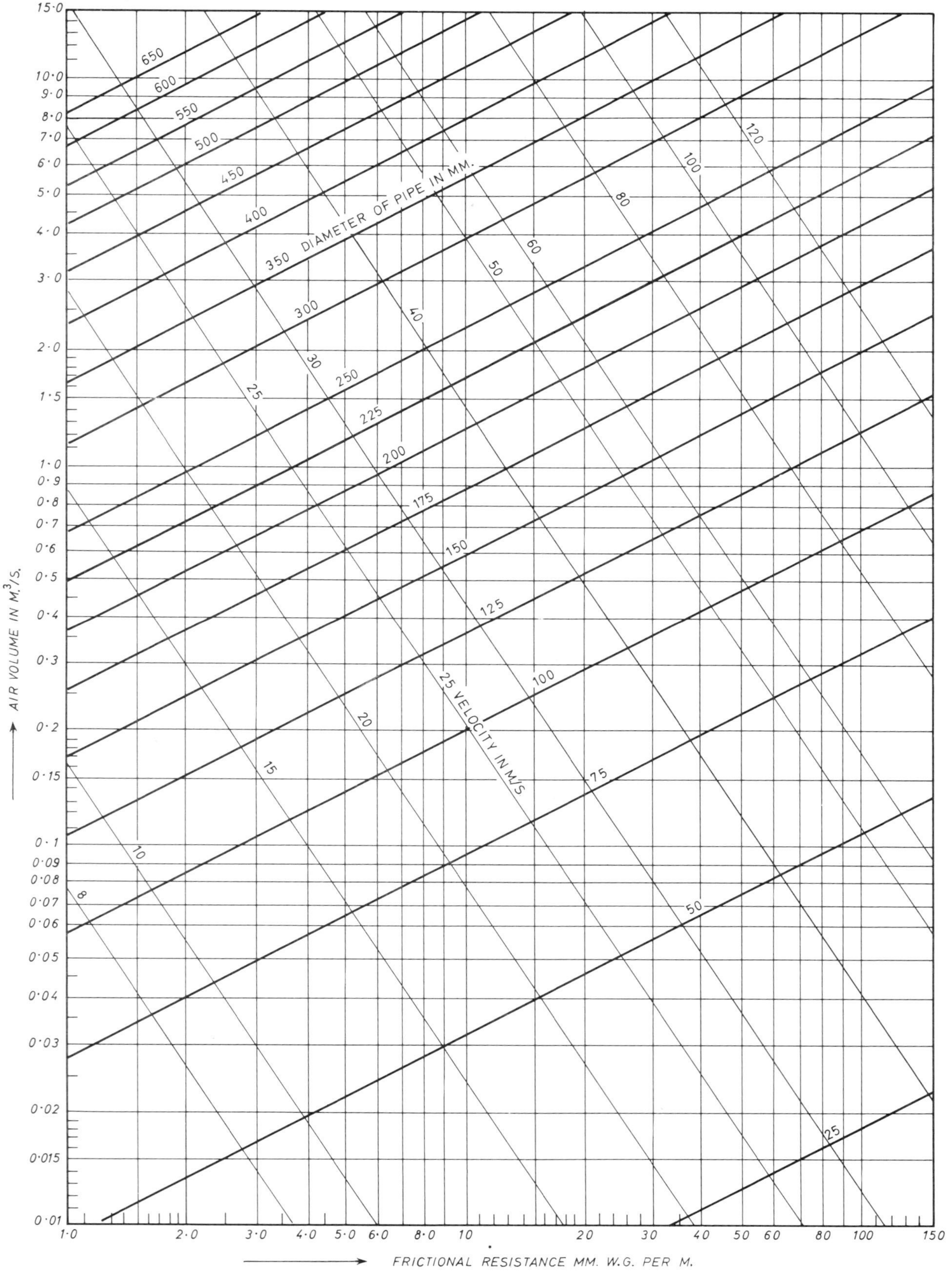

HUMIDITY

Dry-Bulb Thermometer, i.e., an ordinary thermometer, indicates the air temperature without being affected by the moisture in the air.

Wet-Bulb Thermometer has its bulb covered with cloth dipped into water, and it is exposed to a current of air. It indicates the temperature of adiabatic saturation.

The Wet-Bulb Depression is the difference in the readings of Dry-Bulb and Wet-Bulb Temperatures, and is a measure of the relative humidity of the air. (See V. 11 to V. 18.)

DETERMINATION OF HUMIDITY

1. **Dew Point Method.**—A mirror surface is cooled by evaporation of ether, or any other low-boiling solvent, and the temperature at which moisture begins to form is measured.

2. **Wet-Bulb Method.**—Dry Bulb and Wet Bulb Temperatures are measured by a Sling Psychrometer or an Aspiration Psychrometer. Tables V. 17 and V. 18 to be used for determination of the relative humidity.

3. **Hair Hygrometer.**—The sensitive hygroscopic element is a human hair.

4. **Electric Hygrometer** is an instrument, the electrical resistance of which varies with the relative humidity of the surrounding air.

APPARATUS

Wet and Dry Bulb Thermometer: On a wet bulb thermometer the water evaporates from the surface of the cloth dipped into water. The latent heat of evaporation is taken from the surrounding air, and, therefore, the wet bulb thermometer gives a lower reading than the dry bulb thermometer.

The Wet Bulb Depression is the difference in reading on Dry and Wet bulb thermometer and is the measurement for the humidity in the air.

Sling Psychrometer consists of a Dry bulb and a Wet bulb thermometer mounted on a board with a handle for slinging through the air. The peripheral velocity should be not less than 15 ft. per sec.

Aspiration Psychrometer (Assman) consists of a Dry bulb and Wet bulb thermometer the bulbs of which are enclosed in a casing. Air is drawn through the casing by means of a spring operated propeller fan.

Humidity in Heated Rooms: The humidity in heated rooms should be maintained within 35% to 50%, for good practice 40% relative humidity at 68°F. Dry bulb temperature.

DESIGN OF AIR CONDITIONING SYSTEMS

Air Conditioning Cooling Load:

The Air Conditioning Cooling Load is made up from the Sensible Heat Load and the Latent Heat Load.

(*a*) *The Sensible Heat Load* consists of:

1. Heat transfer through walls, windows, ceilings, etc.
2. Heat gained by solar radiation.
3. Heat emission of occupants.
4. Heat introduced by infiltration of outside air, or ventilation.
5. Heat emission of appliances (mechanical, electrical, gas, etc.).

(*b*) *The Latent Heat Load* consists of:

1. Moisture given off by people.
2. Infiltration of outside air, or ventilation.
3. Latent heat from appliances.

(See pages V. 14, VI. 13 to VI. 18.)

Methods of Cooling:

1. Spray-type air conditioners, air washers.
2. Surface type coolers.
3. Adsorption type.
4. Absorption type.

Compression Refrigeration System: The plant consists of a compressor, condenser, receiver, reducing valve and evaporator.

The function is as follows: Hot compressed vapour, called refrigerant, leaving the compressor is liquefied in the condenser by cooling with air or water. The liquid refrigerant then passes from the high pressure side through the expansion valve to the low pressure side and evaporates in the evaporator, extracting the required latent heat of evaporation from the surroundings thus reducing the temperature of the medium to be cooled.

Direct System: The evaporator in which the expansion of the liquid refrigerant takes place is fixed directly in the room to be cooled, e.g. a direct expansion coil in an air washer.

Indirect, or Brine System: Indirect cooling coils are supplied with cold brine which has been previously cooled by a refrigerant.

Absorption System of Refrigeration: The compressor used in the compression system is replaced by a generator where the refrigerant, ammonia, is absorbed and then driven off by applying heat.

AIR WASHER

Air Washers are sheet metal, or sometimes bricked or concrete chambers, where air is drawn through a mist caused by spray nozzles and then through eliminators. The water for the spray nozzles is recirculated by a pump and can be heated or cooled. A tempering heater is installed before, and a reheating battery after, the air washer.

General Data on Air Washers:

Cleaning efficiency: 70 per cent with fine dust.

Cleaning efficiency: 98 per cent with coarse dirt.

Air velocity through washer: 450 to 550 ft./min. = 2·3 to 2·8 m./s.

Resistance: 0·2 to 0·5 ins. W.G. = 5·0 to 14·0 mm. W.G.

Water pressure for sprays: 15 to 25 lb. per sq. in. = 100 to 170 kN/m.^{2}

Water quantity for spray nozzles: { 3 to 3·5 gal. per 1,000 cu. ft. air. / 0·45 to 0·55 l. per m.^{3} }

Humidifying efficiency:

$$E = \frac{t_1 - t_2}{t_1 - t_w} 100 \text{ per cent.}$$

t_1 = initial dry bulb temperature.

t_2 = final dry bulb temperature.

t_w = initial wet bulb temperature.

E = Efficiency of humidification or saturation.

E = 60 to 70 per cent with one bank of nozzles, downstream.
65 to 75 per cent with one bank of nozzles, upstream.
85 to 100 per cent with two banks of nozzles.

Refrigeration Units:

The Standard Term for measuring the capacity of refrigeration equipment is the Ton of Refrigeration. One Ton of Refrigeration is equal to the quantity of Heat in B.t.u. required to melt 1 American ton of pure ice of 32°F. to water 32°F. in 24 hours.

Therefore, one Ton of Refrigeration equals:

2,000 lb. per 24 hrs. × 144 B.t.u./lb. = 288,000 B.t.u. per 24 hrs.
= 12,000 B.t.u. per hour.
= 200 B.t.u. per min.
= 3·517 kW.

SURFACE TYPE AIR CONDITIONERS

Surface Type Air Conditioners consist of plain or gilled tubes containing a heating medium (steam, hot water, etc.) or a cooling medium (cold water, brine, refrigerant). Dehumidification is achieved by cooling the air below dew point causing part of the moisture content to be condensed out.

Required Surface of a Cooler:

$$A = \frac{H}{C\,(t_a - t_m)}$$

A = Area of cooler, or heater in sq. ft.. or m.^{2}

H = Sensible heat loss to be removed in B.t.u. per hr. or W.

$(t_a - t_m)$ = log. mean temperature difference between air and cooling, or heating medium, °F. or °C. (see IV. 14).

C = Heat transfer coefficient, in B.t.u. per sq. ft. per hr. per °F. or W/m.^{2} deg. C.

ADSORPTION SYSTEMS

In Adsorption Systems humidity of the air is reduced by adsorbent materials, such as silica-gel or activated alumina. The adsorbent material can be reactivated by heating, the water adsorbed being evaporated, and used repeatedly. The adsorption system is especially suitable for dehydration at room temperature, and where gas or high pressure steam is available for reactivation.

Temperature for Reactivation: 325 to 350°F. or 160 to 175°C.

Heat required for Reactivation: 2,100 to 2,500 B.t.u./lb. / 4,800 to 5,800 kJ/kg. of water removed.

The action is physical—no chemical change.

Silica-gel, Si O_2 is a hard, hygroscopic crystalline substance; size of a pea; very porous.

Voids are about 50% by volume.

Adsorbs water up to 40% of own weight.

Bulk density: 30 to 45 lbs. per cu. ft. = 480 to 720 kg./m.^{3}

Specific heat: 0·27 B.t.u./lb. °F. = 1·13 kJ/kg. deg. C.

Activated Alumina, about 90% aluminium oxide, Al_2O_3 ; very porous.

Voids about 50 to 70% by volume.

Adsorbs water up to 60% of own weight.

Bulk density: 50 to 54 lb. per cu. ft. = 800 to 870 kg./m.^{3}

Specific heat: 0·24 B.t.u./lb. °F. = 1·0 kJ/kg. deg. C.

ABSORPTION SYSTEMS

Humidity from the air is absorbed by a hygroscopic solution such as calcium chloride solution.

SUMMARY OF AIR CONDITIONING PROCESSES

Air Conditioning Processes are either heating, cooling, humidifying or dehumidifying of the air, and the air temperature and humidity alter by any of these processes. Computations can be carried out by using either algebraic equations or a psychrometric chart. (See V. 10 to V. 21.) Computations are to be based upon unit mass of dry air since the mass remains constant during the process.

SUMMARY OF PROCESSES

Process	*Indication in chart Fig.*	*Final Air Conditions*	*Equipment used*
Heating	A-B	S.H. and D.P. unchanged. R.H. reduced. W.B.T. increased.	Air passes over surface air heater.
Cooling above Dew Point.	A-C	S.H. and D.P. unchanged. R.H. increased. W.B.T. reduced ; no condensation of vapour.	Air passes an air washer using refrigerated water for spray nozzles or air passes a surface cooler.
Cooling to Dew Point Temperature.	A-D	S.H. and D.P. unchanged. R.H. becomes 100 per cent. D.B.T. and W.B.T. coincide; the air is saturated but there is no condensation of vapour.	
Cooling below Dew Point Temperature *Dehumidification*	A-D-E	S.H. and D.P. reduced. The final air is saturated and part of the humidity has condensed.	
Humidification i.e. adiabatic saturation.	A-F	The process is adiabatic, no heat is added or abstracted. W.B.T. unchanged. R.H. increased. D.P. and S.H. increased.	Air passes air washer using recirculated water for spray nozzles.
Mixing two airs	A-G-H	Points A and G representing the initial conditions of the atmospheres to be mixed are to be connected by a straight line. This line is to be divided inversely proportional to the weights of both airs. The point "H" thus found indicates the conditions of the air mixture.	Mixing chamber or two ducts with dampers.

See Chart and Key on page X. 23.

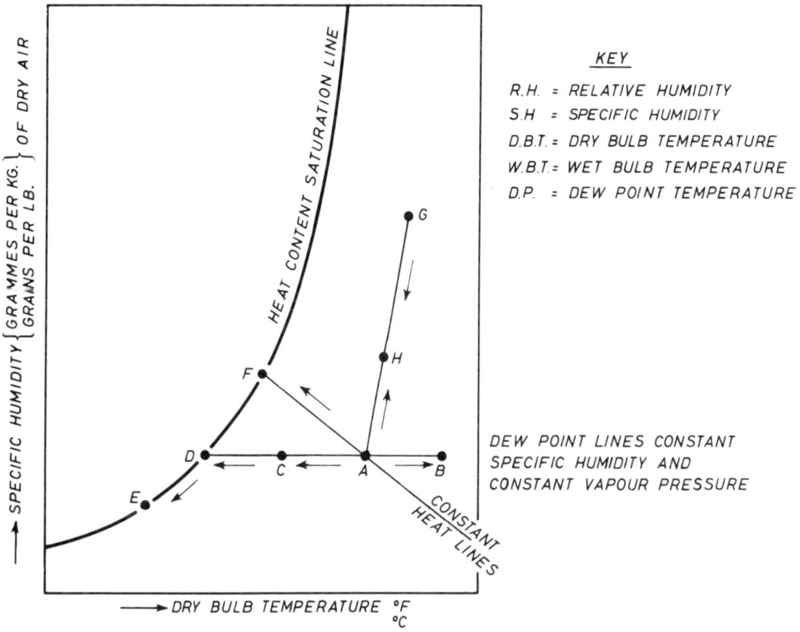

PSYCHROMETRIC CHART

The Psychrometric Chart is plotted with Dry Bulb Temperature as abscissa and Specific Humidity as ordinate. Any point on the chart represents conditions of air-vapour mixture, such as Dry Bulb Temperature, Wet Bulb Temperature, Dew Point, Relative Humidity, Specific Humidity, Specific Volume, and Vapour Pressure. Any two values locate the point representing the state of air, and the remaining values can be read from the Chart.

Example of the use of the Psychrometric Chart:

All conditions to be found for air at 40°C. Dry Bulb Temperature and 60% Relative Humidity.

The Wet Bulb Temperature is found on the intersection of the constant heat line with saturation line: it is 32°C.

The Dew Point is found on the intersection of the horizontal Dew Point line with the saturation line: it is 30·5°C.

The Specific Humidity is found by following the same horizontal dew point line to the scale on the left; it is 28·2 g. per kg.

The Vapour Pressure is found by following the same horizontal dew point line to the scale on the right: it is 43 millibars.

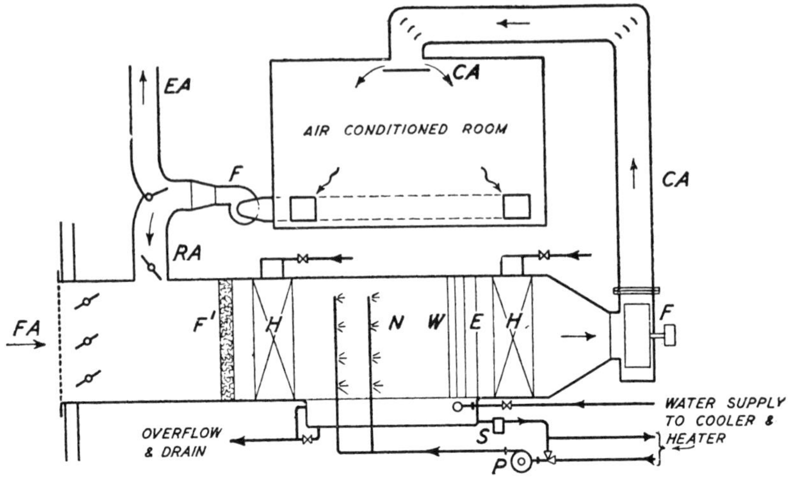

AIR-CONDITIONING LAYOUT WITHOUT BYPASS

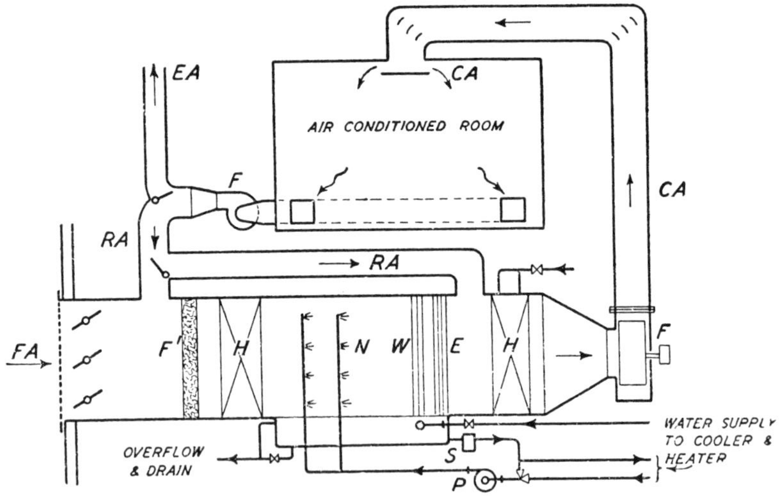

AIR-CONDITIONING LAYOUT WITH BYPASS

REFERENCE

FA = FRESH AIR
CA = CONDITIONED AIR
RA = RECIRCULATED AIR
EA = EXHAUST AIR

F = FAN
F' = FILTER
H = HEATER
N = SPRAY NOZZLES

P = PUMP
S = STRAINER
W = WET SCRUBBER
E = ELIMINATOR PLATES

Refrigerant	*Symbol*	*Boiling Temp.*	*Critical Temp.*	*Properties*	*Inflammability*	*Use*
Ammonia	NH_3	–28°F. –33°C.	271°F. 133°C.	Penetrating odour soluble in water, harmless in concentration up to 1/30 per cent.	Non-inflammable explosive.	Large plants.
Sulphur dioxide	SO_2	14°F. –10°C.	311°F. 155°C.	Colourless vapour, unpleasant to breathe but not poisonous.	Not combustible or corrosive if not in contact with water.	Small plants.
Carbon dioxide	CO_2	–108·4°F. –78°C.	87·8°F. 30°C.	Heavy, colourless odourless gas, harmless to breathe.	Not inflammable.	Large plants.
Ethyl chloride	C_2H_5Cl	55°F. 12·8°C.	—	Colourless, very volatile, sweet taste.	Inflammable; non-corrosive with steel, iron, copper.	—
Methyl chloride	CH_3Cl	–10·7°F. –23·7°C.	289·6°F. 143°C.	Colourless, sweet smelling vapour.	Inflammable, explosive, non-corrosive with steel,iron,copper	Small plants.
Dichloro-difluoro-methane, F-12 "Freon."	CCl_2F_2	–21·7°F. –30°C.	222·7°F. 106°C.	Odourless, non-toxic.	Not inflammable not explosive; non-corrosive.	Small plants.
Carrene, methylene chloride.	CHCl	103·6°F. 39·8°C.	473°F. 245°C.	Colourless liquid at atmospheric conditions, sweet, pleasant odour, similar to chloroform.	Not explosive.	—
Dieline dichloro-ethylene.	$C_2H_2Cl_2$	122°F. 50°C.	470°F. 243°C.	Colourless liquid odourless, non-toxic.	Inflammable non-corrosive with steel, iron, nickel, copper, aluminium.	Small plants.
Trieline Trichlor-ethylene	C_2HCl_3	∸126°F. –87·8°C.	188°F. 86·6°C.	Heavy, colourless liquid, pleasant odour.	Non-inflammable; non-explosive; non-corrosive.	—
Water	H_2O	212°F. 100°C.	706°F. 374°C.	See data, p. IV. 31.	—	Steam jet and centrifugal plants.

EQUIPMENT FOR INDUSTRIAL EXHAUST SYSTEMS

A. Suction hoods, booths, or canopies for fume and dust collection, or suction nozzles, or feed hoppers for pneumatic conveying.

B. Conveying, ducting or tubing.

C. Fan or exhauster to create the necessary pressure or vacuum for pneumatic conveying.

D. Dust separator, for separating the conveyed material from the conveying air.

CLASSIFICATION OF SCHEMES

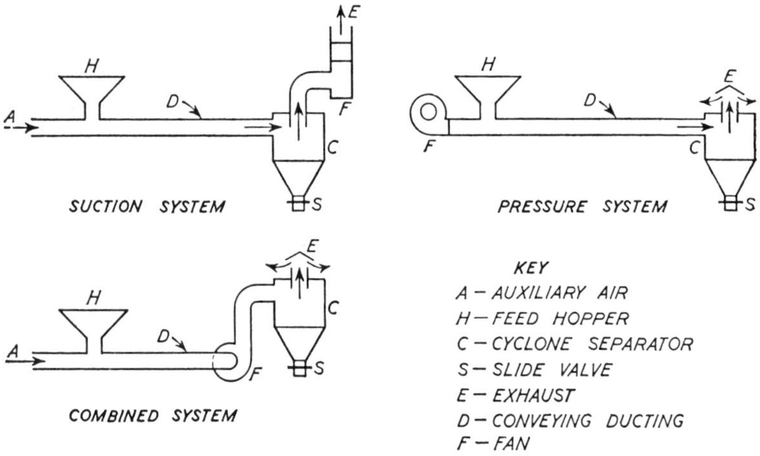

Pneumatic Conveying Plants are suitable for conveyance of material in powdered form or in solids up to 2 in. size, dry: not more than 20 per cent moisture, not sticking.

Efficiency of pneumatic conveying plants is low but compensated by easy handling, free of dust.

Suction Type—Distance of conveying up to 300 m. difference in heights up to 40 m. Required vacuum 200 to 400 mm. mercury.

Pressure Type—Distance of conveying above 300 m. working pressure up to 40 kN/m.^{2} Advantage: possibility of conveying material over long distances by connecting more systems in series.

Working pressure above 40 kN/m.^{2} not suitable, because of high running cost.

TYPES OF HOODS

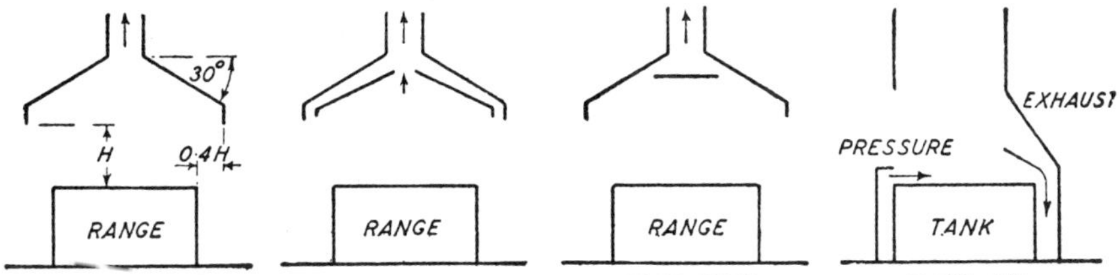

SINGLE HOOD — DOUBLE HOOD — HOOD WITH BAFFLE PLATE — BLOW AND EXHAUST HEAD

Single Hoods—For removing fumes which rise naturally, for ranges, forge fires, vats, kettles, etc. Projection beyond the range approx. 0·4 m. per m. of height above range. 0·25 to 1·5 m./s. entrance velocity. Duct area about one fifth to one-tenth of hood area.

Double Hoods with gap around the perimeter for fume extraction in rooms with cross currents, high velocity of entering air, approximately 5·0 m./s.

Velocity in Ducts—Approximately 10 m./s.

Recommended Velocities through Top Hoods and Booth, subject to cross draughts in m./s.:

Canopy Hood, open	1·0 –1·5	Canopy Hood, double ...	5·0
Canopy Hood, closed 1 side	0·9 –1·0	Booths, through 1 side ...	0·5 –0·75
Canopy Hood, closed 2 sides	0·75–0·9	Laboratory Hoods, through doors	0·25–0·35
Canopy Hood, closed 3 sides	0·5 –0·75		

Coefficients of Entry and Velocity. Pressure Loss of Dust Extraction Hoods:

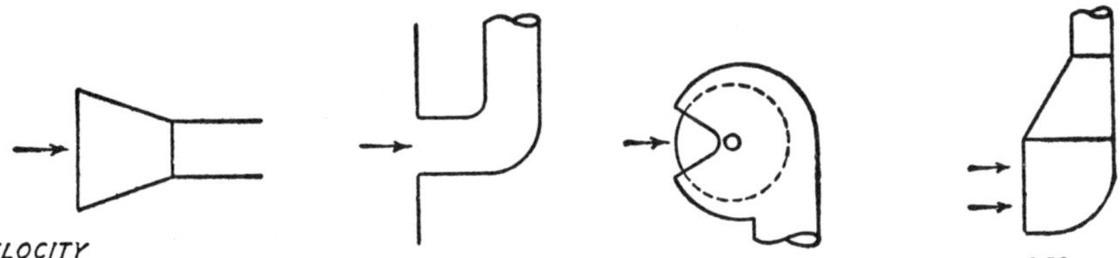

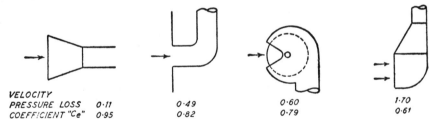

VELOCITY PRESSURE LOSS	0·11	0·49	0·60	1·70
COEFFICIENT "Ce"	0·95	0·82	0·79	0·61

Flow of Air into a Hood:

$Q = 4\,C_e\,A_t\,\sqrt{h_t}$
Q = Air volume m.^{3}/s.
C_e = Entrance coefficient.
A_t = Area of throat, m.^{2}
h_t = Static suction in throat, mm. W.G.

$Q = 4{,}000\,C_e\,A_t\,\sqrt{h_t}$
Q = Air volume ft^3/min
C_e = Entrance coefficient.
A_t = Area of throat, ft.^{2}
h_t = Static suction in throat, inches W.G.

Coefficient of Entry:

$$C_e = \frac{\sqrt{h_v}}{h_t}$$ (h_v = Velocity pressure.)

Entrance Loss into Hood:

$$h_e = \frac{(1 - C_e{}^2)}{C_e{}^2}\,hv.$$

The Transporting Velocity for material varies with the size, specific gravity and shape of the material. (Dalla Valle).

Vertical Lifting Velocity:

$$V = 10{\cdot}7 \frac{s}{s+1} \times d^{0.57} \qquad V = 13{,}300 \frac{s}{s+1} \times d^{0.57}$$

Horizontal Transport Velocity:

$$V = 8{\cdot}4 \frac{s}{s+1} \times d^{0.40} \qquad V = 6{,}000 \frac{s}{s+1} \times d^{0.40}$$

V = Velocity m/s.
s = Specific gravity of material.
d = Average dia. of largest particle in mm.

V = Velocity ft./min.
s = Specific gravity of material.
d = Average dia. of largest particle in inch.

Friction Loss of Mixture:

$$\frac{F_m}{F_a} = 1 + 0{\cdot}32 \frac{W_s}{W_a}$$

where F_m = Friction loss of mixture.
F_a = Friction loss of air.
W_s = Mass of solid.
W_a = Mass of air.

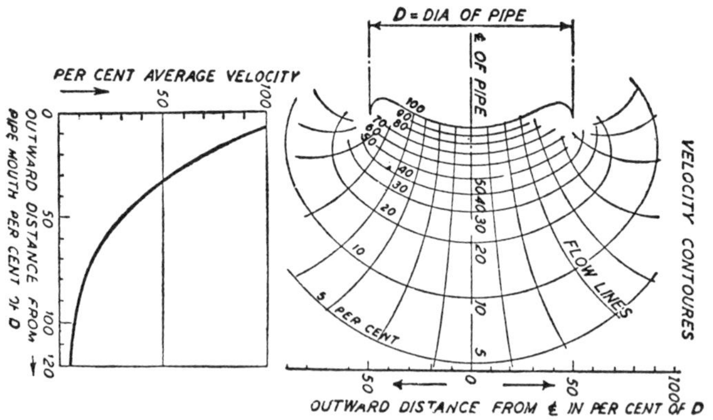

Velocity contours and flow directional lines in radial plane of circular suction pipe.

Velocities for Dust Extraction:

Material	*m/s.*	*ft./min.*
Sawdust and Shavings, light... ...	10 to 15	2,000 to 3,000
Sawdust and Shavings, heavy ...	17 to 23	3,500 – 4,500
Grinding and Foundry Dust... ...	17 to 23	3,500 – 4,500
Sandblast	17 to 23	3,500 – 4,500
Lead Dust	20 to 25	4,000 – 5,000
Cotton Lint Flyings	7·5 to 10	1,500 – 2,000
Grain Dust, Rubber Dust	10 to 15	2,000 – 3,000
Bakelite Moulding Powder	15 to 17	3,000 – 3,500
Bakelite Dust	10 to 12	2,000 – 2,500

Velocities for Pneumatic Conveying:

Material	*m/s.*	*ft./min.*
Coal, Powdered	20 to 28	4,000 to 5,500
Sawdust	20 to 30	4,000 – 6,000
Cork	17 to 28	3,500 – 5,500
Pulp Chips	22 to 36	4,500 – 7,000
Wool, Jute, Cotton	22 to 30	4,500 – 6,000
Coffee Beans	15 to 20	3,000 – 4,000
Ashes, Powdered Clinker	30 to 43	6,000 – 8,500
Sand, Cement	30 to 46	6,000 – 9,000
Lime	25 to 36	5,000 – 7,000
Flour	17 to 30	3,500 – 6,000
Rags	22 to 33	4,500 – 6,500
Corn, Wheat, Rye	25 to 36	5,000 – 7,000
Oats	22 to 30	4,500 – 6,000

USUAL DIMENSIONS OF CYCLONE SEPARATORS FOR DUST COLLECTING.

SEPARATION FACTOR OF CYCLONES:

$$S = \frac{V^2}{gr}$$

V = tangential velocity

r = radius of cyclone = $\frac{A}{2}$

g = gravity of earth

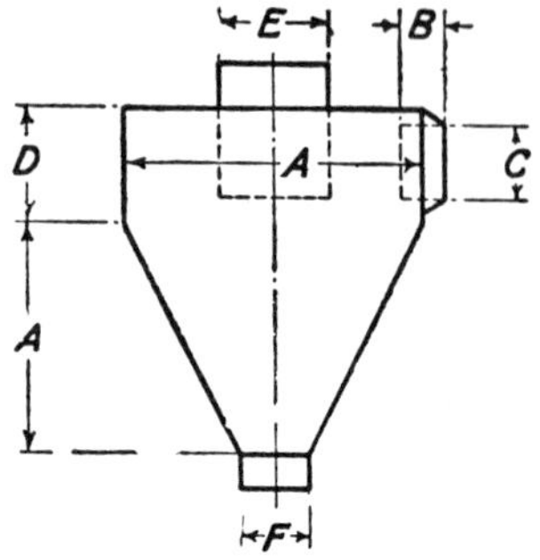

Fan Outlet		Dimensions in inches					
Dia., in.	*Area, sq. in.*	A	B	C	D	E	F
5	20	30	3	9	14	9	5
7–8	35–50	42	4	12	19	13	8
10	78	54	5	18	23	18	8
11–12	95–113	60	5½	21	26	20	12
13–14	133–154	66	6½	24	30	24	12
16–17	201–227	78	8	30	36	28	12
18	254	84	8½	32	39	32	13
19–20	283–314	90	9	34	42	34	13
22	380	98	10	40	48	39	14
23–24	415–452	102	11	43	50	41	14
30	707	126	12	60	66	58	15
34	908	138	13	66	72	68	15

Air Outlet = 4 × inlet area. Leaving Air Velocity = 350 to 450 ft/min.

ONE MICRON = 0.001 mm. is the usual unit for measuring the dimensions of fine particles.

TYLER STANDARD SCREEN SCALE:

Meshes per inch	10	20	35	48	65	100	150	200	325
Micron Scale ...	1650	830	420	300	220	150	110	74	44

Dust = over 10 microns. *Clouds* = 0·1 to 10·0 microns. *Smoke* = below 0·001 microns.

MINIMUM PARTICLE SIZE FOR WHICH VARIOUS SEPARATOR TYPES ARE SUITABLE:

Gravity	200 microns
Inertial	50 to 150
Centrifugal, large dia. cyclone	40 to 60
Centrifugal, small dia. cyclone	20 to 30
Fan Type	15 to 30
Filter	0·5
Scrubber	0·5 to 2·0
Electrical	0·001 to 1·0

Size of Particles:

Outdoor Dust	0·5 microns
Sand Blasting	1·4
Foundry Dust	1·0 to 200
Granite Cutting	1·4
Coal Mining	1·0
Raindrops	500 to 5000
Mist	40 to 500
Fog	1 to 40
Fly Ash	3 to 70
Pulverized Coal	10 to 400

Automatic Control Apparatus or Regulators:

These operate to maintain measured conditions subject to changes within certain limits. They consist of the two following parts:

The Controller or Measuring Device, indicating the change of the required conditions, and the magnitude of the change; for instance, thermostats, humidistats, pressure controls, etc.

The Operating Device, carrying out the corrective action based on the indications of the controller; for instance, magnetic valves, motorized valves, etc.

Classification of Controls according to operation:

(*a*) Two positions, or "off" and "on", or positive acting control.
The operating device is either fully open or fully shut; no intermediate positions.

(*b*) Gradual, intermediate, modulating, or graduated acting control. The motion of the operating mechanism is proportional to the change in required conditions.

Classification of Controls according to operating medium:

(*a*) Self-contained, or direct operated apparatus. Controller and operating mechanism in self-contained unit.

(*b*) Electric control systems.
Electricity is used for operating the controlling devices, either line voltage or low voltage circuits are switched on and off by the controller.

(*c*) Pneumatic control systems.
Compressed air is used for operating the controlling devices. Controller and operating mechanism are connected to compressed air piping, the air pressure being varied by the measuring devices, according to the change in the required conditions.

EQUIPMENT

Controllers:

Thermostats are sensitive to changes of temperature and cause the regulation of the flow of heating or cooling medium (steam, hot water, cold water, brine, etc.)
Room thermostat for installation in rooms.
Immersion thermostat for tubings.
Duct thermostats for air ducts.

Pressure and Vacuum Controls are sensitive to changes of pressure.

Humidistats or Psychrostats are sensitive to the difference between dry bulb and wet bulb temperature.

Clocks and Time Switches are used for maintaining different conditions by day and by night, for starting up and switching off plants, etc.

Operating Mechanisms:

Electric Motors fitted to valves, dampers, i.e., motorized valves, motorized dampers, etc.

Magnetic or Solenoid Valves, dampers, etc.

Diaphragm Valves, operated by fluid pressure on diaphragm.

Basic systems: 1. Single duct system.
2. Dual duct system.
3. Induction system.

SINGLE DUCT SYSTEM

A central plant delivers conditioned air through high velocity ducting to attenuator boxes throughout the building. The distribution boxes have sound attenuators.

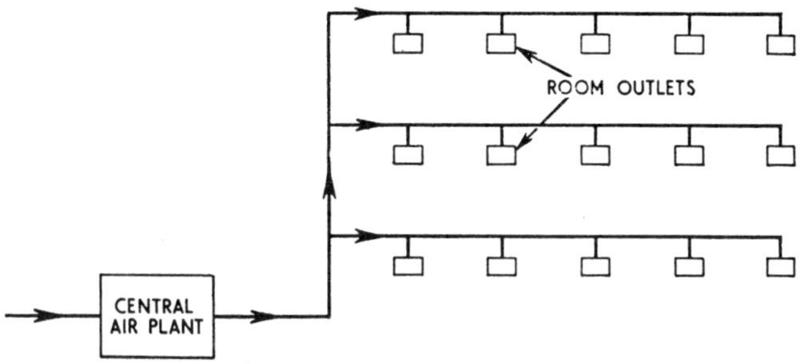

Advantages:

Space saving through use of high velocity small diameter ducts.
Low initial cost.
Zone control can be used.

Disadvantages:

Large volume of air to be treated in central plant.
Individual room control not possible.
Recirculating system necessary.

DUAL DUCT SYSTEM

A central plant delivers two streams of air through two sets of high velocity ducting to attenuator mixing boxes in the various rooms. The two streams are at different temperatures and humidities, and the ratio in which they are mixed can be set at each mixing box independently of all others in the building.

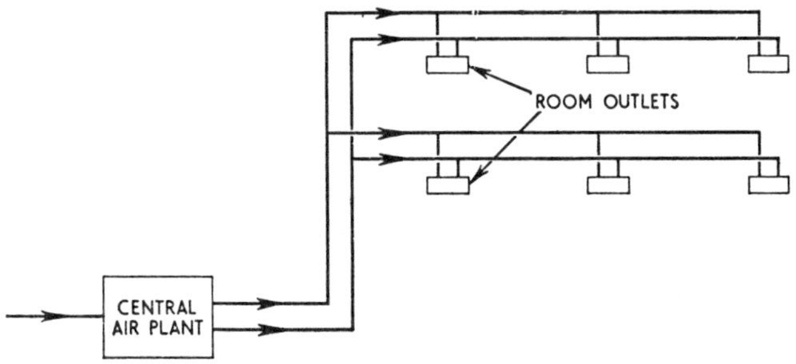

Advantages:

Space saving through use of high velocity small diameter ducts.
Individual room control—zoning not necessary.
Flexible in operation.
Units are available for mounting under window-sills or in ceilings.
Suitable for use where large air volumes are required.

Disadvantages:

Two sets of air ducting are needed, using more space.
More air has to be treated in central plant.
Recirulating system is necessary.

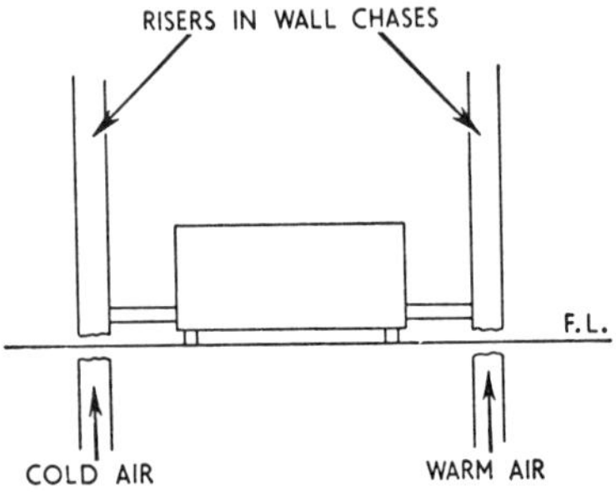

DUAL DUCT SYSTEM—TYPICAL CONNECTIONS TO ROOM UNIT

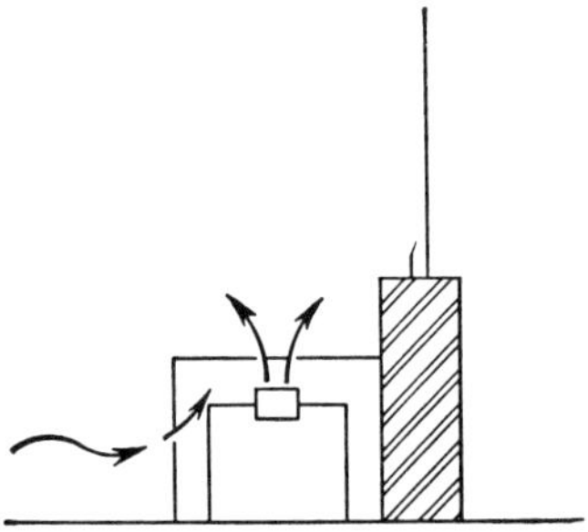

TYPICAL ARRANGEMENT OF UNIT UNDER WINDOW SILL

Control

An individual pneumatic or electric room thermostat adjusts the ratio in which air streams are mixed.

Each stream is kept at constant temperature and humidity by automatic controls in plant room.

INDUCTION SYSTEM

A central air plant delivers conditioned air through high velocity ducting to induction units in the rooms. Water from a central plant is also supplied to the induction units. The conditioned, or primary, air supplied to the units induces room, or secondary, air through the unit. This induced secondary air passes over the water coil and is thus heated or cooled.

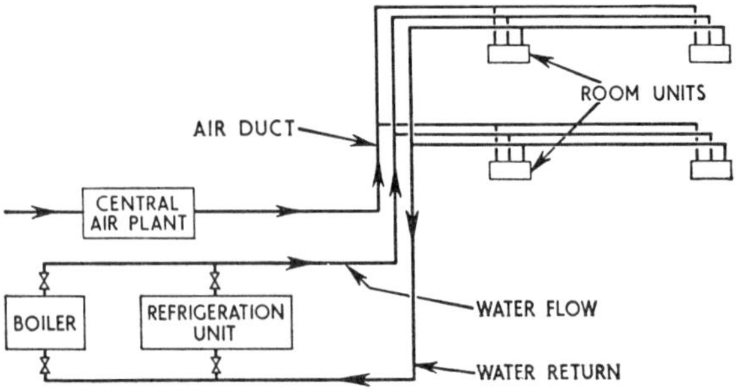

Advantages:

Space saving through use of high velocity small diameter ducts.
Low running costs.
Individual room control.
Very suitable for modular building layouts.
Central air plant need handle only part of the air treated.
Particularly applicable to perimeter zones of large buildings—hotels, hospitals, schools and flats, etc.
Suitable for large heat loads with small air volumes.

Disadvantages:

Higher capital cost.

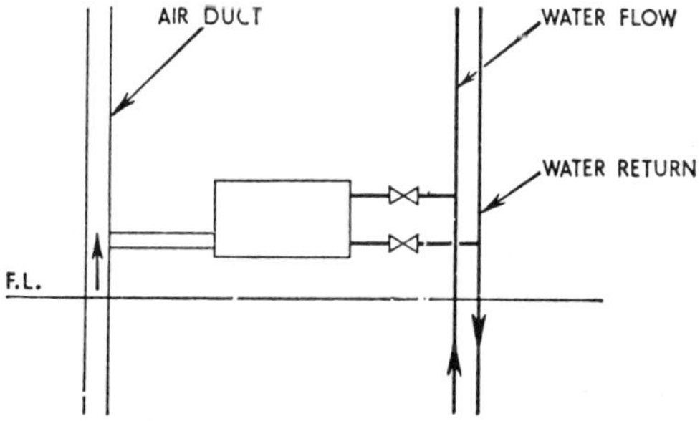

INDUCTION SYSTEM — TYPICAL CONNECTIONS TO ROOM UNIT

PRIMARY AIR PLANT

Central air handling plant consists of:

1. Intake louvres.
2. Air filter.
3. Preheater.
4. Humidifier or dehumidifier.
5. Reheater.
6. Fan.
7. Silencer or sound absorber.

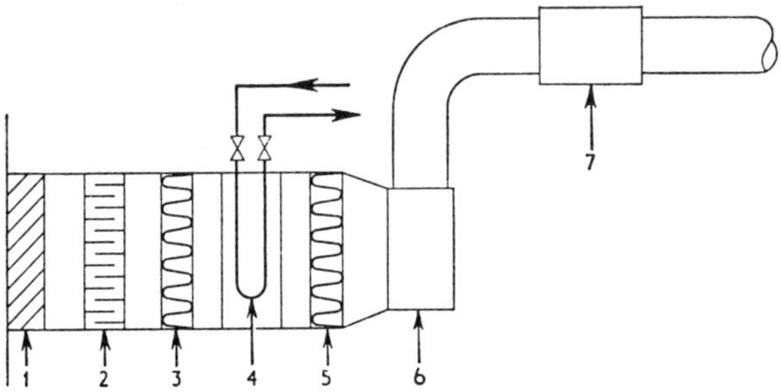

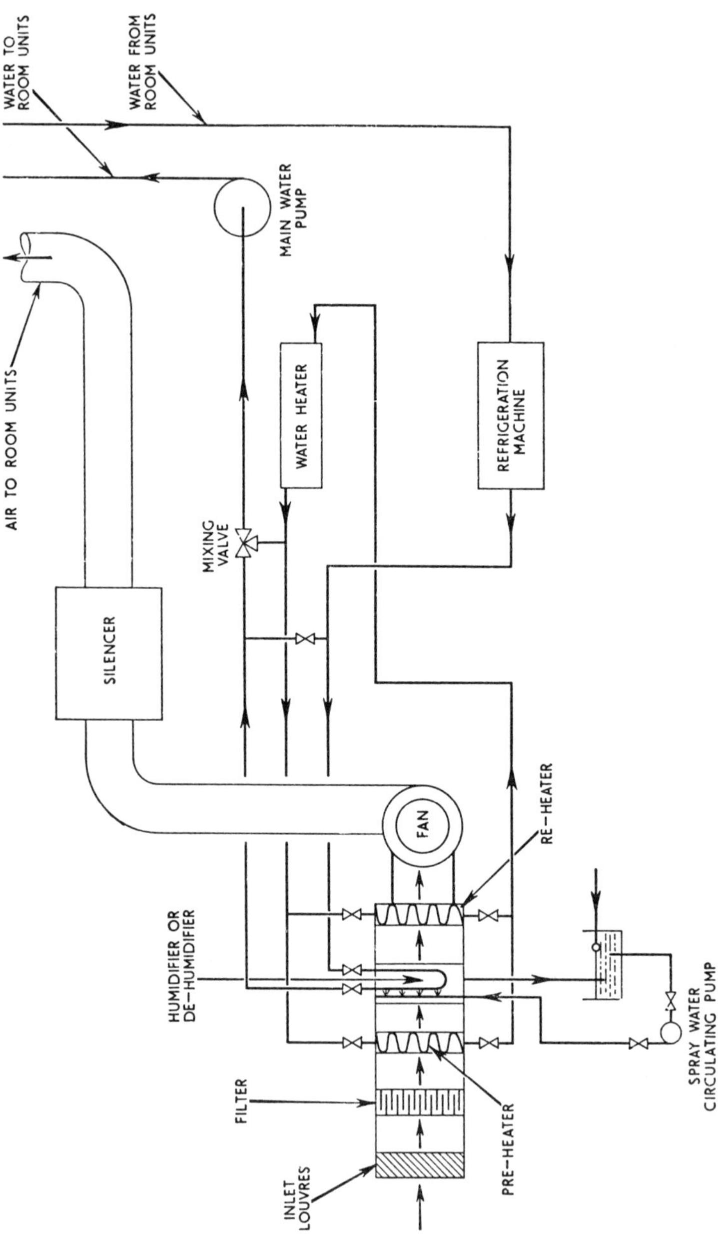

SCHEME OF CENTRAL PLANT FOR INDUCTION SYSTEM

Air-handling plant can be installed either in the basement or on the roof.

To provide heated water and chilled water for the air plant a boiler and a refrigerator unit are provided. For reasons of weight these are usually installed in the basement; they need not necessarily be near the air-handling plant.

AIR DISTRIBUTION FOR HIGH VELOCITY SYSTEMS

High pressures are used to obtain the advantages of high velocities and small ducts.

Pressure at inlet to farthest unit ...	10 to 25 mm. w.g.	½ to 1 in. w.g.
Typical pressure at fan	125 to 150 mm. w.g.	5 to 6 in. w.g.
Air velocities in ducts	15 to 20 m./s.	3,000 to 4,000 ft/min.

Ducts for these systems are always circular. Welded construction is used to avoid leaks at the higher air pressures. Spirally wound ducting is also available.

Alternatively special fittings are obtainable.

Ducts must be insulated.

ROOM DISTRIBUTION UNITS

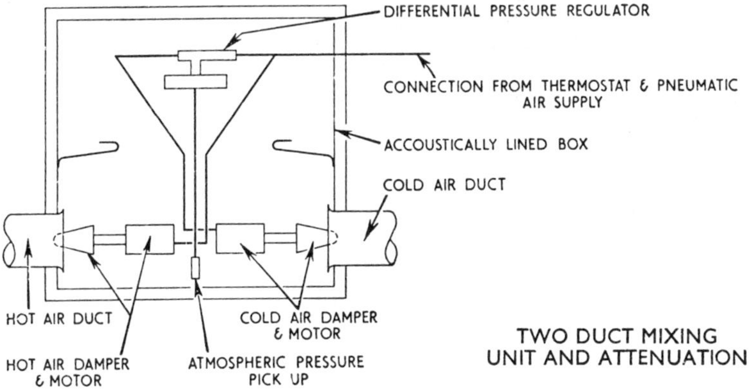

TWO DUCT MIXING UNIT AND ATTENUATION

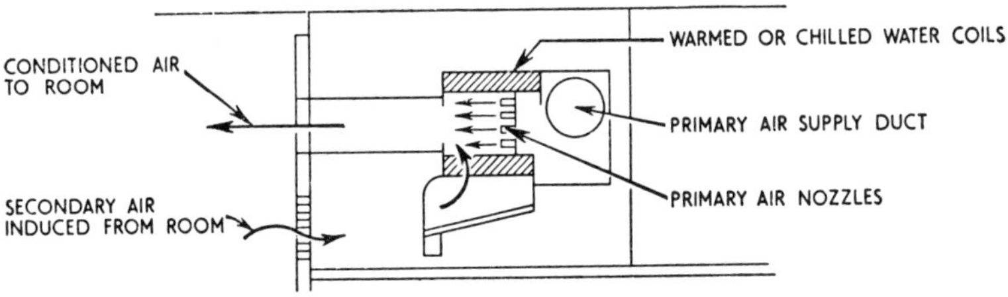

CEILING MOUNTED INDUCTION UNIT

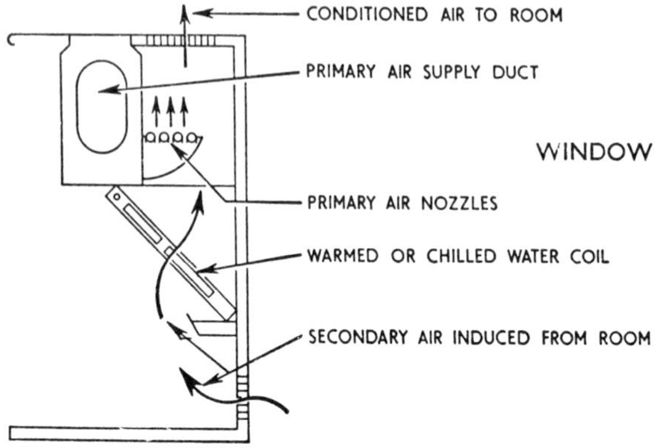

WINDOW SILL INDUCTION UNIT

DESIGN DATA

Air Heater: 800–1,200 ft./min. over free area (4·0–6·0 m./s.).
500–600 ft./min. over total area (2·5–3·0 m./s.).

Supply of primary fresh air per person:
0·010 m.^{3}/s., 1,400 cu. ft./hr.
0·015 m.^{3}/s., 1,800 cu. ft./hr. where heavy smokers are present.

Air velocities in primary ducts:
3,000–4,000 ft./min. (15–20 m./s.).

Induction units:
Secondary air/primary air — 3/1.
Pressure of primary air — ¾ in. w.g. (20 mm. w.g.).

Design sequence for induction units:

1. Calculate heat gain.
2. Select suitable room unit from manufacturer's data.
3. Temperature of chilled water is settled by dew-point of room air.
4. Allow primary air quantity from data above.
5. Design ducting, using "equivalent length" method. Table on page X.39 takes into account static regain.
6. Water pipes are designed and sized exactly as for a conventional radiator system.

FRICTION LOSS THROUGH FITTINGS

EL = Equivalent Length of Pipe.

	FITTING		4 in. 100mm	6 in. 150mm	8 in. 200mm	10 in. 250mm	12 in. 300mm	14 in. 350mm	16 in. 400mm	18 in. 450mm	20 in. 500mm
	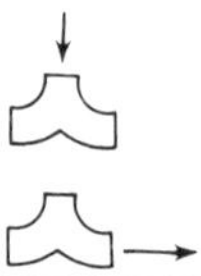	LE ft.	-9	-15							
		LE m.	-3	-5							
		LE ft.	12	21							
		LE m.	4	7							

	FITTING		4 in. 100mm	6 in. 150mm	8 in. 200mm	10 in. 250mm	12 in. 300mm	14 in. 350mm	16 in. 400mm	18 in. 450mm	20 in. 500mm
	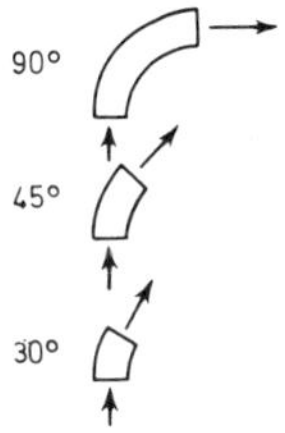	LE ft.	3	4	7	10	12	15	18	21	24
		LE m.	1	1·2	2	3	4	5	6	7	8
		LE ft.	1	2	4	5	6	8	9	10	12
		LE m.	0·3	0·6	1·2	1·5	1·8	2·4	3	3	4
		LE ft.	1	1	2	3	4	5	6	7	8
		LE m.	0·3	0·3	0·6	1	1·2	1·5	1·8	2	2·4
	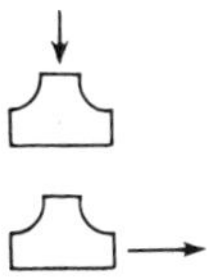	LE ft.	–5	–9	–13	–17	–22	–26	–31	–36	–42
		LE m.	–1·5	–3	–4	–5	–7	–8	–10	–11	–13
		LE ft.	12	21	30	40	52	63	75	87	100
		LE m.	4	6	10	12	16	19	22	25	30
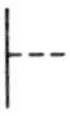	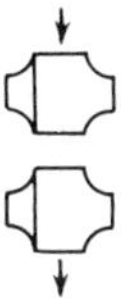	LE ft.	–13	–22	–32	–42	–54	–66	–78	–91	–105
		LE m.	–4	–7	–10	–13	–16	–20	–24	–28	–32
		LE ft.	13	22	32	42	54	66	78	91	105
		LE m.	4	7	10	13	16	20	24	28	32
	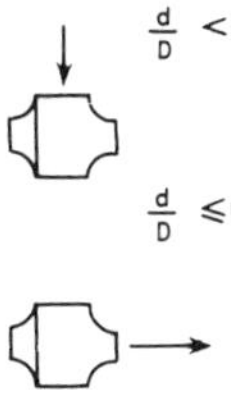 d/D <0·4	LE ft.	–8	–10	–11	–13	–17	–20	–24	–28	–32
		LE m.	–2·4	–3	–3·3	–4	–5	–6	–7	–9	–10
	d/D ⩽0·4	LE ft.					–9	–9	–10	–10	–10
		LE m.					–3	–3	–3	–3	–3
		LE ft.	13	22	32	42	54	56	78	91	105
		LE m.	4	7	10	13	16	20	24	28	32
	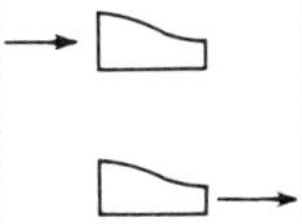	LE ft.		–21	–30	–40	–52	–63	–75	–87	–100
		LE m.		–6	–10	–12	–16	–19	–22	–25	–30
		LE ft.	13	22	32	42	54	66	78	91	105
		LE m.	4	7	10	13	16	20	24	28	32
		LE ft.	14	23	37	49	62	76	90	106	121
		LE m.	4	7	11	15	19	23	27	32	37

CALCULATION OF PRESSURE LOSS

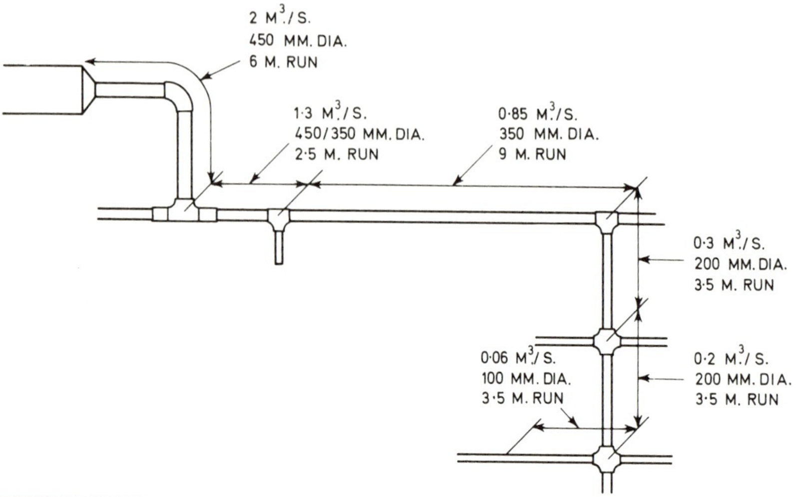

Quantity	*Symbol*	*Dia. mm.*	*Pressure Drop mm./m.*	*Duct M.*	*Length m.*	*Pressure Drop mm. W.G.*	*Air Velocity m./s.*	
2		450	0·4	32				
				6		13.6		
				7	3·4			
				−11				
1·3			0·16	25	0			
				−25				
		350	0·52	20				
				2·5	2·5	1·3		
				−20				
0·85			0·28	+20				
				9	23	6·4		
				−6				
0·3		200	0·71	10				
				3·5	3·5	2·5		
				−10				
0·2			0·35	10				
				3·5	10·2	3·6		
				−3·3				
0·06		100	1·7	4	7·5	12·8		
				3·5				
						40·2		

Heated air is blown across a door opening to prevent or reduce ingress of cold atmospheric air.

Applications: Door-less shop fronts.
Workshop entrances.
Doors of public buildings which are frequently opened.

Temperatures:

Discharge Temperature for small installation		35–50°C.
,, ,, ,,	large installation	25–35°C.
Suction Temperature		5–15°C.

Air Velocity:

Flow from above	5–15 m/s.
,, ,, below	2– 4 m/s.
,, ,, side	10–15 m/s.

Air Quantity: Quantity required depends on too many variable factors for exact calculation to be possible. The quantity should be made as large as possible consistent with practicable heat requirements. Suggested values: 2,000–5,000 $m.^3/m.^2$ hr. of door opening. In very exposed situations or other difficult cases this can be increased to 10,000 $m.^3/m.^2$ hr.

Let V_o = Quantity of air entering in absence of curtain.
V = Quantity blown by curtain.

For one-sided curtain $V = 0{\cdot}45\ V_o$.

For two-sided curtain $V = 0{\cdot}9\ V_o$.

Example: Width of door 4 m. Height of door 2 m. Speed of outdoor air 2 m/s.

$$\therefore\ V_o = 4 \times 2 \times 2 = 16 m.^3/s.$$
$$\therefore\ V = 0{\cdot}45 \times 16 = 7{\cdot}2\ m.^3/s.$$

Discharge velocity, say 10 m/s.

$$\therefore\ \text{Grille area} = \frac{7{\cdot}2}{10} = 0{\cdot}72\ m.^2$$

Height of grille = height of door = 2 m.

$$\therefore\ \text{Width of grille} = \frac{0{\cdot}72}{2} = 0{\cdot}36\ m.$$

Section XI

HYDRAULICS

Pressure due to Head of Water:

$$p = \rho g h.$$

h = head of water, m.
g = acceleration due to gravity = 9·81 m./s.²
ρ = density, kg./m.³
p = pressure, N./m.²

Bernoulli's Theorem: The total energy of unit mass of a perfect fluid (or total head) flowing through any section of a stream is the sum of its kinetic head, pressure head, and elevation head, and is constant along any stream line.

$$\frac{v_1^2}{2g} + \frac{P_1}{\rho g} + Z_1 = \frac{v_2^2}{2g} + \frac{P_2}{\rho g} + Z_2 = H.$$

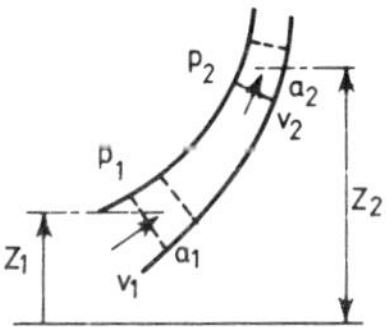

Z_1, Z_2 = Heights.
v_1, v_2 = Velocities.
p_1, p_2 = Pressures.
ρ = Density.

Venturimeter: Instrument for measuring the quantity of water flowing through a pipe.

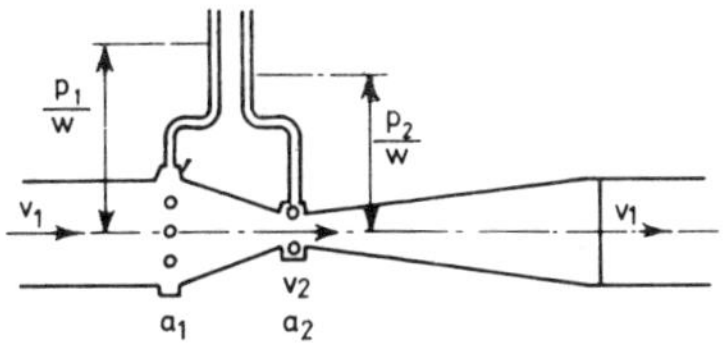

$$Q = \frac{C_v a_2}{\sqrt{1 - \left(\frac{a_2}{a_1}\right)^2}} \times \sqrt{2\frac{P_1 - P_2}{\rho}}$$

Q = Quantity of water in m.³/s.
a_1 a_2 = Areas in m.²
P_1 P_2 = Pressures in N./m.²
C_v = Coefficient of velocity
= 0·96 to 0·99.
ρ = Density in kg./m.³

Discharge of Water through Small Orifices:

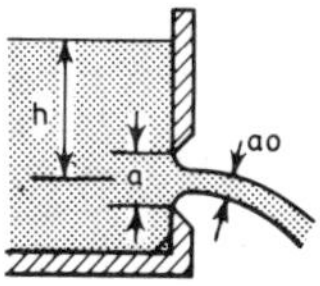

$$v = C_v \sqrt{2gh}.$$

$$C_c = \frac{a_o}{a}.$$

$$C = C_v \times C_c.$$

$$Q = v a_o$$

$$= C a \sqrt{2gh}.$$

where: v = Velocity in m./s.
C_v = Coefficient of velocity
= 0·96 to 0·99.
g = Gravity of earth
= 9·81 m./s.²
h = Height in m.
a = Height of orifice, m.²
a_o = Area of flow, m.²
C_c = Coefficient of contraction.
C = Coefficient of discharge.

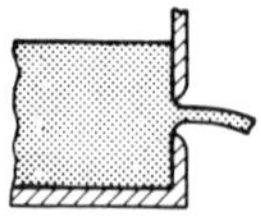

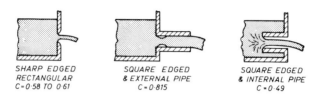

SHARP EDGED
RECTANGULAR
C = 0·58 TO 0·61

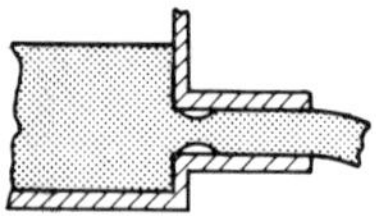

SQUARE EDGED
& EXTERNAL PIPE
C = 0·815

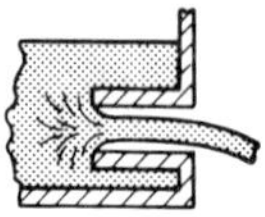

SQUARE EDGED
& INTERNAL PIPE
C = 0·49

DISCHARGE THROUGH NOTCHES

Rectangular Notch:

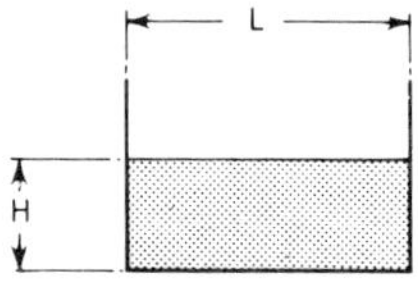

Discharge—$Q = \frac{2}{3}\, Cd\sqrt{2g}\, LH^{\frac{3}{2}}$

Triangular or "V" Notch:

Most satisfactory type for measuring of quantity of flowing water.

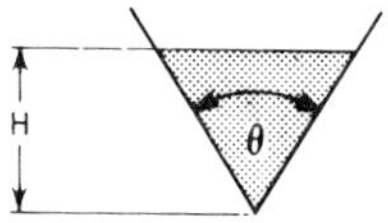

Discharge—

angle of notch

$$Q = \frac{8}{15} C_d \tan\frac{\theta}{2} \sqrt{2g}\, H^{\frac{5}{2}}$$

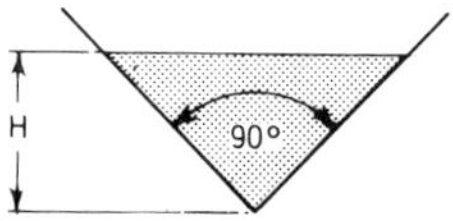

90° notch

$$Q = \frac{8}{15} C_d \sqrt{2g}\, H^{\frac{5}{2}}.$$

Q = Discharge through notch ($m.^3/s.$).
H = Height of water surface above sill (m.).
L = Breadth of notch (m.).
C_d = Coefficient of discharge
= 0·62 average.

The Pitot Tube is an instrument for measuring the velocity head of a flowing fluid.

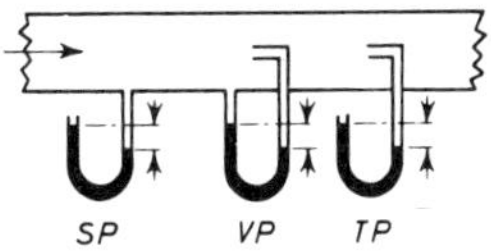

$$v = K\sqrt{2gH}$$

SP = Static pressure.
VP = Velocity pressure.
TP = Total pressure.
(Velocity Head) = (Total Head) — (Static Head)
v = Velocity
H = Head.
K = Coefficient for the instrument.
= 1·0 for British Standard pitot-static-tube.

FLUID FLOW THROUGH PIPES

Total Loss of Head (H) = Loss of head due to friction (H_1).
+ Loss of head due to obstructions (H_2).
(Fittings, Valves, Bends, etc.).

$$H = H_1 + H_2.$$

Loss of Head due to Friction:

$$H_1 = li;\quad i = \frac{H_1}{l}.$$

$$i = \frac{4f}{d} \cdot \frac{v^2}{2g}$$

i = Slope of hydraulic gradient.
= Head lost in resistance per unit length of pipe.
l = Length of pipe (m.).
d = Inside diameter of pipe (m.).
v = Velocity of fluid (m/s.).
f = Friction factor.

Friction factor (f) is a function of the "Reynold's Number"

$$f = \phi\left(\frac{dv\rho}{\mu}\right)$$

$$\left(\frac{dv\rho}{\mu}\right) = R_e.$$

ρ = Density of fluid.
μ = Absolute viscosity of fluid.
R_e = Reynolds' number.

Viscous, or streamline flow—motion solely in axial direction—low velocities and highly viscous fluids.

$$R_e < 2{,}000.$$

Turbulent or eddying flow—radial velocity components, eddies.

$$R_e > 3{,}000$$

Region of Discontinuity (Critical Region) R_e = From 2,000 to 3,000. (Lower and higher critical velocity).

Loss of Head due to Obstructions (Fittings):

$$H_2 = F\frac{v_1^2}{2g}\rho$$

H_2 = Loss of head in m.
v_1 = Actual velocity of water at the point where the resistance is encountered in m/s.
F = Coefficient of resistance (always applied to smaller velocity).
ρ = Density of fluid.

VELOCITY HEADS AND THEORETICAL VELOCITIES OF WATER

$$h = \frac{v^2}{2g}.$$

h = Head in m.
v = Velocity in m./s.
g = Gravity of earth = 9·81 m./s.[2]

v m./s.	h m.	v m./s.	h m.	v m./s.	h m.	v m./s.	h m.
0·01	0·0000051	0·80	0·0326	1·60	0·130	2·40	0·293
0·05	0·000127	0·85	0·0368	1·65	0·139	2·45	0·306
0·10	0·00051	0·90	0·0413	1·70	0·147	2·50	0·318
0·15	0·00115	0·95	0·046	1·75	0·156	2·55	0·331
0·20	0·00204	1·0	0·0510	1·80	0·165	2·60	0·344
0·25	0·00319	1·05	0·0561	1·85	0 174	2 65	0·358
0·30	0·00459	1·10	0·0617	1·90	0·184	2·70	0·371
0·35	0·00624	1·15	0·0674	1·95	0·194	2·75	0·385
0·40	0·00815	1·20	0·0734	2·0	0·204	2·80	0·400
0·45	0·0103	1·25	0·0797	2·05	0·214	2·85	0·414
0·50	0·0127	1·30	0·0862	2·10	0·225	2·90	0·429
0·55	0·0154	1·35	0·0930	2·15	0·236	2·95	0·444
0·60	0·0183	1·40	0·100	2·20	0·246	3·0	0·459
0·65	0·0125	1·45	0·107	2·25	0·258		
0·70	0·0250	1·50	0·115	2·30	0·269		
0·75	0·0287	1·55	0·122	2·35	0·281		

$$h = \frac{v^2}{2g}.$$

h = Head in ft.
v = Velocity in ft./s.
g = Gravity of earth = 32·2 ft./s.[2]

v ft./s.	h ft.	v ft./s.	h ft.	v ft./s.	h ft.	v ft./s.	h ft.
0·1	0·0002	2·1	0·068	4·1	0·261	6·1	0·578
0·2	0·0006	2·2	0·075	4·2	0·274	6·2	0·597
0·3	0·0014	2·3	0·082	4·3	0·289	6·3	0·616
0·4	0·0025	2·4	0·089	4·4	0·301	6·4	0·636
0·5	0·0039	2·5	0·097	4·5	0·314	6·5	0·656
0·6	0·0056	2·6	0·105	4·6	0·329	6·6	0·676
0·7	0·0076	2·7	0·113	4·7	0·343	6·7	0·697
0·8	0·0099	2·8	0·122	4·8	0·358	6·8	0·718
0·9	0·0126	2·9	0·131	4·9	0·373	6·9	0·739
1·0	0·0155	3·0	0·140	5·0	0·388	7·0	0·761
1·1	0·019	3·1	0·149	5·1	0·404	7·1	0·783
1·2	0·022	3·2	0·159	5·2	0·420	7·2	0·805
1·3	0·026	3·3	0·169	5·3	0·436	7·3	0·827
1·4	0·030	3·4	0·179	5·4	0·453	7·4	0·850
1·5	0·035	3·5	0·190	5·5	0·470	7·5	0·874
1·6	0·040	3·6	0·201	5·6	0·487	7·6	0·897
1·7	0·045	3·7	0·212	5·7	0·505	7·7	0·921
1·8	0·050	3·8	0·224	5·8	0·522	7·8	0·945
1·9	0·056	3·9	0·236	5·9	0·541	7·9	0·969
2·0	0·062	4·0	0·248	6·0	0·559	8·0	0·994

RESISTANCE OF VALVES AND FITTINGS TO FLOW OF FLUIDS IN TERMS OF EQUIVALENT LENGTH OF STRAIGHT PIPE

Description of Fitting		*Nominal Diameter*										
		in.	$\frac{1}{2}$	$\frac{3}{4}$	1	$1\frac{1}{4}$	$1\frac{1}{2}$	2	$2\frac{1}{2}$	3	4	5	6
		mm.	15	20	25	32	40	50	65	80	100	125	150
Globe Valve ...		E.L. ft.	13	16	26	35	40	55	65	80			
		m.	4	5	8	11	12	17	20	24·5			
Angle Valve ...		E.L. ft.	8	11	15	18	20	27	32	40			
		m.	2·5	3·5	4·5	5·5	6	8·3	10	12			
Gate Valve ...		E.L. ft.	0·3	0·5	0·5	0·5	1	1	1·5	2	2·5	3	3
		m.	0·09	0·15	0·15	0·15	0·3	0·3	0·45	0·6	0·75	0·9	0·9
Elbow		E.L. ft.	1	2	2	3	4	5	6	8	11	13	17
		m.	0·3	0·6	0·6	0·9	1·2	1·5	1·8	2·5	3·5	4	5·2
Long Sweep Elbow		E.L. ft.	1	1·5	2	2·5	3	3	5	4	6	8	10
		m.	0·3	0·45	0·6	0·75	0·9	1·0	1·2	1·5	1·8	2·5	3·0
Run of Tee		E.L. ft.	1	1·5	2·5	2·5	3	3	4	5	6	8	10
		m.	0·3	0·45	0·75	0·8	0·9	1·0	1·2	1·5	1·8	2·5	3·0
Run of Tee, reduced to $\frac{1}{2}$ ...		E.L. ft.	1	2	2	3	4	5	6	8	11	13	17
		m.	0·3	0·5	0·7	0·9	1·2	1·5	1·8	2·5	3·5	4	5·2
Branch of Tee		E.L. ft.	3·5	5	6	8	10	13	15	18	24	30	35
		m.	1·1	1·5	1·8	2·5	3·0	4	4·5	5·5	7·3	9	11
Sudden Enlargement	$\frac{d}{D}=\frac{1}{4}$	E.L. ft.	1	2	2	3	4	5	6	8	11	13	17
		m.	0·3	0·5	0·7	0·9	1·2	1·5	1·8	2·5	3·5	4	5·2
	$\frac{d}{D}=\frac{1}{2}$	E.L. ft.	1	1·5	2	2·5	3	3·5	4	5	7	9	11
		m.	0·3	0·45	0·6	0·75	0·9	1·1	1·2	1·5	2·1	2·7	3·5
	$\frac{d}{D}=\frac{3}{4}$	E.L. ft.	0·3	0·5	0·5	1	1	1	1·5	2	2·3	3	3
		m.	0·09	0·15	0·15	0·25	0·3	0·35	0·45	0·6	0·75	0·9	1·0
Sudden Contraction	$\frac{d}{D}=\frac{1}{4}$	E.L. ft.	0·8	1·0	1·2	1·5	2·0	2·5	3	4	5	6	8
		m.	0·25	0·3	0·35	0·45	0·6	0·75	0·9	1·2	1·5	1·8	2·5
	$\frac{d}{D}=\frac{1}{2}$	E.L. ft.	0·5	0·8	1·0	1·2	1·5	2	2	3	4	5	6
		m.	0·15	0·25	0·3	0·35	0·45	0·6	0·6	0·9	1·2	1·5	1·8
	$\frac{d}{D}=\frac{3}{4}$	E.L. ft.	0·4	0·5	0·6	1·0	1·0	1·5	1·5	2·0	2·5	3	3·5
		m.	0·12	0·15	0·18	0·3	0·3	0·4	0·5	0·6	0·8	0·9	1·1
Ordinary Entrance		E.L. ft.	1	1	1·5	2	2·5	3	3·5	4·5	6	8	10
		m.	0·3	0·3	0·45	0·6	0·75	0·9	11	1·4	1·8	2·5	3·0
Boiler, Radiator, Tank		E.L. ft.	3	4·5	6	7·5	9	12	15	18	21	24	40
		m.	0·9	1·4	1·8	2·25	2·7	3·7	4·5	5·5	6·5	7·3	12

Notation:

V_1 V_2 = Absolute velocity of entering and leaving water, respectively.
W_1 W_2 = Velocity of water relative to wheel at inlet and outlet, respectively.
U_1 U_2 = Tangential velocity of wheel at inlet and outlet, respectively.
V_{w1} V_{w2} = Velocity of whirl at inlet and outlet, respectively, parallel to direction of motion.
V_{F1} V_{F2} = Velocity of flow at inlet and outlet, respectively, perpendicular to direction of motion.
θ = Angle between relative velocity & direction of motion at inlet.
ϕ = Angle between relative velocity & direction of motion at outlet.

VELOCITY TRIANGLES
(For Inlet and Outlet)

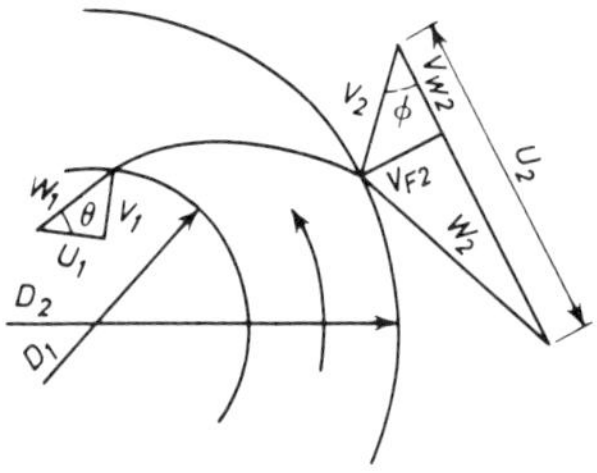

Theoretical Gross Lift $= H + \frac{v_d^2}{2g} = \frac{V_{w2} U_2}{g}$

Gross Lift $= h + h_f + \frac{v_d^2}{2g}$, $h = \frac{v_2^2}{2g} - \frac{v_1^2}{2g}$.

H = Total theoretical lift (m.).
v_d = Discharge velocity (m/s.).
h = Actual height water is lifted (m.).
h_f = Head lost in friction in delivery and suction pipe (m.).

Actual Efficiency $= \frac{M h_a g}{P}$.

Hydraulic Efficiency $= \frac{M h_g g}{P}$

where
M = mass of water pumped, kg./s.
h_a = actual lift, m.
h_g = gross lift, m.
g = acceleration of gravity.
P = input to pump, W.

Manometric Efficiency $= \frac{\text{Gross Lift}}{\text{Theoretical Gross Lift}}$.

The Specific Speed of a centrifugal pump is the speed at which the pump would deliver 1 m.³/s of water under a head of 1 m.

$$n_s = \frac{n \sqrt{Q}}{h^{\frac{3}{4}}}.$$

n_s = Specific speed in revolutions per minute.
n = Speed in revolutions per minute.
Q = Discharge in m^3/s.
h = Total head or lift in m.

Data for Centrifugal Pumps:

Actual lift not more than 140 ft. (or 50 m.) for 1 impeller.
Speed between 1,000 and 2,000 revolutions per minute.
Velocities in suction and delivery pipes 5 to 15 ft. per sec. (= 1·5 to 5 m/s.).
Blade angles θ and ϕ between 12° and 30°.

Characteristic Curves of Pumps for one speed, showing relation between head, discharge, efficiency and brake horse power.

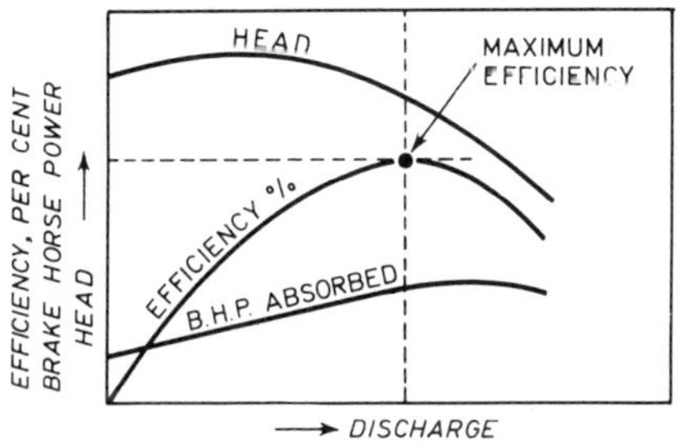

Pump Laws:

1. Water quantity delivered varies directly as speed $\frac{Q_2}{Q_1} = \frac{N_2}{N_1}$
2. Head developed varies as the square of speed. $\frac{H_2}{H_1} = \left(\frac{N_2}{N_1}\right)^2$
3. Horse-power required varies as the cube of speed. $\frac{hp_2}{hp_1} = \left(\frac{N_2}{N_1}\right)^3$

TYPES OF FANS

1. Axial Flow Fans or Propeller Fans:

Static head (or resistance head) up to 1 in. W.G. (30 mm. W.G.).

2. Radial Flow or Centrifugal Fans:

TYPES OF FAN BLADES

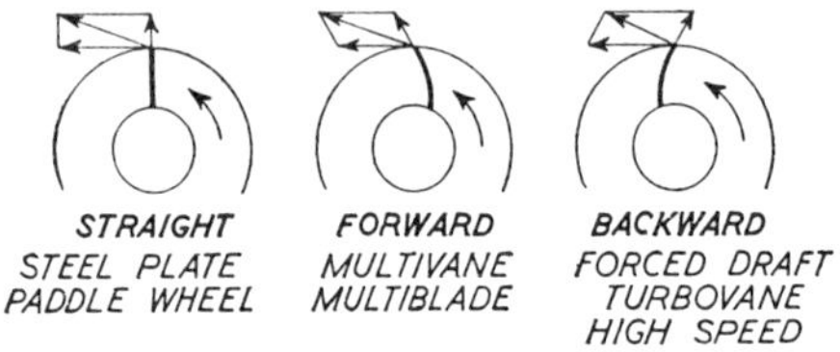

VELOCITY TRIANGLES

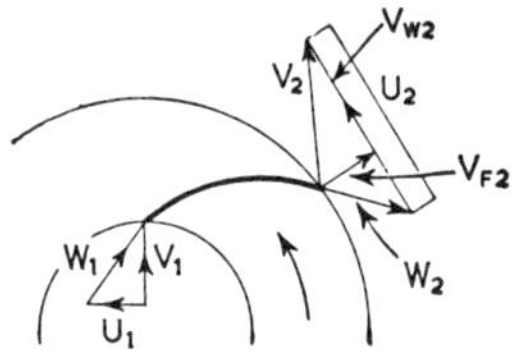

V_1 V_2 = Absolute velocity of entering & leaving air, respectively.
W_1 W_2 = Velocity of air relative to wheel at inlet & outlet, respectively.
U_1 U_2 = Tangential velocity of wheel at inlet & outlet, respectively.
V_{F2} = Velocity of flow at outlet.
V_{w2} = Velocity of whirl at outlet.

Theoretical Head:

$$Hth = \frac{U_2^2 - U_1^2}{2g} + \frac{W_1^2 - W_2^2}{2g} + \frac{V_2^2 - V_1^2}{2g} = \frac{V_2 V_{w2}}{g}.$$

Actual Head:

$$H_A = H_{th} \times E.$$

E = Efficiency = 0·40 for small fans.
0·60 for medium fans.
0·80 for large fans.

The Power:

$$p = \frac{M h g}{E}.$$

where p = Power in W.
M = Mass of air handled in kg./s.
h = Total head of air in m.
g = Acceleration of gravity.
E = Efficiency.

Static Pressure = Pressure of Air to Casing.

Dynamic Pressure = Pressure Head of Velocity $= \frac{v^2}{2g} \rho$ m. of air.

Static Efficiency $= \dfrac{0{\cdot}9982 \times \text{volume in m}^3\text{/s.} \times \text{static pressure in mm. W.G.}}{\text{Power input in Watts}}.$

Total or Mechanical Efficiency $= \dfrac{0{\cdot}9982 \times \text{volume in m}^3\text{/s.} \times \text{total pressure in mm. W.G.}}{\text{Power input in Watts}}.$

LAWS APPLYING TO CENTRIFUGAL FANS

1. Air capacity varies directly as speed. $\dfrac{Q_2}{Q_1} = \dfrac{N_2}{N_1}$
2. Static pressure varies as square of speed. $\dfrac{H_2}{H_1} = \left(\dfrac{N_2}{N_1}\right)^2$
3. Horse-power varies as cube of speed. $\dfrac{HP_2}{HP_1} = \left(\dfrac{N_2}{N_1}\right)^3$

Horse-power and static pressure vary directly as barometric pressure (at constant capacity and speed).

Horse-power and static pressure vary inversely as the absolute temperature.

Characteristic Curves for Centrifugal Fans:

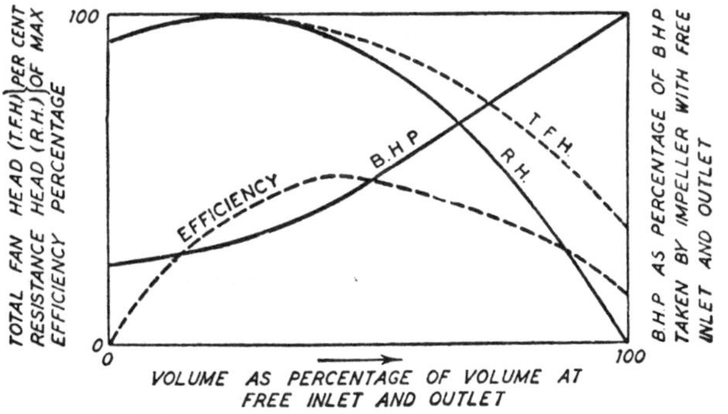

Characteristic Curves for Axial Flow Fans:

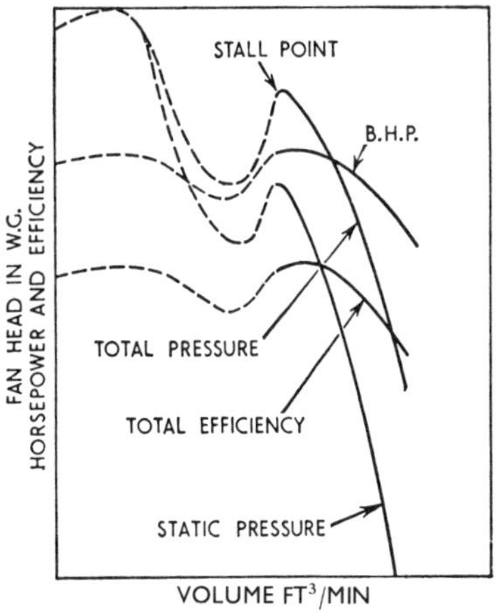

TYPICAL DUTIES OF AXIAL FLOW FANS

Fan Dia. in.	Approximate Volume Range ft³/min		Maximum Static Pressure. High speed fan, in. of water	
	Minimum Low speed fan	Maximum High speed fan	Single Stage	Two Stage
12	200	2,500	0·86	2·6
15	400	5,000	1·5	4·1
19	600	11,000	2·5	6·8
24	1,000	20,000	3·8	12·5
30	2,000	35,000	6·0	19·5
38	3,000	45,000	2·6	7·0
48	6,000	60,000	4·2	11·4
60	10,000	100,000	2·8	7·8
75	20,000	130,000	2·0	—
95	50,000	150,000	1·0	—

(Woods of Colchester Ltd.)

Data for the Selection of Fans:

1. Air volume to be moved.
2. Static Pressure or Resistance.
3. Degree of noise permissible.
4. Motive power available.

Fans should be selected as near as possible to the point of maximum efficiency.

Resistance Heads commonly used for typical Vent-Installations:

For Public Buildings, ventilating only	9–15 mm. W.G.	($\frac{3}{8}$ to $\frac{1}{2}$ in.)
For Public Buildings, heating and ventilating	15–25 mm. W.G.	($\frac{1}{2}$ to 1 in.)
For Public Buildings, heating and ventilating, including air washer	17–30 mm. W.G.	($\frac{3}{4}$ to $1\frac{1}{4}$ in.)
For Factories, heating	17–40 mm. W.G.	($\frac{3}{4}$ to $1\frac{1}{2}$ in.)
For Factories, heating and ventilating, including air washer	30–50 mm. W.G.	($1\frac{1}{4}$ to 2 in.)

FAN OUTLET VELOCITIES FOR SILENT RUNNING

	Inlet		*Extract*	
	m./s.	*ft./min.*	*m./s.*	*ft./min.*
Sound Studios, Churches, Libraries	4 to 5	800 to 1,000	5 to 7	1,000 to 1,400
Cinemas, Theatres, Ballrooms	5 to 7·5	1,000 to 1,500	6 to 8	1,200 to 1,600
Restaurants, Hotels, Offices, Stores	6 to 8	1,200 to 1,600	7 to 9	1,400 to 1,800

Sound is energy travelling as a pressure wave:

$$I = \frac{W}{A} = \frac{p^2}{\rho c}.$$

where I = Intensity of sound.
W = Power.
A = Area.
p = Root means square pressure
ρ = Density.
c = Velocity of sound.

One Decibel is equal to ten times the logarithm to base 10 of the ratio of two quantities.

Sound Power Level: $\text{PWL} = 10 \log_{10} \frac{W}{W_o}.$

Sound Intensity: $\text{IL} = 10 \log_{10} \frac{I}{I_o}.$

Sound Pressure Level: $\text{SPL} = 10 \log_{10} \frac{P^2}{P_o^2}.$

$$= 20 \log_{10} \frac{P}{P_o}.$$

The usual reference levels are:

$W_o = 10^{-12}$ watts.
$I_o = 10^{-16}$ watts/cm.^{2} = 10^{-12} watts/m.^{2}
$P_o = 0{\cdot}0002\ \mu$bar.

At room temperature and at sea level SPL = IL + 0·2 decibels.

METHOD OF ADDING LEVELS EXPRESSED IN DECIBELS

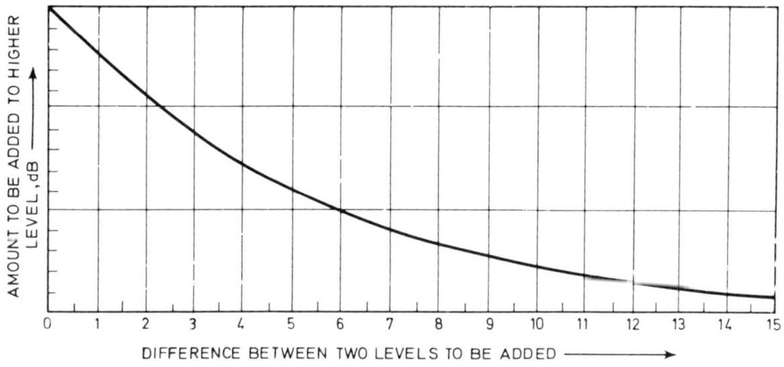

ADDITION OF DECIBELS

Noise Rating: Graphs are plotted of Sound Pressure Level (SPL) v. frequency, to show how the acceptable sound level varies with frequency. What is acceptable depends on the use to which the room will be put, and so a different curve is obtained for each type of use. Each such curve is designated by an NR number.

NR No.	*Application*
NR 25	Concert Halls, Broadcasting & Recording Studios, Churches.
NR 30	Private Dwellings, Hospitals, Theatres, Cinemas, Conference Rooms.
NR 35	Libraries, Museums, Court Rooms, Schools, Hospital Operating Theatres and Wards, Flats, Hotels, Executive Offices.
NR 40	Halls, Corridors, Cloakrooms, Restaurants, Night Clubs, Offices, Shops.
NR 45	Department Stores, Supermarkets, Canteens, General Offices.
NR 50	Typing Pools, Offices with Business Machines.
NR 60	Light Engineering Works.
NR 70	Foundries, Heavy Engineering Works.

NR LEVELS (SPL, dB re 0·00002 N/m.^{2})

Noise Rating	*Octave band mid-frequency, Hz*							
	62·5	125	250	500	1000	2000	4000	8000
NR10	42	32	23	15	10	7	3	2
NR20	51	39	31	24	20	17	14	13
NR30	59	48	40	34	30	27	25	23
NR35	63	52	45	39	35	32	30	28
NR40	67	57	49	44	40	37	35	33
NR45	71	61	54	48	45	42	40	38
NR50	75	65	59	53	50	47	45	43
NR55	79	70	63	58	55	52	50	49
NR60	83	74	68	63	60	57	55	54
NR65	87	78	72	68	65	62	61	59
NR70	91	83	77	73	70	68	66	64
NR75	95	87	82	78	75	73	71	69
NR80	99	91	86	82	80	78	76	74

Sound obeys the Inverse Square Law:

$$p^2 = K\frac{W}{r^2}.$$

where p = Root mean square pressure.
K = Constant.
W = Power.
r = Distance from source.

or $SPL = PWL - 20 \log_{10} r + K'$. $K' = \log_{10} K$ = Constant.

For a continuing source in a room, the sound level is the sum of the direct and the reverberant sound and is given by

$$SPL = PWL + 10 \log_{10}\left[\frac{Q}{4\pi r^2} + \frac{4}{R}\right] + 10 \text{ db.}$$

where

$$Q = \frac{\text{SPL at distance r from actual source}}{\text{SPL at distance r from uniform source of same power}}.$$

$$R = \text{Room constant} = \frac{S\alpha}{1-\alpha}.$$

S = Total Surface Area of Room.
α = Absorption Coefficient of Walls.

COEFFICIENT OF ABSORPTION α
(For range of frequencies usual in fan work.)

Plaster Walls	0·01–0·03	25 mm. wood wool cement on battens ...	0·6–0·7
Unpainted Brickwork ...	0·02–0·05	50 mm. slag wool or glass silk	0·8–0·9
Painted Brick	0·01–0·02	12 mm. acoustic felt ...	0·5–0·6
3-plywood panel	0·01–0·02	Hardboard (Masonite) ...	0·3
6 mm. cork sheet	0·1– 0·2	25 mm. sprayed asbestos	0·6–0·7
6 mm. porous rubber sheet	0·1–0·2	Persons, each	2·0–5·0
12 mm. Fibreboard on battens	0·3 –0·4	Acoustic tiles	0·4–0·8

Sound Insulation of Walls:

$$\text{Transmission coefficient } \tau = \frac{\text{transmitted energy}}{\text{resident energy}}$$

$$\text{Sound Reduction Index} \quad \text{SRI} = 10 \log_{10} \left[\frac{1}{\tau}\right] \text{db.}$$

Empirical Expression is:

$\text{SRI} = 14{\cdot}3 \log_{10} W + 22{\cdot}7$ db. W = Weight of wall in lb. per sq. ft. surface.

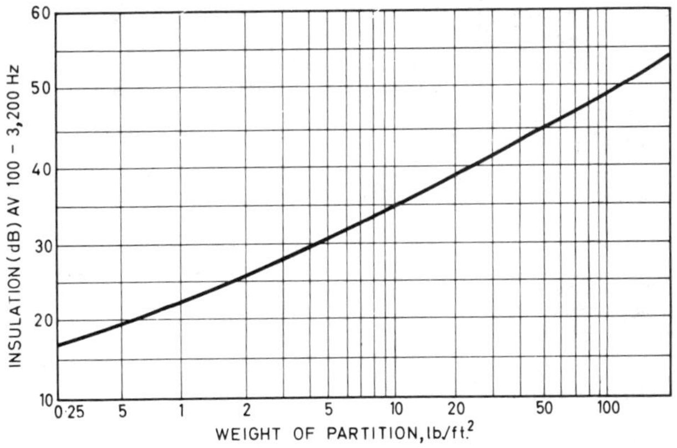

SOUND INSULATION OF SOLID WALLS ACCORDING TO THEIR WEIGHT

Transmission Through Walls:

$$(SPL)_1 - (SPL)_2 = SRI - 10 \log_{10} \frac{S_P}{S_2 \alpha_2} \text{ db.}$$

where $(SPL)_1$ = Sound pressure in sending room.
$(SPL)_2$ = Sound pressure in receiving room.
SRI = Sound Reduction Index (see page XI.13).
$S_2\alpha_2$ = Equivalent Absorption in receiving room.
S_P = Area of partition wall.

SOUND INSULATION OF WINDOWS

Single or Double Window	*Type of Window*	*Type of Glass*	*Sound Reduction in db.*
Single	Opening type (closed)	Any Glass	18–20
Single	Fixed or opening type with air-tight weather strips	24/32 oz. Sheet Glass	23–25
		6 mm. Polished Plate Glass	27
		9 mm. Polished Plate Glass	30
Double	Opening type (closed) plus absorbent material on sides of air space	24/32 oz. Sheet Glass 100 mm. space ...	28
		24/32 oz. Sheet Glass 200 mm. space ...	31
		6 mm. Polished Plate Glass 100 mm. space	30
		6 mm. Polished Plate Glass 200 mm. space	33
Double	Fixed or opening type with air-tight weather strips	24/32 oz. Sheet Glass 100 mm. space ...	34
		24/32 oz. Sheet Glass 200 mm. space ...	40
		6 mm. Polished Plate Glass 100 mm. space	38
		6 mm. Polished Plate Glass 200 mm. space	44

ATTENUATION BY BUILDING STRUCTURE

Structure	*Attenuation db.*	*Structure*	*Attenuation db.*
9 in. brick wall	50	Double window 50 mm. spacing	30
6 in. (150 mm.) concrete wall...	42	12 mm. T. & G. boarded partition	26
Wood joist floor and ceiling ...	40	2·5 mm. glass window ...	23
Lath and plaster partition ...	38		

Transmission through Ducts:

$$\frac{\text{Attenuation}}{\text{Duct Length}} = 1{\cdot}05 \ \alpha^{1{\cdot}4} \ \frac{P}{A} \text{ db. per ft.}$$

$$= 3{\cdot}44 \ \alpha^{1{\cdot}4} \ \frac{P}{A} \text{ db. per m.}$$

where α = Coefficient of absorption.
P = Perimeter of duct.
A = Cross sectional area of duct.

APPROXIMATE ATTENUATION OF ROUND BENDS OR SQUARE BENDS WITH TURNING VANES IN DB.

Frequency Hz.	20–75	75–150	150–300	300–600	600–1,200	1,200–2,400	2,400–4,800	4,800–10,000
Diameter								
5 to 10 in. 125 to 250 250 mm.	0	0	0	0	1	2	3	3
11 to 20 in. 251 to 500 mm.	0	0	0	1	2	3	3	3
21 to 40 in. 501 to 1,000 mm.	0	0	1	2	3	3	3	3
41 to 80 in. 1,001 to 2,000 mm.	0	1	2	3	3	3	3	3

ATTENUATION DUE TO CHANGES IN AREA IN DB.

Ratio of Areas S_2/S_1	*Attenuation db.*	*Ratio of Areas S_2/S_1*	*Attenuation db.*
1	0·0	3	1·3
2	0·5	4	1·9
2·5	0·9	5	2·6

ATTENUATION AT ENTRY TO ROOM (END REFLECTION LOSS)

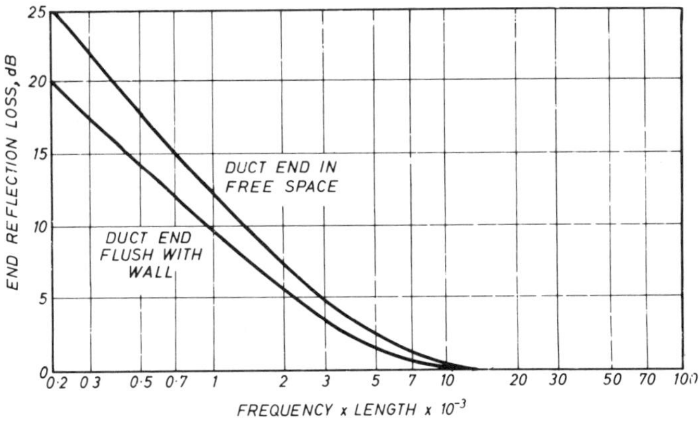

END REFLECTION LOSS
FOR RECTANGULAR OPENING LENGTH = $\sqrt{L_1 \times L_2}$
FOR CIRCULAR OPENING LENGTH = 0·9 x DIAMETER

Sound Power Level (PWL) of Fans: Exact data for any particular fan is to be obtained from the manufacturer. In the absence of this the following approximate expression may be used.

$$\text{PWL} = 90 + 10\log_{10} h_p + 10\log_{10} h.$$
$$\text{PWL} = 55 + 10\log_{10} q + 20\log_{10} h.$$
$$\text{PWL} = 125 + 20\log_{10} h_p - 10\log_{10} q.$$

where h_p = Rated motor horsepower.
h = Fan static pressure in W.G.
q = Discharge in ft.^{3}/min.

$$\text{PWL} = 47 + 10\log_{10} P_j + 10\log_{10} p.$$
$$\text{PWL} = 60 + 10\log_{10} V + 20\log_{10} p.$$
$$\text{PWL} = 34 + 20\log_{10} P_j - 10\log_{10} V.$$

where P_j = rated motor power in watts.
p = Fan static pressure in mm. W.G.
V = Discharge in m.^{3}/s.

TYPICAL CURVES OF FAN FREQUENCY DISTRIBUTION

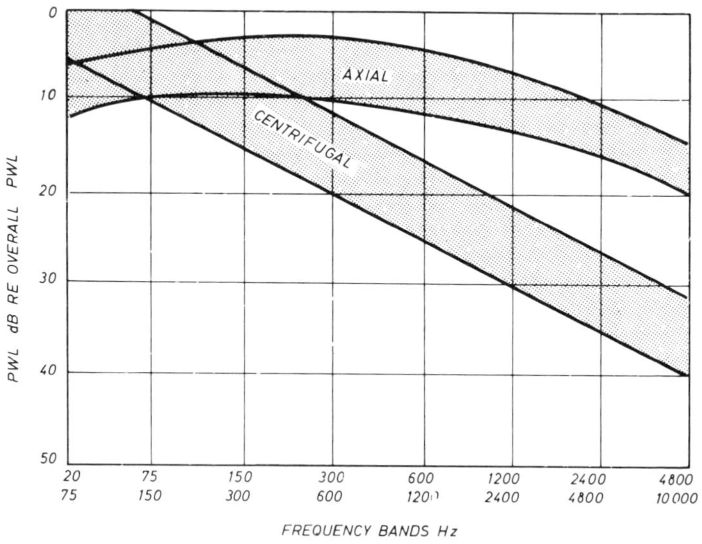

SOUND POWER LEVEL SPECTRA OF FANS

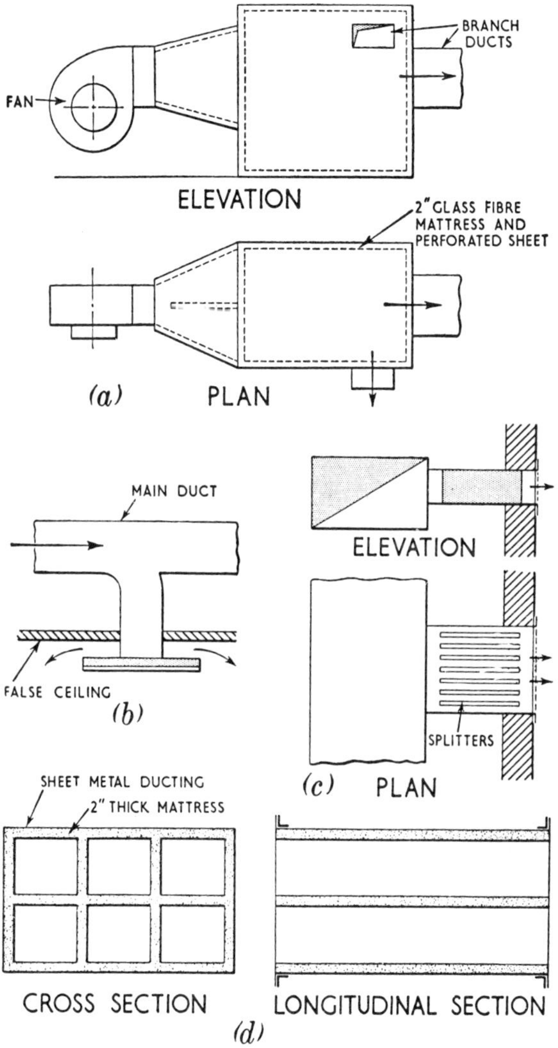

(*a*) Sound absorption by increase of duct area.

(*b*) Ceiling air outlet with sound-absorbing plate.

(*c*) Sound absorption in branch duct with splitter.

(*d*) Arrangement of splitters in main duct.

Section XII

LABOUR RATES FOR INSTALLATION

In the following Tables the times for the erection of pipe lines and heating plants on site are given in hours.

Included: Hauling of all parts into position, erection on site, surveying of builder's work, testing of the plant.

Not included: Delivery to site, travelling time, removing and refixing of radiators for painting, addition for overtime work.

Additions to Total Basic Time:

For jobs up to 1 week	40%
For jobs up to 2 weeks	20%
For jobs up to 3 weeks	8%
For installations in existing buildings, unoccupied ...	5%
For installations in existing buildings, occupied	15%
For installations in existing buildings, with hidden pipes	15–20%

ESTIMATING ERECTION TIME

The working time is stated in normal working hours of one fitter and mate.

	Description of Work	*Time in hours*
BOILERS	**Cast-iron Sectional Boilers for Hot Water Heating,** including Boiler Mountings.	
	Without Insulating Jacket for a duty of:	
	Up to 36 kW.	15
	From 37 to 75 kW.	20
	From 76 to 150 kW.	30
	From 151 to 220 kW.	40
	From 221 to 300 kW.	50
	From 301 to 450 kW.	60
	From 451 to 580 kW.	65
	From 581 to 750 kW.	70
	Cast-iron Sectional Boilers for Low-pressure Steam With Mountings — No Jacket:	
	Up to 36 kW.	28
	From 37 to 75 kW.	30
	From 76 to 150 kW.	40
	From 151 to 220 kW.	50
	From 221 to 300 kW.	65
	From 301 to 450 kW.	75
	From 451 to 580 kW.	80
	From 581 to 750 kW.	90
	Sectional Steel Boilers for Hot Water Heating With Mountings. With Outer Casing:	
	Up to 110 kW.	6
	From 111 to 300 kW.	12
	From 301 to 600 kW.	20

	Description of Work	*Time in Hours*
BOILERS	**Insulating Jackets for Cast-Iron Sectional Boilers:**	
	Up to 150 kW.	6
	From 151 to 300 kW.	8
	From 301 to 580 kW.	10
	From 581 to 750 kW.	12
	Domestic H.W.S. Boilers Non-sectional	
	Up to 15 kW.	8
	From 16 to 30 kW.	10
	From 31 to 45 kW.	12
	Boiler Smoke Pipes:	
	Straight Pipes, 150 mm. dia. per m. ...	1·0
	Elbows, 300 mm. dia., each...	0·3
	Straight Pipes, 300 mm. dia., per m. ...	1·8
	Elbows, 160–300 mm. dia., each	0·6
	Straight Pipes, above 300 mm. dia., per m.	3·0
	Elbows, above 300 mm. dia., each	1·0
BOILER HOUSE EQUIPMENT	**Expansion Tanks,** including Ball Valve and Overflow:	
	Up to 90 litres capacity	6
	From 91 to 225 litres capacity	9
	From 226 to 450 litres capacity	12
	From 451 to 900 litres capacity	15
	Above 900 litres capacity	20
	Centrifugal Pumps, with Motors, Direct Coupled or Belt Drive, with Base Plate or Slide Rails:	
	Duty up to 3·5 l/s. Size: 1½ in. ...	10
	From 3·6 to 7·5 l/s. Size: 1½ to 2 in....	14
	From 7·6 to 20 l/s. Size: 3 to 4 in....	18
	Above 21 l/s. Size: Above 4 in.	24
	Heating Accelerators: Pipe Line Model:	
	Size up to 1½ in.	6
	Size 2 in. to 3 in.	10
	Size 4 in. and above	14
	Starters	2

	Description of Work	Time in Hours
BOILER HOUSE EQUIPMENT	**Hot Water Service Storage Cylinders:**	
	Direct or Indirect, or Calorifiers:	
	Contents up to 250 litres...	9
	From 251 to 550 litres...	12
	From 551 to 900 litres...	20
	From 901 to 1,300 litres...	30
	From 1,301 to 2,250 litres...	35
	From 2,251 to 3,500 litres...	40
	From 3,501 to 5,000 litres...	45
	Centrifugal Fans with Motors:	
	Impeller diameter up to 300 mm. ...	15
	From 301 to 600 mm. ...	20
	From 601 to 1,000 mm. ...	28
	From 1,001 to 1,200 mm. ...	45
	From 1,201 to 1,525 mm.	50
	Electric Motors only, with Slide Rails:	
	Up to 2 kW....	7½
	From 2 to 4 kW....	12
	From 4 to 6 kW....	15
	From 6 to 7·5 kW.	20
	From 7·5 to 12 kW....	25
ROOM HEATERS, RADIATORS, ETC.	**Unit Heaters,** Suspended Type:	
	Duty up to 6 kW.	10
	From 7 to 15 kW.	15
	From 16 to 30 kW.	20
	Radiators with 2 valves:	
	Heating Surface up to 2·5 sq. m.... ...	3
	2·6 to 4·5 sq. m.... ...	5
	4·6 to 10 sq. m.	7
	Radiator Brackets	0·5
	Radiator Top Stays	0·5
	Removing and refixing one radiator ...	1·5
	Towel Rail	8
	Rayrad Panels on Wall:	
	Up to 3 sections	4
	From 4 to 6 sections	6
	From 7 to 10 sections	8

	Description of Work	Time in Hours
ROOM HEATERS, RADIATORS, ETC.	Skirting Heater, per m.	0·75
	Gravity Convector up to 1 m. length ...	6
	1·0 to 1·5 m. length ...	7
	Above 1·5 m. length ...	8
	Rayrad Panels on Ceiling:	
	Up to 3 sections	6
	From 4 to 6 sections	8
	From 7 to 10 sections	10
	Fan Convectors, free standing cabinet type	5
	Fan Convectors, extended type, 2 grilles ...	10
	(Brackets and Top Stays to be marked by Heating Contractor and built in by Builder.)	
	Gas or Electric Water Heaters, up to 3 kW.	4
	Ventilation Plant Equipment, such as Fans, Heater Batteries, Air Filters, Ducting, etc., per 1,000 kg. of equipment ...	90
VALVES, ETC.	Valves, size ½ to 1¼ in.	0·5
	size 1½ to 2 in.	1·0
	size 2½ to 4 in.	2·0
	size above 4 in.	3
	Thermostatic and Motorised Valves:	
	Up to 1¼ in.	2
	1½ to 2 in.	4
	2½ to 4 in.	6
	3-way Mixing Valves and Steam Reducing V Valves:	
	Up to 2 in.	10
	2½ to 4 in.	12½
	5 to 6 in.	15
	Steam Traps:	
	Up to ¾ in.	2
	1 to 1½ in.	3
	Thermometers	1·0
	Pressure Gauge with Cock...	2
	Altitude Gauge with Cock	2
	Dial Distance Thermometer with 2 m. Capillary Tube	4
	Longer Capillary Tube, per metre ...	0·1

	Description of Work		*Time in Hours*
TUBES	**Tubes:** including Brackets and Pipe Fittings such as Bends and Branch Tees		
	½ and ¾ in. n.b.		0·45
	1 and 1¼ in. n.b.	Screwed	0·75
	1½ and 2 in. n.b.	connections	1·0
	2½ and 3 in. n.b.	per metre	1·35
	4 in. n.b.	linear	2·25
	5 in. n.b.		2·70
	1½ and 2 in.		0·60
	2½ and 3 in.		0·75
	4 in.	Welded	0·84
	5 in.	connections	1·56
	6 in. (150 mm.)	per metre	1·80
	7 in. (175 mm.)	linear	2·04
	8 in. (200 mm.)		2·25
	9 in. (225 mm.)		2·46
	10 in. (250 mm.)		2·70
	Lagging:		
	Rigid sections:		
	For tubes up to 6 in.	per metre linear	0·45

OXYGEN WELDING

(Time in hours. Gas in m³.)

Pipe Size Nom. Base in.	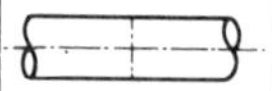		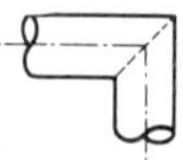		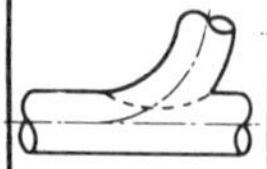		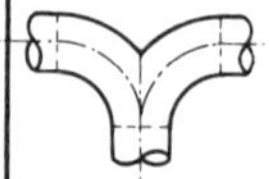		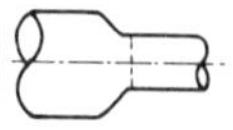	
	Time	*Gas*	*Time*	*Gas*	*Time*	*Gas*	*Time*	*Gas*	*Time*	*Gas*
3/8	0·10	0·062	0·12	0·074	0·3	0·12	0·5	0·14	0·2	0·085
1/2	0·10	0·071	0·13	0·085	0·3	0·12	0·5	0·14	0·2	0·085
3/8	0·15	0·074	0·15	0·088	0·4	0·13	0·6	0·18	0·3	0·099
1	0·15	0·079	0·20	0·093	0·5	0·13	0·6	0·21	0·3	0·099
1¼	0·20	0·093	0·25	0·11	0·5	0·13	0·7	0·24	0·4	0·11
1½	0·25	0·099	0·30	0·12	0·6	0·14	0·7	0·27	0·4	0·11
2	0·30	0·11	0·35	0·13	0·7	0·15	1·1	0·31	0·5	0·13
2½	0·35	0·13	0·4	0·15	1·0	0·17	1·3	0·37	0·6	0·13
3	0·4	0·16	0·5	0·18	1·2	0·20	1·6	0·45	0·7	0·17
4	0·5	0·21	0·6	0·24	1·6	0·27	2·3	0·66	1·0	0·23
5	0·6	0·25	0·7	0·31	1·9	0·34	2·9	0·74	1·2	0·28
6	0·7	0·31	0·8	0·37	2·2	0·40	3·2	0·74	1·3	0·34
7	0·8	0·37	0·9	0·42	2·7	0·48	3·6	1·0	1·6	0·37
8	1·0	0·42	1·1	0·48	3·3	0·54	4·0	1·2	1·8	0·42
9	1·1	0·48	1·3	0·57	3·8	0·62	4·6	1·4	2·1	0·48
10	1·2	0·54	1·4	0·62	4·2	0·71	5·2	1·6	2·2	0·54
11	1·3	0·59	1·6	0·68	4·8	0·76	5·7	1·7	2·5	0·60
12	1·5	0·68	1·8	0·79	5·2	0·88	6·2	1·8	2·8	0·68

LOW PRESSURE WELDING

Thickness of Material mm.	Gas Consumption m.^{3}				Filling Wire		Speed of Welding m./hr.
	Oxygen		Acetylene		Size mm.	Consumption m. per m.	
	per m.	per hr.	per m.	per hr.			
0·4	0·004	0·048	0·003	0·042	—	—	15
0·8	0·0065	0·071	0·006	0·062		0·33	10·5
1·2	0·015	0·14	0·013	0·12	1·6	0·50	10
1·6	0·027	0·23	0·023	0·20	3·2	0·66	8·5
2·4	0·048	0·35	0·043	0·31		0·83	7·25
3·2	0·070	0·42	0·060	0·35		0·92	6·0
4·8	0·14	0·54	0·13	0·37		1·33	3·6
6·4	0·25	0·65	0·20	0·57		1·66	2·7
8·0	0·43	0·85	0·37	0·74	4·8	2·25	2·0
9·6	0·64	1·0	0·56	0·88		2·75	1·6
12·7	1·1	1·3	1·0	1·1		3·5	1·1
16·0	2·2	1·7	2·0	1·5	6·0	4·6	0·75
19·0	4·5	2·4	3·9	2·1	6·0	5·4	0·53
22	7·4	3·4	6·5	3·0	to	7·1	0·45
25–40	7·8	3·5	6·9	3·1	10·0	8·3–12·5	0·45

CAST IRON WELDING

Thickness of Material mm.	Gas Consumption m.^{3}				Filling Rod		Speed of Welding m./hr.
	Oxygen		Acetylene		Size mm.	m. per m. run	
	per m.	per hr.	per m.	per hr.			
6	0·045	0·10	0·042	0·096	6·0	1·8	2·25
12	0·37	0·31	0·33	0·28	6·0	4·5	0·90
20	1·3	0·79	1·2	0·74	6·0	9	0·60
25	3·3	1·3	3·2	1·2	6·0	13	0·40
32	7·9	2·4	7·0	2·1	6·0	18	0·30
40	18	4·1	17	3·9	6·0	23	0·225

NOZZLE SIZES, WORKING PRESSURES AND GAS CONSUMPTIONS FOR HIGH PRESSURE WELDING.

Blowpipe Nozzle (Litres)	*Blowpipe B.O.C. Type*	*Mild Steel Plate, Thickness in inches*	*Regulator Pressure, lb./sq. in.*		*Gas Consumption Cu. ft./hr.*
			Acetylene	*Oxygen*	
1–5	"O"	$\frac{1}{16}$	$1\frac{1}{2}$	$1\frac{1}{2}$	0·2
2–12	"O"	$\frac{1}{16}$	$2\frac{1}{2}$	$2\frac{1}{2}$	0·4
3–31	"O"	$\frac{1}{16}$	3	3	1·1
4–62	"O"	$\frac{1}{16}$	4	4	2·2
5–125	"O"	$\frac{1}{16}$	5	5	4·4
25	A	to $\frac{1}{32}$	2	4	0·9
50	A or B	$\frac{1}{32}$	2	4	1·8
75	A or B	$\frac{3}{64}$	2	4	2·7
100	A or B	$\frac{1}{16}$	2	4	3·3
150	A or B	$\frac{3}{32}$	2	4	5·3
225	A or B	$\frac{1}{8}$	3	5	8·0
350	B	$\frac{5}{32}$	3	5	12·4
500	B	$\frac{1}{4}$	4	8	17·7
750	B	$\frac{3}{8}$	5	10	26·5
1,000	B	$\frac{1}{2}$	6	15	35·3
1,500	B	$\frac{1}{2}$–$\frac{3}{4}$	9	18	53·0
2,000	B	1	12	25	70·7
2,500	B	1 and above	20	30	88·3

FORMS OF WELDED JOINTS

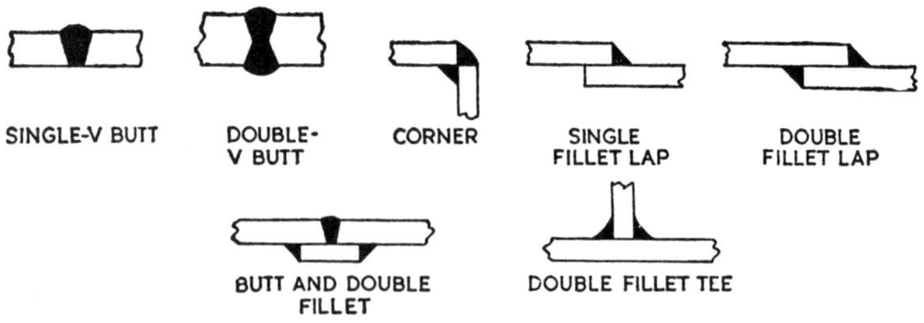

Section XIII

BOILER FEED WATER TREATMENT

HARDNESS OF WATER

A water is said to be hard when it is difficult to obtain lather with soap. Soap dissolves in soft water, but with a hard water it combines with the calcium and magnesium present, and forms an insoluble precipitate. It is necessary to add soap in order to obtain lather from hard water. About $\frac{1}{4}$ oz. of soap is required to remove 1 degree of hardness from 10 gallons of water.

Degrees of Hardness:

English (Clark's) degrees are grains of calcium carbonate per gallon, or parts per 70,000.

French (Continental) degrees are parts of calcium carbonate per 100,000 (by weight).

German degrees are parts of calcium oxide (lime) per 100,000.
1 English degree of hardness = 0·70 French degree of hardness.
1 English degree of hardness = 1·24 German degree of hardness.
1 French degree of hardness = 1·43 English degree of hardness.
1 German degree of hardness = 0·80 English degree of hardness.

In practice hardness varies from zero to 30 degrees, 5 English degrees is generally considered good.

Temporary Hardness is that which is removed by boiling the water. It is due to calcium bicarbonate and/or magnesium bicarbonate.

Permanent Hardness is that which remains after boiling. It is due to calcium and magnesium sulphate, especially $CaSO_4$. Can be softened by chemical methods.

WATER SOFTENING

Removal of Temporary Hardness:

(a) Heating the water to a temperature of 150°F. to 212°F. When heated the soluble bicarbonates decompose into insoluble carbonates which settle to the bottom and can be removed:

$$\underset{\text{Soluble Bicarbonate}}{Ca(HCO_3)_2} \longrightarrow \underset{\text{Insoluble Carbonates}}{CaCO_3} + \underset{\text{Water}}{H_2O} + \underset{\text{Carbon Dioxide}}{CO_2}$$

(b) Adding any alkali soluble in water (lime, caustic soda or baryta) in order to decompose the carbonates.

$$XCO_3 . H_2CO_3 + CaO \longrightarrow XCO_3 + CaCO_3 + H_2O$$

Soluble Bicarbonate + Lime → Insoluble Carbonates + Water

Removing of Permanent Hardness:

(a) By distillation (evaporating and condensing).

(b) Soda-Lime Porter-Clerk Process:

Introducing lime (CaO) to remove temporary hardness. The soluble bicarbonates are converted into insoluble carbonates.

$$Ca(HCO_3)_2 + CaO \longrightarrow 2CaCO_3 + H_2O$$

Introducing soda Na_2Co_3 to remove the permanent hardness due to calcium sulphate $CaSo_4$. The calcium is precipitated as its carbonate

$$CaSO_4 + Na_2CO_3 \longrightarrow CaCO_3 + Na_2SO_4$$

The quantity of the addition depends upon the hardness and other properties of the water and has to be proportioned exactly.

Theoretical addition:

For 1° of Temporary Hardness:

4·4 grains slaked lime $Ca(OH)_2$ for 1 cu. ft. of water.

For 1° of Permanent Hardness:

8·3 grains sodium hydrate, caustic soda ($NaOH$) for 1 cu. ft. of water.

A careful surveying of the process is necessary. A final hardness of 3 to 4 degrees is attainable.

(c) Permutit Process (Dr. R. Gans):

Using Permutit, an artificial, porous, insoluble zeolite, as water filter. The zeolite is converted into its calcium or magnesium derivate and has to be reactivated by treatment with a solution of common salt ($NaCl$).

Softening:

$$Na_2Ze + Ca(HCO_3)_2 \longrightarrow 2NaHCO_3 + CaZe.$$
$$Na_2Ze + CaSO_4 \longrightarrow Na_2SO_4 + CaZe.$$

Re-activating:

$$CaZe + 2NaCl \longrightarrow CaCl_2 + Na_2Ze.$$

Ze = Symbol for a portion of molecule of artificial zeolite.

The process is automatic and no mistake due to adding too much or too little reagent can be made.

A zero hardness is attainable.

Consumption of salt for re-activation is 17·5 to 30·5 grains per cu. ft. of water per 1 degree of hardness.

Time for re-activation:

6 to 8 hours for Natriumpermutit.
1 hour for Neopermutit.

Corrosion:

Internal corrosion in boilers, pipes, etc., may be due to:

1. Acid in the feed water.
2. Acid formed by the decomposition of salts by heating the water.
3. Oxygen dissolved in the water.
4. Electrolysis.

Prevention of Corrosion:

1. Neutralizing the acid by an alkali, chalk milk of lime or sodium carbonate (soda).
2. De-aerating or de-gassing the feed water.
3. Inserting zinc plates connected metallically to the boiler plates, then zinc is corroded instead of iron.
4. Counter-electrolysis. by passing a direct electric current from anode in the boiler to the heating surface.
5. Coating the surface of the boiler or vessel.

Formation of Scale in Boilers:

(a) Soft, non-adherent sludge.
(b) Hard, adherent deposit (scale).

Prevention of Scale:

(a) By water softening.
(b) By adding some substance of a colloidal or gummy nature, causing the formation of sludge instead of scale.
(c) By passing a current of electricity through the water (seldom used).

Removal of Scale:

(a) Mechanically, by chipping with a hammer or chisel.
(b) Chemically, by adding caustic soda or other chemicals to the feed water.
(c) By electrolysis, small bubbles of hydrogen leave the surface of the boiler and loosen the scale.

Priming and Foaming:

Priming is a too violent boiling and projecting of droplets of water into the steam pipes.

Cause: Restricted steam-liberating area or sudden increase of load.

Foaming is the formation of a layer of bubbles on the water.

Cause: Oil, grease, or other organic impurities in water or high concentration of dissolved and suspended mineral salts.

Prevention—Provision of adequate steam place in boiler, no sudden opening of main steam valve, exclusion of oil or grease from boiler feed water.

pH Value

This is an arbitrary symbol adopted to express the degree of acidity or alkalinity of a solution.

The pH number is the common logarithm of the number of litres containing one gram-equivalent of hydrogen ion. A pH of 7 represents a neutral solution, lower values represent acidity, higher values alkalinity.

The alkalinity of boiler water is usually maintained between a pH value of 9·5 and 11·0.

Some Indicators for the Determination of the pH Value.

Thymol blue	1·2–2·8	Red–yellow	Litmus (Azolitmin)	5–8	Red–Blue.
Methyl Orange	2·9–4·6	Orange–red–orange–yellow	Phenolpthalein	8·3–10·0	Colourless–red.
Methyl red	4·4–6·2	Red–yellow	Thymolphthalein	9·3–10.5	Colourless–blue.

Blow-down

A steam boiler has to be blown-down in order to prevent the accumulation of dissolved solids (sludge and contamination of steam).

The theoretical amount of blow-down is $D = \frac{100F}{B}$ per cent,

where D = Blow down in per cent of the evaporation.
F = Dissolved solids in feed water, in parts per 100,000.
B = Permissible total of dissolved solids in boiler water in parts per 100,000.

Types of Blow-down

(*a*) *Intermittent Blow Down*, for less than 25 gallons per hour.

(*b*) *Continuous Blow Down*, by utilizing heat in a heat exchanger or flash vessel.

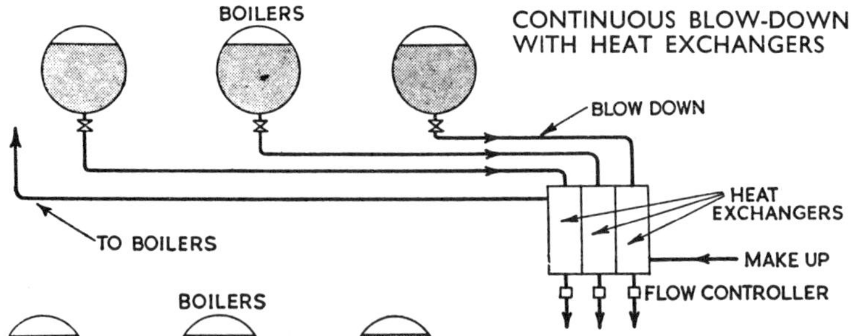

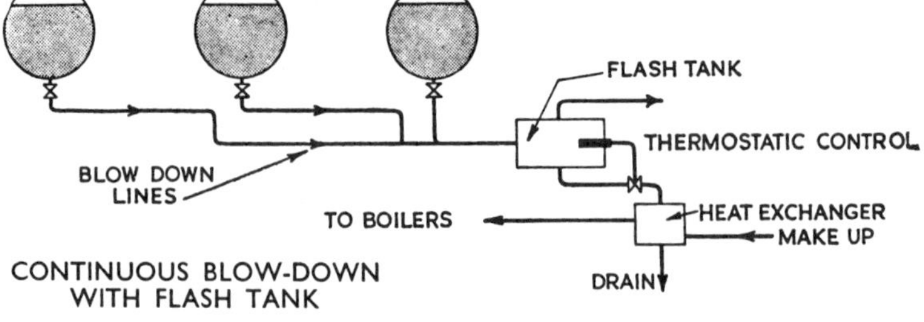

Section XIV

BIBLIOGRAPHY

HANDBOOKS

Guide and Data Book. 3 vols. American Society of Heating, Refrigeration and Air-Conditioning Engineers.

Guide to Current Practice 1970. Institution of Heating and Ventilating Engineers. 49, Cadogan Square, London, S.W.1.

Handbook of Air Conditioning System Design. Carrier Air Conditioning Co., Dingwall Road, Croydon, Surrey.

Kempe's Engineers Year Book. Morgan-Grampian (Publishing) Ltd., 30 Calderwood Street, London S.E.18.

Marks Mechanical Engineers' Handbook. McGraw-Hill, New York and London.

Newnes Engineer's Pocket Book. Newnes-Butterworths, London.

Newnes Electrical Pocket Book. Newnes-Butterworths, London.

Pocket-Book for Mechanical Engineers by David Allan Low, Longmans-Green & Co., London, New York, Toronto.

PERIODICALS

Air Conditioning, Heating and Ventilating. Industrial Press, New York.

Combustion and Flame. American Elsevier, New York.

Fuel. IPC Science and Technology Press, 32 High Street, Guildford, Surrey.

Heating and Air Conditioning Journal. Troup Publications, 76 Oxford Street, London W1.

Heating and Ventilating Engineer and Journal of Air Conditioning. Technitrade Journals Ltd., 11–13 Southampton Row, London WC1.

Heating Piping and Air Conditioning. Reinhold Publishing Corporation, Chicago.

Plumbing. Institute of Plumbing, Southport, Merseyside.

Power. McGraw-Hill, New York and London.

Water Services. 17–19 John Adam Street, London WC2.

BOOKS

HEATING, VENTILATING AND AIR CONDITIONING

Allen, Walker, and James. *Heating and Air Conditioning*. 6th edn. (1946, McGraw-Hill, New York and London).

Barton, J. J. *Heating and Ventilating: Principles and Practice*. (1964, Newnes-Butterworths, London).

Barton, J. J. *Small Bore Heating and Hot Water Supply for Small Dwellings* 2nd edn. (1970, Newnes-Butterworths, London).

Barton, J. J. *Electric Floor Warming* (1967, Newnes-Butterworths, London).

Barton, J. J. *Hot Water Supply: Design and Practice*. 2nd edn. (1968, Newnes-Butterworths, London).

Bedford, T. *Basic Principles of Ventilating and Heating*. 2nd edn. (1964, H. K. Lewis, London).

Blake's *Drainage and Sanitation*. 13th edn., by S. Jennings.

Burkhardt, C. H. *Residential and Commercial Air Conditioning* (1959 McGraw-Hill, New York and London).

Carrier, W. H. *Modern Air Conditioning, Heating and Ventilation* (1959, Pitman, London).

Cornell, R. K. *Heating and Ventilating for Architects and Builders* (1946, Paul Elek (Publishers), London).

Diamant, R. M. E., and McGarry, J. *Space and District Heating* (1968, Newnes-Butterworths, London).

Diamant, R. M. E. *Insulation of Buildings* (1965, Newnes-Butterworths, London).

Dalzell, J. R. *Air Conditioning Furnaces and Unit Heaters* (1938, American Technical Society, Chicago).

Dye, F. W. *Steam Heating* (Spon, London).

The Efficient Use of Fuel. Ministry of Fuel and Power (1944, H.M. Stationery Office, London).

Faber, O. and Kell, J. R. *Heating and Air Conditioning of Buildings*. 5th edn. (1971, Architectural Press, London).

Geiringer, P. L. *High Temperature Water Heating* (1963, Wiley, New York and London).

Goodman, W. *Air Conditioning Analysis* (1943, Macmillan, New York).

Gosling, C. T. *Applied Air Conditioning and Refrigeration* (1974, Applied Science Publishers).

Haines, J. E. *Automatic Control of Heating and Air Conditioning*. 2nd edn. (1961, McGraw-Hill, New York and London).

Harris, N. C. *Modern Air Conditioning Practice* (1959, McGraw-Hill, New York and London).

Heating, Ventilating, Air Conditioning, 5 vols. A home study course (1945, American Technical Society, Chicago).

Hemeon, W. C. L. *Plant and Process Ventilation*, 2nd edn. (1963, Industrial Press, New York).

Holmes, H. W. and Langstaff, W. J. G. *Domestic Small Pipe Heating* (1965, Longmans, London).

Holmes, R. E. *Air Conditioning in Summer and Winter*. 2nd edn. (1951, McGraw-Hill, New York and London).

Hutchinson, F. W. *Design of Air Conditioning Systems* (1958, Industrial Press, New York).

Jones, W. P. *Air Conditioning Engineering*. 2nd edn. (1973, Edward Arnold, London).

Kut, D. *Heating and Hot Water Services in Buildings* (1968, Pergamon Press, Oxford).

Kut, D. *Warm Air Heating* (1970, Pergamon Press, Oxford).

Lyle, O. *The Efficient Use of Steam* (1947, H.M. Stationery Office, London).

Moyer, J. and Fittz, R. U. *Air Conditioning* (1938, McGraw-Hill, New York and London).

Overton, L. J. *Heating and Ventilating*. 5th edn. (1944, Sutherland Publishing Co., Manchester).

Penn and Solley. *Heating and Hot Water Supply* (1960, B. T. Batsford Ltd., London).

Raber, B. F. and Hutchinson, F. W. *Panel Heating and Cooling Analysis* (1947, Wiley, New York, and Chapman & Hall, London).

Rummel, A. J. and Vogelsang, L. D. *Practical Air Conditioning* (1941, Wiley, New York and London).

Sanford, G. C. *Central Heating and Hot Water Supply for Private Houses* (1966, Technical Press, London).

Severns, W. H., and Fellows, J. R. *Air Conditioning and Refrigeration* (1958, Wiley, New York and London).

Shaw, E. W. *Heating and Hot Water Services: Selected Subjects with Worked Examples*. 3rd edn. (1971, Crosby Lockwood, London).

Shaw, W. F. B. *Domestic Heating: A Guide to All Forms* (1960).

Sheridan, N. R. *et al. Air Conditioning* (1963, University of Queensland Press).

Smith, F. C. *Warming Buildings by Electricity* (Spon, London).

Strock, C. and Koral, R. L. *Handbook of Air Conditioning, Heating, and Ventilating*. 2nd edn. (1965, Industrial Press, New York).

Torok, E. *Psychrometric Notes and Tables* (The American Rayon Corporation, New York).

Turpin, F. B. *District Heating* (1966, Newnes-Butterworths, London).

Vernon, H. M. *The Principles of Heating and Ventilation* (Edward Arnold, London).

Wignall, N. *Students' Textbook of Heating and Ventilation*. 11th edn. (1966, Technitrade Journals Ltd., London).

Wilcox, E. A., *Electric Heating* (McGraw-Hill, New York and London).

Wilkes, G. B. *Heat Insulation* (1950, Wiley, New York and London).

HEAT PUMPS

Ambrose, E. R. *Heat Pumps and Electric Heating* (1966, Wiley, New York and London).

Davies, S. J. *Heat Pumps and Thermal Compression* (1950, Constable, London).

Kemble, E. N. and Ogleby, S., jnr. *Heat Pump Applications* (1947, McGraw-Hill, New York and London).

HEAT, HEAT TRANSFER, THERMODYNAMICS

Barton, A. W. *Textbook on Heat* (1963, Longmans, London).

Brown, A. I., and Marco, S. M. *Introduction to Heat Transfer* (1958, McGraw-Hill, New York and London).

Chapman, A. J. *Heat Transfer*. 3rd edn. (1974, Collier-Macmillan, London).

Duncan, J. and Starling, S. G. *Heat, Textbook of Physics*. 6th part 2nd edn. (1948, Macmillan, London).

Everett. *Thermodynamics* (Van Nostrand, New York).

Fishenden, M., and Saunders, O. A. *An Introduction to Heat Transfer* (1950, Oxford University Press, London).

Inchley's *Theory of Heat Engines* (Edited and revised by H. Wright Baker) (Longmans, London).

Jacobs, M. and Hawkins, G. A. *Elements of Heat Transfer and Insulation*. 3rd edn. (1957, Chapman & Hall, London).

Lewitt, E. H. *Thermodynamics Applied to Heat Engines*. 6th edn. (1965, Pitman, London).

McAdams, W. H. *Heat Transmission*. 3rd edn. (1954, McGraw-Hill, New York and London).

Robinson, W., and Dickson, S. M. *Applied Thermodynamics* (1954, Pitman, London).

Rogers, G. F. C. and Mayhew, Y. R. *Engineering Thermodynamics: Work and Heat Transfer, S.I. Units*. 2nd edn. (1967, Longmans, London).

Schack, A. *Industrial Heat Transfer* (Translation from German.) (1965, Chapman & Hall, London).

Schmidt, E. *Thermodynamics: Principles and Applications to Engineering*. Translated from 3rd German edition (Reprinted 1966, Dover, London).

Stoever, H. F. *Applied Heat Transmission* (1941, McGraw-Hill, New York and London).

Walshaw, A. C. *Thermodynamics for Engineers*. 5th edn. (1963, Longmans, London).

Wrangham, D. A. *The Theory and Practice of Heat Engines* (1960, Cambridge University Press, London).

HYDRAULICS

Hansen, A. G. *Fluid Mechanics* (1967, Wiley, New York and London).

Lewitt, E. H. *Hydraulics and Fluid Mechanics* (1966, Engineering Degree Series, Pitman, London).

Massey, B. S. *Mechanics of Fluids*. 2nd edn. (1970, Van Nostrand, London).

FEED WATER TREATMENT

Batley, R. L. and Barber, E. G. *Boiler Plant Technology*. 3rd edn. (1965, Pitman, London).

Boby, W. H. T. and Solt, G. S. *Water Treatment Data* (1965, Hutchinson, London).

I.C.I. *The Use of Sodium Aluminate in Water Softeners.*

I.C.I. *Water Treatment for Industry.*

James, G. V. *Water Treatment.* 4th edn. (1971, Technical Press, London).

Matthews, F. J. *Boiler Feed Water Treatment* (Hutchinson, London).

Ryan, W. J. *Water Treatment and Purification* (McGraw-Hill, New York and London).

COMBUSTION

Batley, R. L., and Barber, E. G. *Boiler Plant Technology*. 3rd edn. (1965, Pitman, London).

Brame and King. *Fuel, Solid, Liquid and Gaseous* (1967, Edward Arnold, London).

Gollin, C. J. *Fuel Oil and Oil Firing* (1952, Reprint from *The Heating and Ventilating Engineer*, London).

Himus, G. W. *Fuel Testing, Laboratory Methods of Fuel Technology* (1953, Leonard Hill, London).

Kent, W. L. R. *et al. Boiler Plant Instrumentation* (1956, George Newnes, London).

Lyle, O. *The Efficient Use of Steam* (1947, H.M. Stationery Office, London).

Steiner, K. *Oil Burners.* 3rd edn. (1954, Heating Publishers Inc., New York).

FANS

Alden, J. L. and Kane, J. M. *Design of Industrial Exhaust Systems.* 4th edn. (1970, Industrial Press, New York and London).

Baumeister, Th. *Fans* (McGraw-Hill, New York and London).

Berry, C. H. *Flow and Fan: Principles of moving air through ducts.* 2nd edn. (1963).

Buffalo Forge Co. *Fan Engineering.* 6th edn. (1961, New York).

Eck, I. B. *Fans* (1973, Pergamon Press, Translated from German.)

De Kovats, A. and Desmur, G. *Pumps, Fans, and Compressors* (Translated from the French by R. S. Eaton, 1958).

Osborne, W. C. *Fans* (1966, Pergamon Press, Oxford).

Stepanoff, A. J. *Turboblowers* (1955, Wiley, New York).

Woods' *Practical Guide to Fan Engineering.* 2nd edn. (1960, Woods of Colchester Ltd.).

PUMPS

Addison, H. *The Pump Users' Handbook* (1958, Pitman, London).

Addison, H. *Centrifugal and other Rotodynamic Pumps* 3rd edn. (1966, Chapman & Hall, London).

De Kovats, A and Desmur, G. *Pumps, Fans, and Compressors* (Translated from the French by R. S. Eaton, 1958).

Hicks, T. G. *Pump Selection and Application* (1957, McGraw-Hill, New York and London).

Hicks, T. G. *Pump Operation and Maintenance* (1958, McGraw-Hill, New York and London).

Karassik, I. J. *Engineers' Guide to Centrifugal Pumps* (1964, McGraw-Hill, New York and London).

MEASURING INSTRUMENTS

Coxon, W. F. *Temperature Measurement and Control* (1960, Heywood, London).

Griffiths, E. *Methods of Measuring Temperatures.*

Kent, W. R. L. *et al. Boiler Plant Instrumentation* (1956, George Newnes, London).

Miles, V. C. *Thermostatic Control. Principles and Practice.* 2nd edn. (1975, Newnes-Butterworths, London).

Ower, E. *Measurement of Flow* (1966, Chapman & Hall, London).

Shell *Flow Meter Engineering Handbook* (Edited by T. S. Preston, 1968).

The Mechanical Properties of Fluids. A Collective Work (Blackie, London).

PIPING

M. W. Kellog Co. *Design of Piping Systems* (1965, Wiley, New York and London).

Littleton, C. T. *Industrial Piping* 2nd edn. (1962, McGraw Hill, New York and London).

Martin, W. L. *Handbook of Industrial Pipework* (1961, Pitman, London).

Pearson, G. H. *Valve Design* (1972, Pitman, London).

WELDING

American Welding Society. *Welding Handbook* 6th edn. (1969–72, Macmillan, London).

Atkins, E. A. and Walker, A. G. *Electric Arc and Oxy-Acetylene Welding.* 4th edn. (1952, Pitman, London).

British Oxygen Company. *Handbook for the Welder.*

Davies, A. C. *The Science and Practice of Welding* 6th edn. (1972, Cambridge University Press, London).

Giachino, J. W. *Welding Skills and Practice* 4th edn. (1971, Technical Press, London).

Giachino, J. W. *Welding Technology* (1968, Technical Press, London).

Kilburn, W. L. *Copper and Bronze Welding.*

Koenigsberger, F. and Adair, J. R. *Welding Technology* 3rd edn. (1966, Macmillan, London).

Lincoln Electric Co. *Procedure Handbook of Arc Welding Design and Practice.* 3rd edn. (1957).

Lincoln Electric Co. *Lessons in Arc Welding.*

Phillips' Practical Welding Course (Phillips Industrial Co., London).

Section XV

BRITISH STANDARDS AND CODES OF PRACTICE APPLYING TO HEATING AND VENTILATING PLANTS

The Standards and Codes of Practice listed in this section are published by:

The British Standards Institution,
2 Park Street,
London W1A 2BS.

BRITISH STANDARDS

No.

10 : 1962 **Flanges and bolting for pipes, valves and fittings**

Flanges in grey cast iron, copper alloy and cast or wrought steel for −328°F (−200°C) to 975°F (524°C) and up to 2800 lb/in^2. Materials and dimensions of flanges, bolts and nuts. Ten tables cover plain, boss, integrally cast or forged, and welding neck types.

21 : 1973 **Pipe threads for tubes and fittings where pressure-tight joints are made on the threads** Metric and Imperial

Specifies a range of jointing threads, size $\frac{1}{16}$ to 6, in metric and inch units where pressure-tight joints are made by mating taper internal and external threads, or taper external and parallel internal threads. Longscrew threads for use with connectors specified in BS 1387 are included. Details of thread forms, dimensions and tolerances are given, together with method of designating each type of thread.

41 : 1973 **Cast iron spigot and socket flue or smoke pipes and fittings** Metric

Material, dimensions and tolerances of pipes, bends and offsets up to 300 mm nominal bore and nominal weight of pipes.

61 :—— **Copper tubes (heavy gauge) for general purposes**

61 : Part 1 : 1947 **Copper tubes (heavy gauge)**

Pressure ranges, (*a*) up to 175 lb/in^2, (*b*) over 175 up to 300 lb/in^2. Dimensions, quality of material; chemical, mechanical, hydraulic tests.

61 : 1969 **Threads for light gauge copper tubes and fittings** Metric and Imperial

Requirements for a series of threads for use with light gauge copper tubes and associated fittings to B.S. 61 : Part 1, B.S. 66, B.S. 99 and B.S. 1306, for which BSP threads to B.S. 21 are not suitable or cannot be utilised due to the thickness being insufficient.

66 & 99 : 1970 **Cast copper alloy pipe fittings for use with screwed copper tubes** Metric & Imperial

Materials and dimensions for banded or beaded cast copper alloy pipe fittings and three piece unions for use with screwed copper tubes suitable for working pressures of up to 125 lbf/in^2 (0.86 MN/m^2) for steam, air, water, gas and oil.

78 :—— **Cast iron spigot and socket pipes (vertically cast) and spigot and socket fittings**

78: Part 1: 1961 **Pipes**

Dimensions and weights of standard pipes for water, gas, sewage, etc. Grades A, B, C and D; internal diameters 3 in. to 48 in. Quality of metal, mode of casting, test requirements, coating, testing facilities.

78: Part 2: 1965 **Fittings**

Dimensions and weights of bends, elbows, tees, angle branches, crosses, condensate receivers, hatchboxes, plugs and caps for water, gas, sewage, etc. Grades A, B, C and D; internal diameters 3 in. to 48 in. Quality of metal, mode of casting, test requirements, coating, testing facilities.

143 & 1256 : 1968 **Malleable cast iron and cast copper alloy screwed pipe fittings for steam, air, water, gas and oil** Metric & Imperial

Materials, tests and dimensions of plain and reinforced fittings suitable for working pressures of up to 200 lbf/in^2 (1·380 MN/m^2) for water and 150 lbf/in^2 (1·035 MN/m^2) for steam, air, gas and oil.

350 :— **Conversion factors and tables**

350: Part 1: 1974 **Basis of tables, Conversion factors** Metric & Imperial

Covers a wide range of subjects of measurement falling under the general headings metrology, mechanics and heat. Extended basic information on units, together with the standard abbreviation for each unit or combination of units considered. Tables of units and conversion factors, showing the relationship between any pair of units used in the measurement of a given physical quantity. Purely electrical units are not dealt with.

350: Part 2: 1962 **Detailed conversion tables** Metric & Imperial

Detailed conversion tables under the general headings metrology, mechanics and heat, with the numerical basis of each table given at the head and tabulated values expressed for the most part to the nearest 6th significant figure. Summary of tables of conversion factors is also included.

Supplement No. 1: 1967 **(PD 6203) Additional tables for SI conversions** Metric & Imperial

Contains 36 additional tables for conversions to and from SI units.

416 : 1973 **Cast iron spigot and socket soil, waste and ventilating pipes (sand cast and spun) and fittings**

One grade of pipe with alternative sockets, and a range of cast iron fittings for use above ground. Quality of materials, finish, testing and marketing. Tables of dimensions to cover pipes, bends, branches, roof-outlets, inspection pieces, holderbats, wire balloons and sanitary connections.

417 :— **Galvanized mild steel cisterns and covers, tanks and cylinders**

417: Part 1: 1964 **Imperial units**

Specifies cold and hot water storage vessels, mainly intended for domestic purposes, as follows:

Cisterns: 20 sizes, from 4 to 740 gallons capacity, in two grades; with covers.

Tanks: 5 sizes, from 21 to 34 gallons capacity, in two grades.

Cylinders: 10 sizes, from 16 to 97 gallons capacity, in three grades.

Materials, workmanship, design and construction, dimensions and tolerances; provision for pipe connections and electric immersion heater and gas circulator connections.

Optional internal coating of cisterns with bitumen, testing and marking requirements, test pressures and working heads.

417: Part 2: 1973 **Metric units**

Specifies cold and hot water storage vessels, mainly intended for domestic purposes. Covers cisterns—20 sizes (from 18 litres to 3364 litres capacity), in 2 grades with covers; tanks—5 sizes (from 95 litres to 155 litres capacity), in 2 grades; cylinders—10 sizes (from 73 litres to 441 litres capacity), in 3 grades. Materials, workmanship, design and construction, dimensions and tolerances; provision for pipe connections and electric immersion heater and gas circulator connections. Optional internal coating of cisterns with bitumen, testing and marking requirements, test pressures and working heads.

499 :—— **Welding terms and symbols**

499: Part 1: 1965 **Welding, brazing and thermal cutting glossary**

Seven sections: terms common to more than one section, terms relating to welding with pressure, fusion welding, brazing, testing, weld imperfections, thermal cutting. Appendices: typical information to appear on welding and cutting procedure sheets.

499: Part 2: 1965 **Symbols for welding**

Scheme for designating on drawing the type, size, location and finish of a weld for a variety of welding processes. Table of standard symbols: illustrations of application of the scheme. Appendices: typical information to be given on welding and cutting procedure sheets.

499C: 1965 **Chart of British Standard welding symbols** (based on B.S. 499 Part 2)

Symbols for denoting type, size and position of welds reproduced from B.S. 499 Part 2 as a wall chart, accompanied by examples of their use.

499: Part 3: 1965 **Terminology of and abbreviations for fusion weld imperfections as revealed by radiography**

Definitions of terms for imperfections, descriptions of their appearance in radiographs, typical radiographs facing each term. Coding system for reporting defects and their location in a radiograph in an abbreviated form.

567 : 1973 **Asbestos-cement flue pipes and fittings, light quality** Metric

Covers flue pipes and fittings of light quality of diameters 50 mm, 75 mm, 100 mm, 125 mm, and 150 mm, intended for use with gas fired appliances of input-rating not exceeding 45 kW. Metricated revision of BS 567: 1968 which it supersedes.

599 : 1966 **Methods of testing pumps**

Describes methods for testing the performance and efficiency of pumps handling fluids which behave as homogeneous liquids.
Covers terms and definitions, symbols and formulae recommended for use in pump testing; standard methods of measuring pressure (head), discharge, power, speed, permissible fluctuations and variations in readings, and limits of error; cavitation testing, tests with vaporizing liquids, tests with viscous liquids and liquids containing solids in suspension.
Provides guidance on estimating the combined limits of error, model testing, and tests at modified speeds.

602 & 1085 : 1970 **Lead and lead alloy pipes for other than chemical purposes** Metric

Specifies requirements for lead pipes in three different compositions in sizes up to and including 125 mm bore (BS 602) and for lead-silver-copper pipes in sizes up to and including 40 mm bore (BS 1085). Suitable for use as service and distributing pipes laid underground; service, cold water and hot water pipes fixed above ground; soil, waste and soil-and-water ventilating, flushing and warning pipes; gas pipes in heavier and lighter weights. See CP 310 for further information on use.

659 : 1967 **Light gauge copper tubes (light drawn)**

Dimensions of tubes of nominal size $\frac{1}{8}$ in to 6 in; tolerances on outside diameters and thicknesses; selection of test samples, mechanical tests, marking, chemical compositions and mechanical properties.

699 : 1972 **Copper cylinders for domestic purposes** Metric

Copper cylinders for heating and storage of domestic hot water. Grades 1, 2 and 3 for maximum working heads of 25 mm, 15 mm and 10 mm; 14 sizes (from 74 litres to 450 litres). Dimensions range from 350 mm to 600 mm external diameter, 675 mm to 1800 mm height. Materials, manufacture, dimensions, connections for pipes and heaters. Tests and marking requirements.

715 : 1970 **Sheet metal flue pipes and accessories for gas fired appliances** Metric

Single- and twin-wall flue pipes, fittings and accessories with welded or folded seams, finished as necessary to resist heat and minimise corrosion. Tabulated dimensions with appendices on testing certain coatings, measurement of *U* value and typical assembly.

749 : 1969 **Underfeed stokers** Metric

Stokers rated up to 550 kg of coal per hour for all applications except metallurgical or other high temperature furnaces. Covers requirements, installation and maintenance; rating by heat output and suitable fuels are appended. (See also CP 3000).

759 : 1967 **Valves, gauges and other safety fittings for application to boilers and to piping installations for and in connection with boilers**

Safety valves, high and low water alarms, stop valves, feed valves, blow-down mountings, water gauges, pressure gauges, test connections and fusible plugs for application to boilers and to main and auxiliary steam pipes and to associated feed and blow-down piping. Fittings for boilers and calorifiers used for low pressure steam heating, hot water central heating and hot water supply. Materials, temperature and pressure limits and methods of construction are specified, together with formulae for determining the minimum aggregate area of safety valves.

778 : 1966 **Steel pipes and joints for hydraulic purposes**

Covers steel pipes and joints for hydraulic fluids such as water and oil as follows:
Part 1: Screwed flange joints and hot finished, cold drawn and electric resistance welded pipes in the size range $\frac{17}{32}$ in. to $5\frac{1}{2}$ in. outside diameter for design pressures of:
up to and including 1500 lbf/in^2
above 1500 lbf/in^2 up to and including 2500 lbf/in^2
above 2500 lbf/in^2 up to and including 5000 lbf/in^2
Part 2: Hot finished and cold drawn seamless pipes for use with welding neck flanges in the size range 4 in. to $12\frac{3}{4}$ in. outside diameter for design pressures of 2500, 5000 lbf/in^2 and 6000 lbf/in^2
Part 3: Cold drawn seamless and electric resistance welded pipes for use with compression pipe fittings in the size range 0·250 in to 2·0 in outside diameter for design pressures of 2500, 5000 lbf/in^2 and 6000 lbf/in^2
The pipes specified are in accordance with B.S. 3601, grades 22 and 27, and thicknesses are given for water and oil service fluids. An appendix gives the design basis for calculating the thickness of pipes.

779 : 1961 **Cast iron boilers for central heating and hot water supply**

Steam boilers of over 150,000 Btu per hour rating for an operating pressure not exceeding 15 lbf/in^2, hot water central heating boilers for use in open systems designed to operate at a pressure not exceeding 120 ft head or 52 lbf/in^2 at a maximum working temperature of 212°F, hot water central heating boilers of over 150,000 Btu per hour rating for use in open systems designed to operate at a pressure not exceeding 140 ft. head or 60 lbf/in^2 and a maximum working temperature of 212°F, hot water central heating boilers of over 150,000 Btu per hour rating for use in pressurized (medium pressure) systems designed to operate at a pressure not exceeding 50 lbf/in^2 at a maximum working temperature of 250°F, boilers for direct hot water supply of 55,000–2,000,000 Btu per hour rating designed to operate at a pressure of 120 ft. head or 52 lbf/in^2 at a maximum working temperature of 250°F.

Efficiencies; fuels (solid, gaseous, liquid); rating on performance; materials; workmanship and construction; inspection, testing and marking; mountings and appliances. Bacharach or Shell smoke scale appended.

799 :—— **Oil burning equipment**

799: Part 1: 1962 **Atomizing burners and Associated equipment**

Fully automatic, semi-automatic and hand-controlled atomizing oil burners and associated equipment for boilers, heaters, furnace, ovens and other similar static plant. Does not apply to steam boilers of a capacity over 25,000 lb. of steam per hour per boiler or to marine and mobile installations. Timings for controls and safety devices on ignition failure and flame failure for both the fully automatic and semi-automatic types, construction and plate thickness of oil storage tanks, filling pipes and connections, oil heaters, filters and other items of equipment forming part of the oil burner assembly. Oil storage tank capacity and overfilling hazards are appended.

799: Part 2: 1964 **Vaporizing burners and Associated equipment**

Fully automatic, semi-automatic and hand-controlled vaporizing oil burners and associated equipment for boilers, heaters, furnaces, ovens and other similar static flued plant (free standing space-heating appliances for single family dwellings) suitable for oil to class C and D of BS. 2869.
Performance requirements, timings for controls and safety devices on ignition failure and flame failure for both the fully automatic and semi-automatic types, construction and plate thickness of oil storage tanks up to 750 gallons capacity, integral tanks and barometric tanks, filling pipes and connections, filters and other items of equipment forming part of the oil burner assembly. Smoke test appended.

799: Part 3: 1970 **Automatic and semi-automatic atomizing burners up to 36 litres per hour and associated equipment** Metric

Requirements for automatic atomizing burners, including pressure jet burners and semi-automatic atomizing burners for boilers, space heating and cooking appliances, and similar static and mobile plant having a total burning rate not exceeding 36 litres/hour and suitable for burning Classes C and D fuel complying with BS 2869.

835 : 1973 **Asbestos-cement flue pipes and fittings, heavy quality** Metric

Covers flue pipes and fittings of heavy quality of diameters from 75 mm to 600 mm inclusive intended for use with solid fuel and oil burning appliances of output rating not exceeding 45 kW, for gas fired appliances, and for incinerators not exceeding 0·09 m^3 capacity. Metricated revision of BS 835 : 1967 which it supersedes.

845 : 1972 **Acceptance tests for industrial type boilers and steam generators** Metric

Methods to be adopted and data required when carrying out a simple efficiency test at minimum cost on hot water and steam raising plants using solid, liquid or gaseous fuel. Not for large power station trials.

848 :—— **Methods of testing fans for general purposes, including mine fans**

848: Part 1: 1963 Performance

Nine methods for testing the performance and efficiency of axial flow, centrifugal and propeller fans, including the site testing of fans in mines and tunnels. Terms, definitions, symbols and formulae recommended for use in fan testing; procedure and instruments for measuring pressure, air velocity and air volume under test conditions; tolerances on fan performance. Selection of test methods for particular fans; illustrations of layouts for each test.

848: Part 2: 1966 Fan noise testing

In-duct, free-field and reverberant field test methods from which the sound output of a fan may be derived; also semi-reverberant test method and methods for site testing. Terms, definitions, symbols and formulae for use in fan noise testing. Procedure and instruments for measuring fan sound output. Illustrations of modification to test airway (in BS 848: Part 1) for the purpose of sound measurement.

853 :—— **Calorifiers for central heating and hot water supply**

853: Part 1: 1960 Mild Steel and cast iron

Calorifiers for central heating and hot water supply for operating pressures and temperatures in the shell not exceeding 65 lb/in^2 and 250°F respectively and for design pressures in the tube battery not exceeding 250 lb/in^2; applies to steam heated calorifiers of any output and also to water-to-water and electrically heated calorifiers exceeding 50 gallons capacity.

853: Part 2: 1960 Copper

Calorifiers for central heating and hot water supply for operating pressures and temperatures in the shell not exceeding 65 lb/in^2 and 250°F respectively and for design pressures in the tube battery not exceeding 250 lb/in^2; applies to steam heated calorifiers of any output and also to water-to-water and electrically heated calorifiers exceeding 50 gallons capacity.

855 : 1961 **Welded steel boilers for central heating and hot water supply**

Steam boilers of 15,000–5,000,000 Btu per hour rating for an operating pressure not exceeding 30 lb/in^2; hot water central heating boilers of 150,000–5,000,000 Btu per hour for use in pressurized (medium pressure) systems designed to operate at pressures not exceeding 65 lb/in^2 and a maximum temperature of 270°F; hot water central heating boilers of up to 5,000,000 Btu per hour for use in open systems designed to operate at pressures not exceeding 150 ft. head or 65 lb/in^2 and a maximum temperature of 212°F; boilers for direct hot water supply of 55,000–2,000,000 Btu per hour rating designed to operate at a pressure not exceeding 150 ft. head or 65 lb/in^2.
Efficiencies, fuels (solid, gaseous, liquid) rating of boilers on performance. Materials, construction, workmanship (including welding requirements), design, inspection, testing and marking, mountings and appliances.
Appendices cover weld preparation for butt welds, attachment of stand pipes and pads (secured by welding alone), Bacharach or Shell smoke scale.

864 : 1953 **Capillary and compression tube fittings of copper and copper alloy**

Dimensions and requirements essential for satisfactory installation and performance; covers the most commonly used types of fittings of nominal sizes $\frac{1}{8}$ in to $2\frac{1}{2}$ in inclusive. Design, construction, workmanship; hydraulic test, porosity test (cast fittings), working pressures and temperatures for various installations.

864: Part 2: 1971 Metric

Specifies requirements for capillary and compression fittings for use with copper tubes complying with BS 2871 : Part 1 Tables X, Y and Z. Applies to the most commonly used type of fittings of nominal sizes ranging from 6 mm to 54 mm inclusive. Specifies design, construction, hydraulic test, porosity test, maximum cooling temperatures and pressures.

1010 :—— **Draw-off taps and stop valves for water services (screwdown pattern)**

1010: Part 1: 1959 **Imperial units**

Requirements for sizes from $\frac{1}{4}$ in to 2 in, and also for rising spindle/rising top, rising spindle/non-rising top, and non-rising spindle types. Material and workmanship for casting and hot pressing. Component parts; waterways, screw threads, spindle and head threads, tap bodies, washers and washer plates, gland packing and union connections. Hydraulic test. Provision is made for ends suitable for direct connection to copper tubing or to lead or iron piping. 'Easy clean' taps and valves ($\frac{1}{2}$ in and $\frac{3}{4}$ in, non-rising spindles with 'O' ring seals) are covered in Amendment No. 7 (design, dimensions, performance tests for ring materials).

1010: Part 2: 1973 **Draw-off taps and above-ground stopvalves** Metric

Specifies, in metric terms, the dimensions and test requirements for screw-down pattern draw-off taps and above-ground stopvalves from $\frac{1}{4}$ in to 2 in nominal sizes. Material requirements (mainly metal), design details and dimensions of components and union ends are also included. BS 1010: 1959 has become 'Part 1'.

1042 :—— **Methods for the measurement of fluid flow in pipes**

1042: Part 1: 1964 **Orifice plates, nozzles and venturi tubes**

Geometrical shape, relative dimensions, constructional requirements and accuracy of manufacture of square-edged, conical-entrance and quarter-circle orifice-plates, nozzles (of ISA 1932 profile), venturi nozzles and venturi tubes. Calculation of rate of flow of homogeneous fluids in pipes of inside diameter not less than one inch; calculation of dimensions of a device for metering a given rate of flow. Applicable to both compressible and incompressible fluids, to the flow of viscous liquids at low Reynolds number and to flow at high pressure differences (initial flow); not to fluids exhibiting non-Newtonian behaviour, to suspensions of solids in liquids or gases, to non-steady or pulsating flow or to flow in partially filled pipelines.

1042: Part 2A: 1973 **Pitot tubes. Class A accuracy** Metric

Describes methods of measuring the rate of flow of a liquid or gas in a pipe, within a specified range of velocities, using a pitot-static tube (or any other tube whose coefficient is known). The conditions for which a total-pressure tube may be used with a wall-static pressure tapping in place of a combined pitot-static tube are also described. Specifies the shape, relative proportions, limits or size and constructional requirements of a family of pitot-static tubes. Restricted to measurements of fluid flow in pipes of circular cross section.

1042: Part 3: 1965 **Guide to the effects of departure from the methods in Part 1**

Effects of non-compliance with the requirements laid down in Part 1. Additional tolerances or corrections which cannot, however, compensate for deviation from Part 1.

1192 : 1969 **Building drawing practice** Metric

Deals with sizes, layouts of drawing sheets, scales, dimension of drawings, drawing materials and reproduction, various types of projection, symbols and graphic representation, together with examples of schedules. Drawings are also included.

1211 : 1958 **Centrifugally cast (spun) iron pressure pipes for water, gas and sewage**

Pipes with spigot and socket joints cast in either metal or sand moulds. Three classes of pipe: Class B, field test pressure 400 ft. head of water; Class C, field test pressure 600 ft. head of water; Class D, field test pressure 800 ft. head of water. Standard lengths, internal diameters, hydraulic tests and tests for straightness.

1212 :—— **Ballvalves (excluding floats)**

1212: Part 1: 1953 **Piston type**

Seven sizes, for the attachment of floats to BS 1968 and BS 2456. Nominal sizes $\frac{3}{8}$, $\frac{1}{2}$, $\frac{3}{4}$, 1, $1\frac{1}{4}$ and 2 in. Provision for removable seats; these have different sized orifices to suit particular requirements of pressure and flow. Materials, quality, workmanship, dimensions, tests for performance, test for the mechanical strength of levers. An appendix gives information on computed flow through seat orifices at various heads of water.

1212: Part 2: 1970 **Diaphragm type (brass body)** Metric

Material, workmanship, design, construction, sizes, tolerances and performance of diaphragm operated ballvalves in two nominal sizes ($\frac{3}{8}$ in and $\frac{1}{2}$ in), having four seat sizes ($\frac{1}{8}$ in, $\frac{3}{16}$ in, $\frac{1}{4}$ in, and $\frac{3}{8}$ in bore). (N.B. The nominal sizes and the seat sizes remain in imperial units until BS 1212: 1953, to be renumbered BS 1212: Part 1, has been metricated.)

1250 :—— **Domestic appliances burning town gas**

1250: Part 1: 1966 **General requirements**

General requirements for construction and performance. Information on test gases and on conditions and methods of test common to all appliances is given in appendices.

1250: Part 3: 1963 **Water-heating appliances**

Specifies design and performance requirements for water-heating appliances (other than laundering appliances); methods of test are described in appendices.

1250: Part 4: 1965 **Space heating appliances**

Design and performance requirements for space heating appliances. Methods of test.

1289 : 1945 **Pre-cast concrete flue blocks for gas fires (of the domestic type) and ventilation**

Provides for two types of flue way, type 1 for use with gasfires of not more than 15,000 Btu and type 2 suitable for use with gasfires over 15,000 Btu. Specifies the material, the surface texture and the dimensions of the flue together with the terminals and joints. It also specifies the compressive strength and drying shrinkage together with methods of test for determining these properties.

1294 : 1946 **Soot doors for domestic buildings**

A range of five sizes of soot doors for use in concrete and brickwork flues of domestic buildings, such as flats, houses and schools, is provided.

1306 :—— **Non-ferrous pipes and tubes for steam services**

1306: Part 1: 1955 **Non-ferrous pipes and piping installations for and in connection with land boilers**

Applies to the design and construction of non-ferrous pipework connecting a land steam boiler to engine, turbine or industrial plant and all auxiliary pipework, together with the individual pipes and pipe fittings forming parts of such installations. It lays down maximum design pressures and temperatures for copper pipes and gunmetal castings used in these installations. It is similar and complementary to B.S. 806.

1334 : 1969 **The use of thermal insulating materials for central heating and hot and cold water supply installations** Metric

Minimum thicknesses of insulating material for heat conservation or for frost protection on installations working at up to *c.* 100°C, based on manufacturer's declared thermal conductivity values. Appendices describe methods of calculation.

1339 : 1965 **Definitions, formulae and constants relating to the humidity of the air** Metric

Gives definitions, etc., with tables of saturation vapour pressures and bibliography.

1387 : 1967 **Steel tubes and tubulars suitable for screwing to B.S. 21 pipe threads** Metric & Imperial

Materials, dimensions and testing of welded and seamless screwed and socketed steel tubes and tubulars, plain end steel tubes suitable for screwing to B.S. 21 pipe threads, nominal bores $\frac{1}{8}$ in. (16 mm) to 6 in. (150 mm) in medium and heavy thickness and up to 4 in. (100 mm) in light thickness. Dimensions, weight per foot and per metre run. Marking. Testing for galvanizing.

1394 :—— **Power driven circulators**

1394: Part 1: 1971 **Glanded and glandless pumps** Metric

Covers centrifugal glanded pumps and glandless pumps above 200 W and not exceeding 2000 W input at maximum voltage (single and three phase electrical supply) used on heating and domestic hot water supply piping systems. Design, construction, performance and testing; specific electrical requirements for glandless pumps.

1394: Part 2: 1971 **Domestic glandless pumps** Metric

Covers domestic glandless pumps with power input not exceeding 200 W at maximum rated voltage (single phase electrical supply) used on small bore piping systems of domestic heating and domestic hot water supply installations. Design, construction, performance, testing, electrical requirements.

1415 : 1955 **Mixing valves (manually operated) for ablutionary and domestic purposes**

Gives manufacturing and test requirements for $\frac{1}{2}$ in., $\frac{3}{4}$ in. and 1 in. manually operated mixing valves. The method of operation, permissible head loss, and methods of connection to the service pipes are given and a suitable testing rig is shown.

1427 : 1962 **Routine control methods of testing water used in industry** Metric

Covers methods of test for the following:

1. Alkalinities
2. Aluminium
3. Ammonia (free and saline)
4. Calcium
5. Chloride
6. Residual chlorine (free and total)
7. Colour
8. Dissolved oxygen
9. Dissolved solids
10. Electrical conductivity
11. Free carbon dioxide
12. Hardness
13. Hydrazine
14. Iron
15. Magnesium
16. Nitrite
17. pH value
18. Phosphates
19. Silica
20. Specific gravity
21. Sulphate
22. Sulphite
23. Fluoride

1563 : 1949 **Cast-iron sectional tanks (rectangular)**

Deals with bolted sectional cast iron tanks up to 40 ft. square and 12 ft. deep excluding supporting structures not subject to pressure other than static head. The tanks specified are with either internal or external flanges and open or closed tops, made from unit plates 2, 3 or 4 ft. square. Details are included of mild steel tie rod ends and turnbuckles, and sizes and thicknesses of unit plates. Four half-tone illustrations of typical tanks, four tables of scantlings and seven tables of approximate weights and nominal capacities are provided.

1564 : 1949 **Pressed steel sectional tanks (rectangular)**

Deals with bolted sectional pressed steel tanks up to 52 ft. square and 16 ft. deep excluding supporting structures not subject to pressure other than static head. The tanks specified are with internal or external flanges, with open or closed tops made from unit plates 4 ft. square. Details are included of the thicknesses of unit plates, and staying and welded connections. Ten drawings, and four tables of approximate weights and nominal capacities for tanks with external flanges are provided.

1565 :—— **Galvanized mild steel indirect cylinders, annular or saddle-back type**

1565: Part 1: 1949 **Imperial units**

One of a series of standards relating to domestic hot water storage vessels, the others being BS 417, BS 699 and BS 1566. It provides for a range of seven sizes of indirect cylinder in two classes, suitable for maximum permissible working heads of 60 ft. and 30 ft. respectively.

The standard deals with minimum thickness of material, minimum heating surfaces, methods of manufacture, radius of curvature, bolted ends and hand-holes, method of galvanizing, screwed connections for pipes and screwed connections for auxiliary electric heating. The latter provision gives a variety of methods so that purchasers may choose that most suitable for the operating conditions. The test requirements deal with primary heaters and complete cylinders. A marking clause gives the various marks to be added in all cases, and recommends the incorporation of further information likely to assist in installation.

All the main dimensions are set out in a table and drawings illustrate the method of measuring them.

1565: Part 2: 1973 **Metric units**

Specifies materials, manufacture, design, dimensions, testing and marking requirements for indirect cylinders in eight sizes (109 litres to 455 litres) and two grades (max. permissible working heads of 18 m and 9 m). Screwed connections for pipes, electric immersion heaters, thermostats; hand holes, draining taps, gas circulators.

1566 :—— **Copper indirect cylinders for domestic purposes**

1566: Part 1: 1972 **Double feed indirect cylinders** Metric

Traditional indirect copper cylinders for heating and storage of hot water. Grades 2 and 3 for maximum working heads of 15 m and 10 m, 14 sizes (from 72 litres to 440 litres). Dimensions range from 350 mm to 600 mm external diameter, 675 mm to 1800 mm height. Materials, manufacture, dimensions, connections for pipes and heaters. Tests and marking requirements.

1566: Part 2: 1972 **Single feed indirect cylinders** Metric

Self-venting type indirect copper cylinders for heating and storage of hot water. Grades 2 and 3 for maximum working heads of 15 m and 10 m, 6 sizes (from 86 litres to 180 litres). Dimensions range from 400 mm to 500 mm external diameters, 750 mm to 1200 mm height. Materials, manufacture, dimensions, connections for pipes and heaters. Tests and marking requirements.

1710 : 1971 **Identification of pipelines** Metric

Defines the meaning and application of a limited number of colours for the identification of pipes conveying fluids in liquid or gaseous condition in land installations and on board ships.

1737 : 1951 **Jointing materials and compounds for water, town gas and low-pressure steam installations**

Relates to the materials, composition strength and sizes of compressed asbestos fibre, plain rubber, rubber insertion and corrugated metal joint rings. The properties of jointing compounds are covered in a special section while methods of testing are given in a range of appendices.

1740 :—— **Wrought steel pipe fittings (screwed BSP thread)**

1740: Part 1: 1971 **Metric units**

Specifies requirements for welded and seamless wrought pipe fittings of nominal size 6 mm to 150 mm inclusive, for use with steel tubes to BS 1387, screwed BSP thread to BS 21.

1740: Part 2: 1971 **Imperial units**

Specifies requirements for welded and seamless wrought pipe fittings of nominal size $\frac{1}{8}$ in to 6 in inclusive, for use with steel tubes to BS 1387, screwed BSP thread to BS 21.

1756 :—— **Methods for sampling and analysis of flue gases**

1756: Part 1: 1971 **Methods of sampling** Metric

Deals with the general principles to be adopted in order to obtain a representative flue gas sample for analysis.

1756: Part 2: 1971 **Analysis by the Orsat apparatus** Metric

Apparatus, reagents, method, sample, analysis, calculation and reporting of results.

1756: Part 3: 1971 **Analysis by the Haldane apparatus** Metric

Apparatus, reagents, method, sample analysis, calculation and reporting of results.

1756: Part 4: 1965 **Miscellaneous analysis** Metric

Deals with the determination of moisture, sulphur dioxide and sulphur trioxide, sulphuric acid dew-point, carbon monoxide (less than 0·5 per cent) and nitrogen oxides.

1756: Part 5: 1971 **Semi-routine analyses** Metric

Describes semi-routine methods for the determination of carbon dioxide, carbon monoxide and total oxides of sulphur, mainly employed for the assessment of the combustion performance of domestic gas appliances.

1894 : 1952 **Electrode boilers of riveted, seamless, welded and cast iron construction for water heating and steam generating**

Specifies requirements for the following types of electrode boiler:
1. Steel water-heating boilers.
2. Steel steam boilers.
3. Cast iron water-heating boilers.

Applies to boilers for design temperatures not exceeding 650°F and is concerned solely with boilers for water-heating or for the generation of steam, in which the water is heated by its resistance to the passage of an alternating electric current. It deals with materials, construction and workmanship, scantlings, inspection and testing of electrode boilers of riveted, seamless, welded, and cast iron construction. Safety valves have been dealt with at length in this standard and electrical safety devices are also specified. The formulae in this standard give in all cases the minimum scantlings and apply to boilers constructed throughout under competent supervision.

1952 : 1964 **Copper alloy gate valves for general purposes**

Rating, design and manufacture, materials, dimensions, tests and marking for Classes 100, 125, 150, 200 and 250, with flanged or screwed ends; size range: $\frac{1}{4}$ in. to 3 in., the gate operated by inside screw, rising or non-rising stem or by outside screw, rising stem.

a. Wedge gate, either solid or split.
b. Double disk.
c. Parallel slide.

1965 :— **Butt-welding pipe fittings for pressure purposes**

1965: Part 1: 1963 **Carbon steel**

Specifies leading dimensions, tolerances and materials for 90° and 45° elbows, 180° return bends, equal tees, concentric and eccentric reducers and caps, in carbon steel, for butt welding on to pipes. Details of testing and marking are also given.

1965: Part 2: 1964 **Austenitic stainless steel**

Dimensions tolerances and materials for 90° and 45° elbows, 180° return bends, concentric and eccentric reducers, equal and reducing tees and caps, in wrought and cast austenitic stainless steel, for butt-welding on to pipes. Details of testing and marking.

1972 : 1967 **Polythene pipe (Type 32) for cold water services** Metric

Requirements for extrusion compound, pipe material, classification and dimensions of pipes, physical and mechanical characteristics, sampling, marking, and stocking and transport. Test methods for toluene extract of carbon black and antioxidant content in appendices.

2060 : 1964 **Copper alloy screw-down stop valves for general purposes**

Rating, design and manufacture, materials, dimensions, tests and marking for globe, angle and oblique (or 'Y') types with rising stem, inside or outside screw, and with flanged or screwed ends; Classes 100, 125, 150, 200 and 250; size range: $\frac{1}{4}$ in. to 3 in.

2456 : 1954 **Floats for ballvalves (plastics) for cold water**

Gives sizes and dimensions, lifting effort required, types of material and tests for floats manufactured from polythene, P.V.C. and expanded ebonite which are suitable for use with B.S. 1212 ballvalves and for general use.

2594 : 1955 **Horizontal mild steel welded storage tanks**

Relates to the design and construction of mild steel fusion welded horizontal storage tanks with dished and flanged ends complying with Table 4 of B.S. 1966, 'Domed ends for tanks and pressure vessels', and also to tanks with unflanged outwardly dished ends. Requirements in respect of materials, manholes, weld preparations, pressure test and inspection are also given. The sizes of tanks range from 3 ft. 6 in. diameter by 4 ft. nominal length up to 9 ft. diameter by 30 ft. nominal length.

These tanks are classified by dimensions instead of by capacities which has hitherto been the practice of the trade. As a guide to the user the gross capacities of these tanks are given in the tables. Tanks with flat ends and also tanks with inwardly dished ends are not provided for in this standard. The standard does not relate to tanks on transport vehicles or to lined storage tanks.

2619 : 1955 **Method of test and rating for steam-heated air-heater batteries**

Specifies a method of test for steam-heated air-heater batteries. It describes the equipment required, gives instructions for calculation and interpretation of results, and has a section on 'type testing'.

2740 : 1969 **Simple smoke alarms and alarm metering devices** Metric

Requirements for the construction and operation of instruments designed to give an alarm when smoke emission from a chimney exceeds a chosen Ringelmann shade.

2742 : 1969 **Notes on the use of the Ringelmann and miniature smoke charts** Metric

Explains the purpose and method of use of these charts for the visual assessment of the darkness of smoke emitted from chimneys.

2767 : 1972 **Valves and unions for hot water radiators** Metric and Imperial

The revision now covers the inclusion of non-rising stem angle valves; the extension of the service conditions for which certain valves and unions, when specially tested, may be used; the incorporation of Y-pattern straight valves: the inclusion of requirements for materials for toroidal sealing rings and the alignment of minimum stem diameters, and body and bonnet wall thickness with the once-designated Class 100 valves in the latest revisions of BS 1952 and BS 2060.

2777 : 1974 **Asbestos-cement cisterns** Metric

Specifies requirements for twelve types of cistern with waterline capacities from 17 to 701 litres. Gives sizes and approximate weights together with sizes of co-ordinating spaces. Includes a system of referencing and test requirements.

2811 : 1969 **Smoke density indicators and recorders** Metric

Describes instruments for measuring the emission of black or grey smoke from a chimney in terms of optical density or percentage obscuration.

2831 : 1971 **Methods of test for air filters used in air conditioning and general ventilation**

Two tests are included: the first, applicable to any filter, deals with the determination of methylene blue efficiency; the second, not applicable to designs precluding accurate before and after weighing, concerns the determination of the dust-holding capacity. Does not specify performance standards.

2852 : 1970 **Rating and testing room air-conditioners** Metric

Prescribes three sets of standard rating conditions for the statement of performance of room air-conditioners employing air-cooled or water-cooled condensers, when used for cooling duties only and gives the corresponding test conditions and test procedures based on the use of a room calorimeter.

2869 : 1970 **Petroleum fuels for oil engines and burners** Metric

Four classes of fuel for internal combustion engines and seven classes of fuel for oil burners including a class (identified by a distinctive symbol) intended for use with free-standing domestic appliances not connected to flues. Minimum temperatures for storage and for outflow from storage and for handling are given for those burner fuels requiring heating facilities. Includes a viscosity conversion chart and a nomograph for calculated cetane index.

2871 :—— **Copper and copper alloys, tubes**

2871: Part 1: 1971 **Copper tubes for water, gas and sanitation** Metric

Specifies requirements in metric terms for copper tubes for water, gas and sanitation. Supersedes BS 1386 and BS 3931.

2871: Part 2: 1972 **Tubes for general purposes** Metric

Specifies requirements for copper and copper alloy tubes for general purposes including chemical composition, condition, dimensions, mechanical properties and non-destructive tests. Supersedes BS 61: Part 1, BS 885, BS 1306: Part 2 and BS 2017.

2879 : 1957 **Drainage taps (screw-down pattern)**

Specifies the design, materials and dimensions of, and tests for, the screw-down pattern non-ferrous taps used for draining hot and cold water and heating systems. It covers $\frac{1}{2}$ in., $\frac{3}{4}$ in., and 1 in. nominal size taps, operated by loose keys, and gives a range of loose key square sizes.

3048 : 1958 **Code for the continuous sampling and automatic analysis of flue gases. Indicators and recorders**

Deals with automatic instruments which give a direct indication or record of the composition of flue gases from industrial plant, and is complementary to B.S. 1756, which covers manually operated instruments.

3063 : 1965 **Dimensions of gaskets for pipe flanges**

'Plan' dimensions for 'inside bolt circle' and 'full face' gaskets for pipe flanges in accordance with B.S. 10 and B.S. 2035, and 'full face' gaskets for flanges to B.S. 1770. The tables giving dimensions bear the same designation as the flange tables in the appropriate pipe flange standard. Marking of gaskets for purchasing and identification.

3198 : 1960 **Combination hot water storage units (copper) for domestic purposes**

Gives general requirements and tests for both direct and indirect types of hot water storage units of 25 gallons capacity, made in copper.

3208 : 1960 Methods of test and rating for hot-water air-heater batteries

Specifies a method of test for hot-water air-heater batteries. It describes the equipment required, gives instructions for the calculation and interpretation of results, and includes procedures for dealing with special cases.

3276 : 1960 Thermometers for measuring air cooling power Metric

Specifies essential requirements of a series of thermometers for measuring low wind speeds or the efficiency of ventilation in ships, factories, hospitals, mines, etc. Standard charts are included, from which the 'cooling power' of the atmosphere and hence the air speed can be read off.

3284 : 1967 Polythene pipe (Type 50) for cold water services Metric

Requirements for extrusion compound, pipe material, classification and dimensions of pipes, physical and mechanical characteristics, sampling, marking and stocking and transport. Test methods for toluene extract of carbon black and antioxidant content in appendices.

3416 : 1961 Black bitumen coating solutions for cold application

Specifies two types of black bitumen coating solutions. Types I and II. Type I is a material for general purposes for the protection of iron and steel and Type II a material for the coating of drinking water tanks.

Details of the requirements of the coatings are given and methods of test for application properties, drying time, finish, protection against corrosion, flexibility, volatile matter, flash point and effect on water are given in the appendices. Special notes on the uses of the material are also given as appendices.

3505 : 1968 Unplasticized PVC pipe for cold water services Metric

Requirements for material, classification and dimensions of pipes, physical and mechanical characteristics, sampling, marking and stocking and transport. Test methods for heat reversion, resistance to acetone and sulphuric acid, opacity, effect on water, impact strength at 0°C and at 20°C, short term and long term hydraulic tests, and for the determination of organotin as tin in aqueous solution.

3561 : 1962 Non-domestic space heaters burning town gas

Specifies constructional and performance requiments for non-domestic space heaters burning town gas; includes general requirements and those specific to fan-assisted air heaters, radiant type overhead heaters, flued convector heaters and room-sealed heaters. Test gases and methods of test are described in appendices.

3899 : 1965 Refrigerated room air-conditioners

Room air-conditioners for all climates defined as 'encased assemblies primarily for mounting in a window or through a wall, or as a console'. Prescribes the construction requirements for cooling capacities generally up to 3 ton—refrigeration (36,000 Btu/h). Specifies a type test to be carried out by the room calorimeter method described in B.S. 2852, and production tests for the electrical equipment.

4118 : 1967 Glossary of sanitation terms

Water supply, from connection with water undertaker's main or from private natural supply; storage and distribution within the curtilage, including hot water supply but not hot water central heating. Sanitary appliances and their associated water fittings and waste fittings. Above-ground drainage of sanitary appliances, roofs and yards. Below-ground drainage, including small private sewage treatment and disposal works and connection to a local authority's sewerage system. Index.

4256 :—— **Oil-burning air heaters** Metric

4256: Part 1: 1972 **Non-domestic, transportable, fan-assisted heaters**

Construction, operation, performance and safety requirements for flued and unflued heaters designed for use with distillate oils such as kerosine, gas oil and domestic fuel oil.

4256: Part 2: 1972 **Fixed, flued, fan-assisted heaters**

Construction, operation, performance and safety requirements for heaters designed for use with distillate oils such as kerosine, gas oil and domestic fuel oil.

4256: Part 3: 1972 **Fixed, flued, convector heaters**

Construction, operation, performance and safety requirements for heaters designed for use with distillate oils such as kerosine, gas oil and domestic fuel oil.

4433 :—— **Solid smokeless fuel boilers with rated outputs up to 45kW**

4433: Part 1: 1973 **Boilers with undergrate ash removal** Metric

Deals with constructional requirements, controllability tests, performance at rated output and at low load, ability to withstand accidental over-run.

4433: Part 2: 1969 **Gravity feed boilers designed to burn small anthracite** Metric

Deals with constructional requirements and testing for controllability performance at rated output and at low load, ability to withstand accidental over-run.

4485 :—— **Water cooling towers**

4485: Part 1: 1969 **Glossary of terms**

Defines general terms, together with specific terms used in the testing of natural draught and mechanical draught water cooling towers.

4856 :—— **Methods for testing and rating fan coil units—unit heaters and unit coolers**

4856: Part 1: 1972 **Thermal and volumetric performance for heating duties without additional ducting** Metric

Deals with methods of carrying out thermal and volumetric tests on forced convection units containing fluid-to-air heat exchangers and incorporating their own fans. The units are for heating applications and the tests are to be carried out on units in an essentially clean condition.

CODES OF PRACTICE

CP 3 :—— **Code of basic data for the design of buildings**

CP 3: Chapter 1(C): 1950 **Ventilation**

This chapter deals with the ventilation of buildings for human habitation. The recommended rate of fresh air supply for different types of occupation are tabulated. The Appendix advises on the choice between natural and mechanical ventilation to meet individual circumstances. For natural ventilation, formulae are given for calculating the rate of air flow due to wind through openings and to temperature differences. Mechanical ventilation is advised when a satisfactory standard cannot be obtained by natural means. The different types are referred to.

CP 3: Chapter II: 1970 **Thermal insulation in relation to the control of environment**

Deals with the use of material to control internal environment of buildings and structures. Does not deal with use for fire protection or for construction of cold stores.

CP 3: Chapter VIII: 1949 **Heating and thermal insulation**

This chapter points out that heating and insulation should be considered together in the early stages of design. It examines the conditions affecting the temperature in dwellings, and recommends standards of warmth for rooms and for indoor places of public assembly. It indicates a method of calculating the degree of insulation appropriate to a building in terms of costs of structure, of heating and of expenditure on fuel, and sets out maximum permissible thermal transmittances for external parts of structure.

CP 99 : 1972 **Frost precautions for water services**

Includes recommendations for minimizing frost effects on water services generally, methods of locating pipes and fittings to obtain maximum frost protection (both for water supply pipes and for soil and waste pipes). Types and efficiency of insulating materials and advice on draining facilities.

CP 131 : 1974 **Chimneys and flues for domestic appliances burning solid fuel**

This code deals with flues depending for their operation upon natural draughts. It is restricted to appliances having a maximum heat output of 100,000 Btu per hour where the temperature of the flue gases leaving the appliance does not exceed 850°F. Recommendations are made regarding sizes and heights of flues for various appliances, the general constructions of chimneys of different materials and the height and position of chimneys and outlets above the roofs in relation to fire hazard and wind effects. An appendix sets out the causes of condensation in flues and suggests methods for reducing the formation of condensate and for protecting the chimney walling from the effects of attack.

CP 310 : 1965 **Water supply**

Deals with the supply of water to houses, schools, offices, public buildings and industrial buildings; the scope extends from the source of supply to the point where the water is drawn off for use. In addition to the statutory water undertakings, other sources of supply such as wells, springs, rivers, ponds and lakes are discussed. Also covers water treatment; distribution; storage; installation of all parts of the water system; inspection, testing and maintenance. An appendix and four figures deal with pipe sizing.

CP 331 :—— **Installation of pipes and meters for town gas**

CP 331: Part 3: 1965 **Installation pipes** (revision of CP 331.103)

Materials and installation for internal low pressure system (installation pipes) applicable to all types of buildings. Materials for pipes and fittings, sizing, jointing, protection, control, inspection and testing and identification. In appendix A provisions are included for gas installation pipes in large premises and in appendix B recommendations on temporary continuity bonding are given. Tables on the interval between the pipe supports, discharge rates in straight horizontal steel and copper pipes for both a $\frac{3}{10}$ in. w.g. and $\frac{5}{10}$ in. w.g. pressure drop together with allowances for elbows, tees and bends.

CP 332 :—— **Selection and installation of town gas space heating**

CP 332: Part 2: 1964 **Central heating boilers for domestic premises**

This code offers guidance on the selection, siting and installation of gas-fired boilers not exceeding 150,000 Btu/h. output utilizing town gas for heating of domestic premises by low pressure hot water and also installations in non-domestic premises requiring boilers of similar outputs. It deals in detail with selection of boilers by type and size; boiler position; boiler accessories; valves; air supply for combustion and ventilation; connections to both gas and electricity supply; connections to a flue and essential controls. Provides information on inspection, testing and servicing. Also included is an appendix giving extracts from CP 337 and data on pipe sizing.

CP 332: Part 3: 1970 **Boilers of more than 150,000 Btu/h (44kW) and up to 2,000,000 Btu/h (586kW) output**

Deals with the selection, siting and installation of gas-fired boilers for central heating and/or hot water supply for commercial and industrial premises.

CP 332: Part 4: 1966 **Ducted warm air systems**

Air heaters not exceeding 150,000 Btu/h. output for warming domestic premises. Guidance on selection of air heaters, heater accommodation, supply of combustion and ventilation air to the heater compartment, room ventilation mounting, return air arrangements, ducting and duct connections, positioning of warm air registers, gas supply, electrical controls and connections, controls, flues, inspection, testing, commissioning and servicing. Appendices contain notes on flues, tables of pipe sizing data and notes on noise.

CP 333 :—— **Selection and installation of town gas hot water supplies**

CP 333: Part 2: 1948 **Schools**

This code deals with systems of heating water for domestic use in schools by means of gas appliances, including auxiliary heating for clothes-drying when central heating is not in use. Consideration is given in the code to the selection of systems and of types of appliances suitable to meet the varying purposes and demand rates required in schools. Necessary requirements are also set out for provision of alternative heating by gas in central heating systems using solid fuel.

CP 337 : 1963 **Flues for gas appliances up to 150,000 Btu/h rating**

Deals with the choice and installation of flues forming part of installations for domestic and commercial purposes, but excluding industrial and specialist applications. Branched flue systems, balanced flue appliances, SE-Ducts and U-Ducts are dealt with and recommendations are included for terminations, flue materials and flue and duct sizes.

CP 341.300-307 : 1956 **Central heating by low pressure hot water**

The code comprises a head-code and seven sub-codes incorporated in one document, the sub-codes having the following titles: Boilers and calorifiers; Storage vessels; Pipework, fittings, valves, taps and cocks; Appliances (column radiators, surface panels, convectors); Unit heaters; Power driven circulating pumps for low pressure hot water heating installations; and Thermal insulation. The head-code 341.300 deals with the general aspects of central heating by low pressure hot water, whilst the sub-codes cover in detail the subjects indicated in their titles. Each sub-code includes a list of the relevant British Standards.

CP 342 : 1950 **Centralized domestic hot water supply**

This code deals with the installation of boilers, calorifiers, storage vessels, pipework and electricity driven circulators in central systems of domestic hot water supply for buildings ranging from small dwelling houses to hotels.

CP 342: Part 1: 1970 **Individual dwellings**

Deals with the planning, designing, installation and commissioning of centralized hot water supply systems in individual dwellings.

CP 352 : 1958 Mechanical ventilation and air conditioning in buildings

This code deals with the work involved in the general design, planning, installation, testing and maintenance of mechanical ventilating and air conditioning installations whereby air is forced into or extracted from buildings. It consists of a head code and eight sub-codes having the following titles: Fans, motors and starting gear; Air heaters; Air distribution system; Air cleaning devices; Thermal insulation; Sound-proofing and anti-vibration devices; Temperature and humidity controls; Cooling and dehumidification. The head code deals with the general aspects of the subject while the sub-codes cover in detail the subject indicated by their title. The code carries fourteen tables and ten illustrations.

CP 413 : 1973 Ducts for building services Metric

Deals with the design and construction of subways, crawlways, trenches, casings and chases for the accommodation of services within buildings and external ducts. Position, dimensions, safety precautions and control of hazards. Fire precautions in builders' or structural ducts and, as an appendix, fire precautions for ventilation ductwork. Supersedes CP 413 : 1951 which is therefore withdrawn.

CP 3000 : 1955 Installation and maintenance of underfeed stokers

In addition to the installation, care and maintenance of underfeed stokers used with domestic, sectional, shell and small-tube boilers, this code makes recommendations on the design of boiler room, fuel storage, flues and chimneys, and on the provision of automatic safety controls.
This Code replaces Appendix A of B.S. 749: 1952 'Underfeed stokers (ram or screw type)'.

CP 3002 :—— Oil firing

CP 3002: Part 1: 1961 Installations burning class D fuel oil and C.T.F. 50

This part of the code deals with the installation of oil burning equipment designed to burn C.T.F. 50 or Class D Fuel Oils in heating boilers, air heaters and cooking equipment connected to flues. It sets out specification data for Class D petroleum oil fuel and for coal tar fuel 50 extracted from current British Standards and other relevant data and also gives in tabulated form approximate design data for all fuels available in the U.K.
Information is given on oil burning and ancillary equipment for small and medium sized installations, including the influence of boiler design, selection of burners, fuel feed to burners, new burner installations, conversion of existing installations, etc. Recommendations are given on the selection and installation of oil storage tanks and ancillary equipment as well as the installa-ation of service tanks. Design and construction of accommodation for boilers and tanks is fully dealt with as well as requirements for chimneys and flues in single family dwellings and other buildings.
Tabulated recommendations are given on the standards of fire resistance for boiler rooms and tank chambers, with examples of the necessary types of construction.
The code concludes with diagrammatic illustrations of the methods of connecting oil storage tanks and of the protection necessary for externally situated tanks.

CP 3002: Part 2: 1964 Installations burning Class C and D fuel oils for vaporizing burners

This code deals with the installation of oil burning equipments embodying vaporizing burners and ancillary equipments designed for domestic heating and hot water installations. It sets out specification data for the Class C and

Class D petroleum oil fuels with approximate design data. Information is given on oil burning and ancillary equipments for installations in new and existing premises and includes the influence of boiler design, selection of burners, fuel feed to burners, new burner installations, conversion of existing installations etc. Recommendations are also given on the selection and installation of oil storage tanks, and ancillary equipments. The design construction of the accommodation for boilers and tanks is fully dealt with as well as requirements for chimneys and flues both in single family dwellings and other buildings.
Tabulated recommendations are set out on the standards of fire resistance for boiler rooms and tank chambers with examples of the necessary types of construction to meet these standards.

CP 3002: Part 3: 1965 **Installations burning pre-heated fuels. Class E, F and G fuel oils and C.T.F. 100 to 250.**

Installation of oil burning equipment designed to burn Class E, F and G fuel oils or C.T.F. 100, 200 or 250 in boilers and air-heaters connected to flues. Data for petroleum oil fuels and the coal tar fuels with approximate design data. Oil burning and ancillary equipment for installation to both new and existing boilers or air-heaters; includes influence on choice of grade of fuel, influence of boiler design and draught control, types of boilers, combustion chamber draught control, air-heaters, selection of burners, oil pre-heaters, feed to burners, sizing of oil pipelines, oil handling systems, ring mains, fillers, pumps, valves, tracing of oil pipelines, thermal insulation of pipelines, inspection and testing, painting and identification, refractory lining and insulation of combustion chamber, materials, appliances and components, fixing of electrical controls and wiring, inspection and testing on site, and maintenance. Selection and installation of oil storage tanks, design and construction of accommodation for boilers and tanks, requirements for chimneys and flues. Fire resistance standards for boiler rooms and tank chambers, examples of constructions which meet these standards.

CP 3006 :—— **Central heating for domestic premises**

CP 3006: Part 1: 1969 **Low pressure forced circulation hot water (small bore) systems**

Deals with planning, designing and installation of heating systems of 25,000 Btu/h to 150,000 Btu/h (7.5kW to 44kW) capacity, but excludes systems heated by back boilers.

CP 3009 : 1970 **Thermally insulated underground piping systems** Metric

Deals with the work involved in the general designing, planning, pre-fabrication, transport, installation, maintenance and testing of thermally insulated underground piping systems for conveying or circulating steam, hot or chilled water, heated oil and other fluids.